Critical Aspects of Safety and Loss Prevention

Critical Aspects of Safety and Loss Prevention

Trevor A. Kletz DSc, FEng, FIChemE, FRSC

Butterworths
London Boston Singapore Sydney Toronto Wellington

PART OF REED INTERNATIONAL P.L.C.

All rights reserved. No part of this publication may be reproduced in any material form (including photocopying or storing it in any medium by electronic means and whether or not transiently or incidentally to some other use of this publication) without the written permission of the copyright owner except in accordance with the provisions of the Copyright, Designs and Patents Act 1988 or under the terms of a licence issued by the Copyright Licensing Agency Ltd, 33–34 Alfred Place, London, England WC1E 7DP. Applications for the copyright owner's written permission to reproduce any part of this publication should be addressed to the Publishers.

Warning: The doing of an unauthorised act in relation to a copyright work may result in both a civil claim for damages and criminal prosecution.

This book is sold subject to the Standard Conditions of Sale of Net Books and may not be re-sold in the UK below the net price given by the Publishers in their current price list.

First published 1990

© **Butterworth & Co (Publishers) Ltd 1990**

British Library Cataloguing in Publication Data

Kletz, Trevor A.
   Critical aspects of safety and loss prevention.
   1. Industries. Accidents. Prevention
   I. Title
   363.1'1

ISBN 0-408-04429-2

Library of Congress Cataloging in Publication Data

Kletz, Trevor A.
   Critical aspects of safety and loss prevention/
   Trevor A. Kletz.
      p.   cm.
   Includes bibliographical references
   ISBN 0-408-04429-2
   1. Petroleum chemicals industry—Great Britain—
Safety measures
I. Title.
TP690.6.K54   1990
65.5'0289—dc20

Composition by Genesis Typesetting, Borough Green, Sevenoaks, Kent
Printed and bound in Great Britain by Courier International Ltd, Tiptree, Essex

Do not try to satisfy your vanity by teaching a great many things. Awaken people's curiosity. It is enough to open minds; do not overload them. Put there just a spark. If there is some good inflammable stuff it will catch fire.

Anatole France

Introduction

This book is a collection of nearly 400 thoughts and observations on *safety* and *loss prevention*, illustrated by accounts of accidents. The items, mostly short, are arranged alphabetically and cross-references are provided.

In 1968, after many years' experience in plant operation, I was appointed the first technical safety adviser to the Heavy Organic Chemicals (later Petrochemicals) Division of *ICI*. My appointment followed a number of serious fires and I was mainly concerned with process safety. However, my views reflect my managerial experience – when safety was one of my line responsibilities – as well as my full-time safety experience and the book will therefore, I hope, be found useful by all who work in design, operations and maintenance as well as by *safety professionals*.

Because I have spent my career in the chemical industry there is a bias towards process safety but nevertheless most of the book will, I hope, interest people in other industries and those concerned with safety generally.

The items are arranged in alphabetical order and the book is self-indexing, that is, cross-references are provided in the body of the text. The titles of other items are printed in italic the first time they appear in each item, if the cross-reference is relevant. (I have not printed words such as 'accident' and 'fire' in italic every time they occur.) Some of the titles are the names of items of equipment such as *valves* and *tanks*; some are the names of places, such as *Flixborough* and *Seveso*, where accidents have occurred; some are the names of hazardous substances such as *asbestos* and *benzene*; others are the names of abstractions such as *management* and *inspection* and a few are the names of people or organizations, such as *Du Pont* or *ICI*.

To describe further the scope of the book it is easier to say what it is not than what it is:

- It is not a comprehensive treatment of the subject. For that see *Loss Prevention in the Process Industries* by F P Lees[1] referred to herein as *Lees*. In fact, this book can be considered a series of footnotes to *Lees*, with the emphasis on practice rather than theory and illustrated by accounts of accidents that have occurred.

- It is neither an encyclopedia nor a dictionary. It is intended to be read, or dipped into, and not just used as a work of reference. Many important topics, such as *fire* and *explosions*, adequately covered elsewhere, are considered briefly or not at all (the brief comments made try to bring out points not usually covered) and instead I have discussed topics, such as *contamination* and *(accidental) purification*, that are neglected elsewhere. I have tried to illuminate neglected corners rather than the whole structure.

  Industrial hygiene is considered only briefly. See *asbestos, cancer* and *fugitive emissions*.

- The book is not a guide to the law, and references to it are concerned with principles rather than details. See, for example, *'reasonably practicable'*.

Several common themes run through the book which tries to express a consistent philosophy. These themes are:

- A commitment to the loss prevention rather than the traditional safety approach.
- An irreverent attitude towards some of the established views and customs of many safety professionals. See, for example *lost-time accident rate* and *human failing*. In safety, as everywhere else, it is no bad thing to question the accepted wisdom from time to time. Most of what we have learned is good, sound stuff but there is some chaff amongst the wheat.
- A belief that we know the answers to most of our problems but lack the will to apply the knowledge we have. See, for example, *philosophers' stone* and *production or safety?*
- A belief that many of our problems are older than we think and that solutions have been known for a long time.
- A preference for specific rather than general advice. See *attitude* and *policy*. The historian Barbara Tuchman writes[2], 'When I come across a generalisation unsupported by illustration I am instantly on my guard; my reaction is show me.' I agree, and when I can do so, without making the accounts too long, I have described accidents to support my recommendations.
- A belief that we cannot do everything at once and therefore need a rational system for setting *priorities*, something better than giving the most to those who shout the loudest.

I have tried to avoid being solemn but that does not mean that I am not serious. When others have put things better than I can I have not hesitated to quote them and I have used quotations from writers on other subjects when they throw light on our own. See, for example, *management* and *persistence*.

American readers should note that I have used the term manager, as it is used in the UK, to include those professionally qualified (or equivalent) people who would be called supervisors or superintendents in most US companies.

Thanks are due to the many colleagues, past and present, who have suggested ideas for inclusion or commented on my contributions. In particular I would like to thank:

- The hundreds of colleagues in ICI with whom I discussed almost every topic in the book during my 14 years (1968–1982) as a safety adviser, and especially the staff of Petrochemicals Division Safety and Loss Prevention Group.
- The staff of the Department of Chemical Engineering at Loughborough University of Technology, and especially Professor F P Lees, for giving me the opportunity of continuing my work after my retirement from ICI.
- The many people in other companies and organizations who have given freely of their knowledge and experience and especially those who have allowed me to describe their companies' mistakes.
- The editors of *Health and Safety at Work* and *The Chemical Engineer* for allowing me to quote from my articles in those journals.
- Mr S Coulson for drawing most of the Figures.
- The Leverhulme Trust for financial support.
- All who have encouraged me, by their support, to write and lecture on safety, as without the opportunity to communicate one's knowledge, it remains vague and ill-defined. The man who said, 'I don't know what I think until I've heard what I've said or read what I've written' was not as foolish as he might at first seem.

I have been unable to trace the owners of the copyright of a few of the Figures and I hope that, in the interests of safety, they will not object to the inclusion of their work.

The advice in the book is given in good faith but without warranty and readers should satisfy themselves that it is applicable to their circumstances.

The following abbreviations are used for those reference works most frequently mentioned:

EVHE: T A Kletz, *An Engineer's View of Human Error*, Institution of Chemical Engineers, Rugby, 1985

LFA: T A Kletz, *Learning from Accidents in Industry*, Butterworths, 1988

Lees: F P Lees, *Loss Prevention in the Process Industries*, Butterworths, 1980

Myths: T A Kletz, *Myths of the Chemical Industry or 44 Things a Chemical Engineer ought NOT to Know*, Institution of Chemical Engineers, 1984. 2nd edition, entitled *Improving Chemical Engineering Practices: A New Look at Old Myths of the Chemical Industry*, Hemisphere Publishing, New York, 1990

WWW: T A Kletz, *What Went Wrong? – Case Histories of Process Plant Disasters*, Gulf Publishing Company, Houston, Texas, 2nd edition, 1988

References to the above are given to Chapters or Sections (§) rather than pages as these differ less between editions.

To avoid the clumsy phrases 'he or she' and 'him or her' I have used 'he' and 'him'. Though there has been a welcome increase in the number of women employed in the process industries the designer, manager or accident victim is usually a man.

The pages that follow, especially the accident reports, record much ignorance, incompetence and folly but also originality and inventiveness in the cause of accident prevention.

Trevor Kletz

1. F P Lees, *Loss Prevention in the Process Industries,* Butterworths, 1980
2. Barbara Tuchman, *Practicing History*, Ballantine Books, New York, 1982, p. 36

Abbeystead

In 1984 an explosion in a *water* pumping station at Abbeystead, Lancashire killed 16 people, most of them local residents who were visiting the plant. Water was pumped from one river to another through a tunnel. When pumping was stopped some water was allowed to drain out of the tunnel leaving a void. Methane from the rocks below accumulated in the void and, when pumping was restarted, was pushed through vent valves into a pumphouse where it exploded.

If the presence of methane had been suspected, or even considered possible, it would have been easy to prevent the explosion by keeping the tunnel full of water or by discharging the gas from the vent valves into the open air. In addition, smoking, the probable source of *ignition*, could have been prohibited in the pumping station (though we should not rely on this alone). None of these things was done because no-one realized that methane might be present. Although there were references to dissolved methane in water supply systems in published papers, they were not known to engineers concerned with water supply schemes.

The official report[1] recommended that the hazards of methane in water supplies should be more widely known, but this will prevent the last accident rather than the next. Many more accidents have occurred because information on the hazards, though well known to some people, was not known to those concerned; the knowledge was in the wrong place. See *lost knowledge, need to know* and *LFA*, Chapter 14.

The Courts ruled that the consulting engineers were responsible for damages as they should have foreseen that methane might be present. However, one judge said that an ordinary, competent engineer could not have foreseen the danger[2].

1. Health and Safety Executive, *The Abbeystead Explosion*, HMSO, London, 1985
2. *Daily Telegraph*, 19 February 1988, p. 4

Abdication, management by

We have heard of *management* by delegation, management by participation and management by exception. More common but less often mentioned is management by abdication. This is illustrated by accident reports which say that the accident was due to *human failing* and that the injured man should take more care; this does nothing to prevent the accident happening again and is merely an abdication of management responsibility.

In some factories over 50%, sometimes over 80%, of the accidents that occur are said to be due to human failing. In other factories it is 10%. There is no difference in the accidents, only in the managers. See *EVHE*.

Another example of management by abdication is turning a *blind-eye*.

Aberfan

This village in South Wales was the scene of one of Britain's worst industrial accidents. In 1966 a colliery waste tip collapsed, a school lay in its path and the 166 people killed were mainly children. The immediate cause was the construction of the tip over a stream but the underlying causes were:

- A failure to learn from the past. Forty years earlier the causes of tip instability were recognized and warned against but the warning went unheeded, as none of the earlier collapses caused any loss of life.
- A failure to inspect adequately. There were no regular *inspections* of the tip and when it was inspected only the tipping equipment was looked at, not the tip itself.
- A failure to employ competent and well-trained people. Tips were the reponsibility of mechanical, not civil, engineers and they received no training on choice of sites or inspection. The official report[1] said, 'it was the blind leading the blind in a system inherited from the blind'.

See *alertness* and *LFA,* Chapter 13.

1. *Report of the Tribunal appointed to inquire into the Disaster at Aberfan on October 21st 1966*, HMSO, London, 1967

Absolute requirements

Under UK safety legislation employers are not required to do everything possible to prevent an accident, only what is *'reasonably practicable'*. However, there are some absolute requirements. For example, dangerous machinery must be securely fenced (guarded) even if the chance of anyone being injured is low and the cost of *fencing* is high.

The difference between the two approaches is not as great as it seems at first sight. To quote from a Factory Inspector, '. . . inspectors have been reluctant to press for an "absolute" standard of fencing as required by the statute where . . . the consequences of achieving that standard would mean that the machine became unworkable. . . This "Nelsonian" approach to the realities of industrial life, relying as it does on the experience and judgement of inspectors, has generally proved satisfactory. Indeed, so successful have been these informal (and in some cases formal) arrangements that there has been little desire to amend the primary legislation to reflect the need for a more pragmatic approach to machinery safety[1].'

1. G W Watson in *Her Majesty's Inspectors of Factories 1833–1983 – Essays to Commemorate 150 Years of Health and Safety Inspection*, HMSO, London, 1983, p. 53

Abstractions

For children, abstractions do not exist. Concrete things like chairs and tables and trees and gardens are real, and so are actions like running or shouting, but abstractions are not.

As we get older we learn to use abstract thought, but we often forget that children are right: abstractions do not really exist but are only a convenient shorthand to simplify our thoughts and conversations.

Thus, it is often convenient to talk about *attitudes* or *policies*. We say that someone's attitude to safety is wrong, meaning that he deals with safety matters in what we think is the wrong way. Instead of trying to change his attitude – difficult because it does not exist – let us discuss his problems with him and try to persuade him to deal with them in a different way. If we are successful we may say that his attitude has changed; so it has, but not as the result of a direct, head-on attack.

Attitude and policy are examples of what philosophers call an epiphenomenon, something that does not exist on its own but only as a sort of glow or halo around other things. If you want to get a halo you don't try to make one or buy one; instead you behave in a saint-like way and hope that a halo will appear and so it goes with attitudes.

Generalizations such as the chemical industry, technology or modern youth do not really exist. Critics blame the chemical industry (or technology) for causing pollution or for making chemical weapons but the industry (or technology) does not have a mind of its own. There are only individual companies, made up of individual people, who have different aims, morals, etc.

The headings in this book include many abstractions (*cause, perception of risk, perspective* and so on) as they provide convenient headings for discussing related subjects.

Acceptable risk

A phrase used, mainly by engineers, to describe a risk which is so small, compared with all the other risks to which we are exposed, that we would not be justified in using our resources of time and money to reduce it even further.

However, the phrase is best avoided when talking to a wider audience as it may cause them to switch off. 'What right have you', they may say, 'to decide what risk is acceptable to me?' and, of course, we have no right to decide what risk is acceptable to others. We should never knowingly fail to act when someone else is at risk but we cannot do everything at once. We have to do some things first and others later so we should talk about priorities rather than acceptable risks. Everybody has problems with priorities and if we talk about them we are more likely to keep the attention of our audience.

The phrase 'tolerable risk' may be more acceptable than 'acceptable risk' and is used in the official UK publication, *The Tolerability of Risk from Nuclear Power Stations*[1].

See *Canvey Island, risk criteria, fatal accident rate* and *unlikely hazards*.

1. *The Tolerability of Risk from Nuclear Power Stations*, HMSO, London, 1988

Access

'As he climbed down he held on to what he thought was a fixed part of the unit. It was not'[1].

This quotation reminds us that accidents can occur because people think about the hazards of carrying out a job but not about the hazards of getting to the site and back. For example, a railwayman was seriously injured while walking down the track at night to the site of an engineering job. In their report the Railway Inspectorate said that 'no thought appears to have been given to the safe means of access for train crews required to reach the site of engineering work at night'.

A repair had to be carried out to a pipeline on a large pipebridge. There was a walkway on one side of the pipebridge but the line to be repaired was on the far side. Scaffolding was erected so that the repair could be carried out safely but access to the scaffolding was a problem. A ladder would have blocked a roadway and traffic would have had to be diverted so instead the two men who were carrying out the repair were asked to crawl, on planks, between the pipes, to reach the scaffolding. They got there without trouble, but the job went wrong, there was an unexpected release of carbon monoxide gas from the pipe and one of the men was overcome. The rescuers had a difficult job dragging him through the pipebridge to the walkway; fortunately he made a complete recovery.

People have been overcome inside vessels. When authorizing entry to a vessel or other confined space always ask how the person entering will be rescued if he collapses inside. If a vessel is entered from the top through a manhole a rope is not usually adequate. One man cannot pull another out of a manhole on a rope; a hoist is usually needed.

According to the UK Factories Act, Section 30 the diameter of a manhole must be at least 18 inches if dangerous fumes are liable to be present. (If the manhole is oval the minimum size is 18 inches by 16 inches.) In fact, it is difficult to get through an 18 inch manhole wearing breathing apparatus or protective clothing and many companies specify a minimum diameter of 24 inches nominal (22.5 inches actual).

1. Health and Safety Executive, *Dangerous Maintenance*, HMSO, London, 1987, p. 24

Accident

An accident is often defined as something that happens by chance and is beyond control. If this is so, then there are very few, if any, accidents in industry (or on the roads or in the home). Most accidents are predictable. I know that during the coming year, in every large chemical works and in many small ones, a *tank* will be sucked in, a *tanker* will be overfilled and another will drive away before the filling or emptying *hose* has been disconnected. A man will be injured while disconnecting a hose and someone will open up the wrong pipeline[1]. (See *identification of equipment*.) I do not know exactly when or where but I am sure these accidents will occur. More serious accidents are also predictable but they occur less often[2].

Most accidents are preventable. We know how to prevent them – or someone does if we do not – but we lack the will to do so, or to find out how to do so. We cannot make gold from lead because we do not know how. In contrast, accidents do not occur because we lack knowledge but because we lack energy, drive and commitment.

In 1884 the Board of Trade's first report on the working of the *Boiler Explosions Act* said, 'The terms "inevitable accident" and "accident" are entirely inapplicable to these explosions . . . So far from the explosions being accidental, the only accidental thing about them is that the explosions should have been so long deferred'.

The Shorter Oxford Dictionary defines an accident as 'Anything that happens' which is a better definition than the usual one but too broad. I prefer 'An undesired event that results in harm to people, damage to property or loss to process' (F Bird).

Some companies call an incident an *accident* only when someone is injured. If no one is injured they call it a *dangerous occurrence* or dangerous incident. It is good practice to call them all accidents in order to draw attention to the fact that it is often a matter of chance whether or not injuries occur and that the investigation and follow up should be the same in each case.

See *mechanical accidents, repeated accidents, (visit accident) sites* and *triangles*.

1. T A Kletz, 'Accidents that will occur during the coming year', *Loss Prevention*, Vol. 10, 1976, p. 151
2. T A Kletz, 'Organisations have no memory', *Loss Prevention*, Vol. 13, 1980, p. 1

Accident chains

See *chains*.

Accident investigation

Craven,[1] an experienced investigator of fires and explosions, suggests that accident investigation should be split into seven steps, to which I have added an eighth. Some of the comments are my own.

Remit

It should be made clear to the investigator or investigating team that it is not their job to allocate blame (unless there has been arson, horseplay or reckless indifference to the safety of others). If witnesses feel that they, or their fellow-workers, may be blamed they will keep quiet, we will never know what happened and we will be unable to prevent it happening again. A tolerant attitude towards errors of judgement or omission is a price worth paying to find out the facts. See *Australia* and *LFA*, Introduction.

Make sure that the relevant authorities, and the *insurance* company if appropriate, have been informed. (In the UK many accidents have to be reported to the Health and Safety Executive.) The authorities and the insurers may wish to carry out their own investigations. If possible the various investigators should work together but do not delay if the others are slow to arrive.

6   Accident investigation

A brief survey

To see what there is to see.

The state of the plant before the accident

Was it normal or abnormal? If the investigator is not familiar with the plant and process he should look at plant descriptions, photographs, drawings and similar plants.

Examination of damage

Nothing should be moved, unless essential to make the plant safe, until the investigators, and any experts they wish to call in, have seen it and photographs have been taken. (See *(visit accident) sites*).

Interviews with witnesses

Do not put ideas into their minds. Avoid questions to which the answer is 'yes' or 'no'. It is easier for a witness to say 'yes' or 'no' than describe what they think happened, especially when they are tired or shocked. (See *falsification* and *memory*.)

Research and analysis

It may be necessary to ask for metallurgical, chemical or other tests to be carried out or to look for information in libraries or company files. In particular a search should be made for reports of similar accidents that have occurred before, on the same plant or elsewhere. Computerized data bases make this searching easier than in the past but the memories of people with long experience in the plant and industry are often better. 'Old boy' relationships with people in other companies can be valuable.

The recommendations

Craven says that these are not part of the investigation but a later stage, possibly carried out by different people. Most of us consider the recommendations to be the most important part of the investigation; if we cannot make any, the whole exercise is pointless.

Investigators should look beyond the immediate technical recommendations and see if they can find ways of avoiding the hazard and of improving the management system. Accident investigation is like peeling an onion: beneath each layer of causes and recommendations there are other, deeper layers. For example, consider the fire described under *amateurism*. The immediate cause was a missing slip-plate and the obvious recommendations concern improved procedures for controlling slip-plating. However, the hazard could have been reduced by a more spacious layout, by installing underground *drains* and by avoiding a mixture of series and parallel units. The investigators should also ask if it was essential to have so much flammable material in the plant ('What you don't have, can't leak' –

see *intensification*) and if it had to be so hot. Even if it is impossible to carry out these recommendations on the existing plant they should be noted for the future.

Finally, why were slip-plating procedures so poor? Did the managers not know what was going on or did they know but not realize that the procedures were poor? Why was the design so poor? Why did those concerned not learn from the other companies in the group? (See *insularity*.)

LFA analyses some accidents in detail and shows that we can learn many lessons, many more than we usually do, from the known facts. We pay a high price for accidents and then fail to use all the gold that is in the mine. See *chains, Challenger, dangerous occurrence* and *Zeebrugge*.

The report

See *accident reports*.

Finally, plan ahead. Draw up a procedure for accident investigation so that everyone knows what to do when one occurs.

1. A D Craven in *Safety and Accident Prevention in Chemical Operations*, edited by H H Fawcett and W S Wood, Wiley, New York, 2nd edition, 1982, Chapter 30

Accident-prone

This term is used to describe people who, as a result of their personal failings, have more than their fair share of accidents. We all know clumsy people who are always dropping things but in industry, especially the process industries, it is generally agreed that accident-prone people are responsible for only a small proportion of the total number of accidents. See *EVHE*, §4.3.

Some people will have more than their fair share of accidents by chance. Suppose that in a factory employing 200 people there are 100 accidents in a year. There are not enough to go round and many people will have no accident. If the accidents are distributed at random then the Poisson equation shows that:

- 121 people will have no accidents
- 61 will have one accident
- 15 will have two accidents
- 3 will have three or more accidents.

The last three people are not accident-prone, just unlucky. To call them accident-prone would be like calling a die loaded because three sixes came up in a row. To prove that a group of people is accident-prone we have to show that they have more accidents than we would expect by chance[1].

If people are really accident-prone this may be due to physical problems such as poor sense–muscle coordination, or to personality. Accident-prone people are often insubordinate, excitable and extrovert and are often absent due to sickness. In some cases they may be so unsuitable for their occupation that they may have to be moved. However, 'Let us beware lest the concept of the accident-prone person be stretched beyond the limits within which it can be a fruitful idea. We should indeed be guilty of a grave error if for any reason we discouraged the manufacture of safe machinery'[2].

If it is not very helpful to talk about accident-prone people; it is helpful to try to identify accident-prone plants or systems of work. Many plants and systems contain *traps* for those who work on them. It is no use telling people to take more care and avoid the traps. A trap is, by definition, something people fall into. We should accept people as we find them and try to change plant designs and methods of working so as to remove opportunities for error (or to protect against the consequences). See *friendly plants*, *human failing* and *EVHE*.

It is easier to match the job to the man than the man to the job.

1. L V Rigby, *The Nature of Human Error*, 4th Annual Technical Conference of the American Society for Quality Control, Pittsburgh, Pennsylvania, May 1970
2. D Hunter, *The Diseases of Occupations*, English Universities Press, 5th edition, 1975, p. 1064

Accident reports

An accident report should tell us:

1. What happened.
2. Why it happened.
3. What we should do differently in future to prevent it happening again, on other plants as well as on the plant where it actually occurred.
4. Who should make the changes.
5. When the changes will be complete. The report can then be brought forward at this time.
6. What the changes will cost.

Item 3 is, of course, the most important but it is surprising how many reports fail to give this vital information or tell us only some of the actions we ought to take. Many reports discuss only the immediate technical causes of an accident but not ways of avoiding the hazards or ways of improving the management system. See *accident investigation* and *chains*.

When we investigate an accident we often find a lot wrong, faults that could cause an accident though they did not do so. Recommendations should cover these.

Reports should be made available to all those who have similar hazards and may be able to profit from our misfortunes, in other companies as well as our own. Do not send them more information than they need or the essential message may not be recognized amongst a mass of detail. The ideal is two reports, one giving the full story, for those who want to see it and the other drawing attention to the essentials. However, do not make them too brief. Include enough detail to make the story convincing or, as W S Gilbert says, to give 'verisimilitude to an otherwise bald and unconvincing narrative'.

See *carelessness*, *communication*, *falsification*, *honesty*, *publication*, *smoke screens* and *LFA*.

Accident statistics

In many companies the safety officer spends a lot of his time preparing complex tables of accident data. For example, one annual report contains 12 pages of tables in which nine types of accident (e.g. number of lost-time accidents, number of alternative work accidents, number of minor accidents, number of fires, lost-time accident frequency rate etc.) are tabulated against 11 variables: location, age, service, occupation, nature of accidents (e.g. falls on level, falls from structure, falls from stairways – 34 headings), cause, type of injury (19 headings), part of body injured etc. In some cases separate figures are quoted for each quarter and one page compares the previous five years.

No conclusions are drawn from all these data and no recommendations are made. Nevertheless returns of this sort are not unusual. I have seen a 39 page report from one large oil company. Who studies these data and why are they produced?

The theory is that a detailed anaysis of the data will show if there has been an increase in, say, accidents involving ladders, especially to electricians, or an increase in hand injuries, especially to men with short service, allowing us to take action to reverse the increase. In practice, conclusions of this sort rarely, if ever, come out of the data. If there has been an increase in ladder accidents the managers and the safety officer should have picked this up from the accident reports as they come in and by talking to people as they go round the site. No magic remedies for accidents will ever come from analysing figures. Most of us already know what we need to do. We just lack the will to get on with it. See *Grimaldi* and *old-timer*.

I suspect that these accident statistics, prepared with so much labour, are given only a quick glance and then filed, never to see daylight again. The safety officer may be preparing them because he thinks the managers want them, while they wonder why he spends so much time in the office instead of getting out on the site.

Obviously we do need a few figures to show us whether or not the accident record is improving and how it compares with other companies (see *comparisons*) and it is true that no single figure such as the *lost-time accident rate* will suffice. Figures that I suggest should be reported are:

- Lost-time accident rate.
- *Fatal accident rate* (even in large companies only a five-year moving average is meaningful).
- Minor accident rate.
- Damage (insured and uninsured) and consequential loss.
- A measure of the results of safety *audits*, such as that provided by the five star grading system. This sets out the objectives or elements of a successful safety programme and awards marks for the standard attained. (See *criteria*.)

Do not congratulate people on an improvement, or blame them for a worsening performance, unless it is statistically significant. For example, suppose a plant averages seven accidents per year; the number would have to fall to two before we could be 90% confident that the accident rate had fallen. See *Lees*, §27.5 and *Myths*, §34.

Another pitfall is the assumption that all accidents are reported. A map of the distribution of bats in Yorkshire showed a large number of sightings near Helmsley but it is probably bat-watchers rather than bats that are common in the area. If one factory reports more dangerous occurrences than other factories, perhaps they have more or perhaps they are more honest. See *under-reporting*.

A senior manager who looks only at accident statistics will not know what is going on. He should also look at the detailed reports on at least some of the accidents that occur. See *management*.

Accuracy

See *confidence limits*.

Acids

See *corrosive chemicals*.

Action

The job of the *safety professional* is not complete when he has made his recommendations, or even when they have been accepted. It is not complete until they have been carried out and shown to be effective. He should not lose interest in the problem until this stage is reached.

Accident reports should be brought forward after an appropriate time to see if their recommendations for action have been carried out. *Hazard and operability study* reports should be brought forward after two or three months to check that team members have carried out the actions they agreed to take.

What instigates action in loss prevention? Undoubtedly the commonest and most effective motive is a serious incident. The nearer the incident the greater its effect. A fire in our own company has more effect than one in another company which in turn has more effect than one in another country.

Nevertheless in many cases awareness of a problem has resulted in action before an accident occurred. I have described elsewhere[1] the 'springs of action' of four major changes in which I was involved:

1. Improvements in the system for the preparation of equipment for maintenance: this followed a serious fire in the company concerned.
2. Improvements in the methods used for testing trips and alarms: this change resulted from recognition of a problem, our increasing dependence on trips and alarms.
3. Improvements in the equipment used for storing liquefied flammable gases: this followed the serious fire at Feyzin in France in 1966. (See *BLEVE.*)
4. High integrity protective systems: they were first designed after recognition of a problem, that some plants, for example, *oxidation*

plants operating close to the explosive range, require far more reliable control equipment than is normally supplied. See *protective systems*.

When there has been a fire or explosion it is easy to convince people of the need for change, in designs and/or procedures. Sometimes the safety professional has to dissuade his colleagues from going too far. (See *over-reaction*.) When the hazard has been recognized but no accident has occurred the safety professional has a harder job persuading his colleagues, but it can be done.

The four changes described above took place fairly quickly but a fifth change, the adoption of *inherently safer designs*, has been slower, probably because it is more far-reaching in its scope, requiring almost a change in *culture*. The chemical industry has been reluctant to admit that large inventories of hazardous materials are dangerous. The view has been that we know how to handle them, so don't worry. When they have not been kept under control, this shows that other companies have lower standards. If an incident occurs in our own company, it was an isolated occurrence. See *cognitive dissonance*.

See *decisions, evangelicalism, persuation* and *selling safety*.

1. T A Kletz, *Proceedings of the Second International Symposium on Loss Prevention and Safety Promotion in the Process Industries*, Dechema, Frankfurt, 1977. p. 1

Action replays

On several occasions people have failed to understand the reason for an accident and allowed it to happen again immediately afterwards.

An *explosion* occurred in a large *tank* 40 minutes after the start of a blending operation in which one grade of naphtha was being added to another. The fire was soon put out and the naphtha moved to another tank. The next day blending was resumed and 40 minutes later another explosion occurred.

The tanks were not blanketed with *nitrogen* and there was an explosive mixture of naphtha vapour and air above the liquid. The pumping rate was rather high and the naphtha became charged with *static electricity*. A spark passed between the naphtha and the roof of the tank and ignited the vapour. The tank was earthed but the naphtha was non-conducting and retained its charge. These explosions led to extensive investigations into the formation of static electricity[1].

A road *tanker* arrived with a load of 65% oleum. The *hose* on the tanker could not be used as the vent cock on it was missing so the operators used one of their own hoses normally used for 20% oleum. After 45 minutes it leaked and there was a large spillage. The operators assumed that the hose was in poor condition and replaced it with another of the same type. This also leaked and there was another spillage.

In the winter of 1952 a large new oil tank was being tested by filling it with water at 4.5°C. A horizontal crack, 0.6 m long, appeared across the first horizontal seam.

The tank was repaired and the test repeated. The tank then split from top to bottom in two places. The fracture started at a repair weld and propagated rapidly, presumably as a result of the low temperature[2]. See *pressure tests*.

1. A Klinkenburg and J L van der Minne, *Electrostatics in the Petroleum Industry*, Elsevier, Amsterdam, 1958
2. *Loss Prevention Bulletin*, No. 086, April 1989, p. 14

Acts of God

This phrase is used to describe natural phenomena such as lightning, earthquakes and floods. It is an unfortunate expression which we should avoid, as it implies that we can do nothing about them. However, to quote from a book on natural disasters[1]:

> Some disasters (flood, drought, famine) are caused more by environmental and resource mismanagement than by too much or too little rainfall. The impact of other disasters, which are triggered by acts of nature (earthquake, volcano, hurricane) are magnified by unwise human actions.
> Humans can make land prone to flooding by removing the trees and other vegetation which absorb this water.
> Humans can make land more drought-prone by removing the vegetation and soil systems which absorb and store water . . .
> In other disasters, such as cyclones and tsunamis, humans can increase their vulnerability by destroying bits of their natural environment which may act as buffers to these extreme natural forces. Such acts include destroying reefs, cutting mangrove forest and clearing inland forests.

The Ethiopian famines of the 1980s were due to mismanagement, not drought. Israel, a food exporting country, has a lower rainfall than Ethiopia but in that country over-grazing and deforestation have resulted in a loss of topsoil and there is little irrigation.

Similarly, suppose a storage *tank* containing petrol or other flammable liquid has been blown up by lightning. We cannot prevent the lightning but we can prevent the results by using a floating-roof tank or *nitrogen* blanketing or installing a *flame trap* in the vent line and by seeing that this *protective equipment* is kept in working order. The explosion could have been prevented by better *design* and/or better *methods of working*.

The use of phrases such as 'cruel fate' implies helplessness and inevitability and discourages a search for ways of preventing the accident or disaster.

Wrigglesworth writes[2]:

> Accidental death is now the only source of morbidity for which explanations that are essentially non-rational are still socially acceptable. Suggestions that 'it was an Act of God', or 'it was a pure accident' or 'it was just bad luck' are still advanced as comprehensive explanations of the causes of accidents and are accepted as such even by professionally trained persons who, in their own fields of activity, are accustomed to applying precise analytical thinking. This approach is typical of the

fables, fictions and fairy tales that mark our community approach to the accident problem.

See *fate*.

1. A Wijkman and L Timberlake, *Natural Disasters – Acts of God or Acts of Man?*, International Institute for Research and Development, London, 1984, pp. 6, 29 and 30
2. E C Wrigglesworth, *Occupational Safety and Health*, April 1972, p. 10

Add-ons

The traditional method of making a plant safer is to add on to it equipment for controlling the hazards. The alternative is to avoid the hazard by *inherently safer design*. However, this is not always possible, particularly on existing plants, and added-on safety features, such as *emergency isolation valves*, will continue to be necessary. See *defence in depth*.

Add-ons have to be tested or inspected regularly or they will not give us the protection they are designed to give. It is easy to buy the *protective equipment*, all we need is money and if we make enough fuss we get it in the end. It is much more difficult to make sure it is tested and inspected regularly. Passive protection such as fire protection (*insulation*) needs *inspection* just as much as active protection such as trips and *relief devices*. If a piece of fire insulation is missing the rest of it is ineffective.

Addresses

The addresses of about 150 sources of information on loss prevention throughout the world are given in *Lees*, Appendix 6. The Health and Safety Executive also publishes a list.

Adequacy

Under the UK *Health and Safety at Work Act 1974* the employer's duties include:

2.(1)(a) The provision and maintenance of plant and systems of work that are, so far as is *reasonably practicable*, safe and without risk to health:
2.(1)(c) The provision of such information, instruction, training and supervision as is necessary to ensure, so far as is reasonably practicable, the health and safety at work of his employees.

Factory Inspectors have sometimes been asked what is a safe system of work, adequate training and so on.

The Factory Inspectorate does give a lot of advice, in speech and writing. (You can get its publications list from the Health and Safety Executive Inquiry Point, Stanley Precinct, Bootle L20 3QY.) However, it is primarily the duty of the employer to decide what is a safe system of work, adequate training and so on. If the Factory Inspector does not agree he will say so, and in extreme cases issue an Improvement or Prohibition Notice, but in the first place it is up to the employer to decide. Those who create a hazard know, or should know, more about it than anyone else.

In deciding how far to go an employer is usually in the clear if he follows recognized *codes of practice* or accepted *good practice*. However, if he has special knowledge and knows that the hazards are greater than is generally believed, he will have to go further in removing them. (See *'reasonable care'*.)

The UK system is unusual. In most countries employers can find out what they have to do by looking it up in a book of rules. Some employers prefer this system, even though they may have to follow rules that are out of date or do not really apply to their plant, as they do not have the responsibility of deciding how far they ought to go. However, the UK system gives greater safety as rules are slow to change and are usually behind the times. The UK system is also flexible; as hazards change or new techniques become available, employers have to do more.

Advice

I was once asked to give just one word of advice on accident prevention to various people. Here are my replies:

'Walk' to plant and works managers

Walking round the plant to see what is happening will improve output and efficiency as well as safety. A plant cannot be managed from an office.

On the way round, keep your eyes open and ask questions. Look for anything unusual and anything that has changed since your last visit. Pick a repair job and ask to see the permit-to-work. Ask why an alarm light is lit up. Look where others do not, behind and underneath equipment. See *magic charm*.

'Talk' to safety professionals

Half the job of the *safety professional* is deciding what advice to give; the other half is giving it. Unless it can be communicated, the best advice is worthless.

The best way of converting people to your view is to talk to them, singly or in groups. You will also have to write when time is limited, but talking is more effective. You will get a response from your audience and you will know whether or not your message is getting through.

'Modesty' to design engineers

The days are long gone when one engineer, George Stephenson, could survey and construct a railway, *design* and construct an engine and drive it on the opening day. Design engineers are now dependent on the advice of experts in materials, piping, machinery, control and so on.

The design engineer needs to know when to ask the expert. Accidents have occurred because an engineer tried to design himself something that should have been left to an expert. See *Flixborough* and *knowledge of what we don't know*.

'Obey' to construction managers

I hope I am not lynched for advocating an old-fashioned virtue, obedience, to the design and to *good practice* when details are not specified in the design. Many accidents have occurred because *construction* teams failed to follow the design or did not construct well, in accordance with good engineering practice, details that were left to their discretion. See *leaks*.

'Ahead' to research and development chemists

Look ahead to the full-scale application of the your process. It may be safe in the laboratory but will it be safe after *scale-up*? Can you avoid the use of hazardous materials or use less of them? See *inherently safer designs*.

'Question' to general managers

It is hardly possible for general managers, responsible for a number of factories, to spend much time walking round them but they can ask searching questions on their visits: what are the problems that prevent an improvement in the safety record?, what action is needed?, what progress has been made? They can also read some *accident reports* picked at random to see if anything has been glossed over. This will be far more effective than the exhortations to do better produced by so many senior managers. See *management* and *platitudes*.

Aerosol cans

At one time the propellant in most aerosol cans was a chlorofluorocarbon. As a result of concern that it might affect the ozone layer of the atmosphere many manufacturers changed to butane during the 1970s. They did not know how to handle liquefied flammable gases and this resulted in a number of *fires* and *explosions* in both factories and warehouses; a good example of failure to foresee the results of technical change and of *exchanging one problem for another*.

The United States' biggest single-building fire loss, in a warehouse in Morrisville, Pennsylvania in 1982, started in a stack of aerosol cans. Bursting cans spread the fire to the rest of the building[1]. Also in 1982, several people were killed and 17–30 injured (reports differ) when a fire occurred in an aerosol factory in West Germany[2]. Altogether I know of seven incidents in the period 1971–85.

See *BLEVE*.

1. *Record* (published by Factory Mutual Insurance), Vol. 6, No. 3, Fall 1983
2. *Daily Telegraph* and *Guardian*, 10 February 1982

Air

Compressed air has been involved in many accidents. Operators find it hard to grasp the power of compressed air (and other compressed gases). Two men were killed when the end was blown off a pressure vessel by compressed air because the vent was choked. The operators found it hard to believe that a 'puff of air' could destroy a steel vessel, and explosion experts had to be brought in to convince them that there had not been a chemical explosion. See *force, pressure* and *LFA*, Chapter 7.

People have been injured when compressed air has been used to remove dust from workbenches or clothing and it has blown into cuts or eyes. A man was killed when, as a prank, a compressed air hose was pushed up his rectum.

Many fires have occurred on air compressors because the delivery temperature was allowed to get too hot (over 140°C) and deposits in the pipelines ignited. Dust in compressed air can contaminate products and cause wear of tools. See *WWW*, §12.1.

Most of the materials handled in the oil and chemical industries will not burn or explode by themselves but only when their vapour is mixed with air (or *oxygen*) in certain proportions. If we keep the materials in the plant and the air out of the plant a fire or explosion cannot occur. The air is just as much responsible for the fire or explosion as the oils or chemicals. See *chokes* and *oxidation*.

Air coolers

See *fin-fans*.

Airlines

Which airline should I fly? Assuming you wish to choose the one most likely to get you there alive, rather than the one with the best food or prettiest girls, then the answer can be found in a *Flight International* article[1]. Ranked by fatal accidents per million flights, including those caused by terrorism, the countries with the best records in the period 1973–84 were:

- Australia   0.3
- Scandinavia   0.45
- Japan   0.6
- USA, France   1.2
- UK, W Germany   1.3

Half the UK accidents occurred on Dan-Air flights.

Some small countries (Austria, Finland, Ireland, Israel and South Africa) had few or no accidents but their airlines are so small that the figures are not statistically significant.

At the other end of the scale, the large airlines with the worst records were:

- Egypt   13
- Turkey   17
- Colombia   27

The average for the World was 2. In 1934 the UK figure was 67.
Why are some airlines so much better than others? The safest may be those in which the crew can argue with the captain. See *Australia*.

These figures, like all failure data, describe past, not future performance and an airline's record can change: Air France was once known as Air Chance and in August 1985, after the *Flight International* article was published, 520 people were killed in a Japan Airlines crash.

Finally, do not worry too much. Even if you travel on a Colombian flight, the chance of a crash is only one in 36 000. On an Australian flight it is 84 times lower.

Air taxis have a poor safety record: in the 1970s they killed on average one ICI employee per year, more than any other item of equipment.

See *failures* and *falsification*.

1. J M Ramsden, *Flight International*, 26 January 1985, p. 29

ALARA

See *'reasonably practicable'*.

Alarms

The advantages of alarms, which tell us when a temperature, pressure, level, concentration or other measurement is approaching a dangerous level, are obvious but there are some pitfalls:

1. It is easy, particularly if *computers* are used, to overload the operator with too much information so that he becomes confused and does not know what action to take. (See *EVHE*, §4.2.1.) On the other hand computers make it possible to analyse a number of more or less simultaneous alarms and pick out the triggering event. The first alarm to sound may not indicate the first event as response times may differ.
2. All alarms can develop fail-danger faults which prevent them operating and should therefore be tested regularly, say, every month, or they may not work when required. An alarm which is not tested regularly is worse than no alarm as operators may rely on it. See *poor equipment* and *tests*.
3. Suppose a *tank* is filled every day. If an alarm is installed operators may not watch the level but instead wait until the alarm sounds. Sooner or later the alarm will fail and the tank will overflow. Testing will not help if demands are more frequent than tests as any failure of the alarm is more likely to be followed by a demand rather than a test.
4. When an alarm sounds unexpectedly, operators often assume it is faulty and do not check the plant. Many tanks have been overfilled for this

reason. (See *instruments*.) It is also a problem on the railways. Hall writes[1], 'It is very easy for the signalman to assume that, when something unforeseen occurs, it is due to technical failure, rather than the particular piece of equipment doing the job for which it was designed . . .'

Plant instructions should always list in a simple, summarized form the action to be taken when an alarm sounds. It is a good plan to mark on record sheets alarm settings and the levels at which actions should be taken.

Computer-generated audio and visual fire alarms are said to be more effective than conventionial firebells.

See *false alarms*.

1. S Hall, *Danger Signals*, Allen, London, 1987, p. 76

ALARP

See *'reasonably practicable'*.

Alcohol

According to the British Medical Association[1] the relationship of alcohol to work injury is a matter of serious and increasing concern. No system exists in the UK for the systematic collection of data on the relationship between alcohol use and industrial accidents but it is believed that the importance of alcohol is considerably under-rated. Some guidance on ways of dealing with the problem is given[2,3] but is not very specific.

Alcohol is known to play a large part in other accidents. Half the people admitted for head injury at a Glasgow hospital were found to be under the influence of alcohol, with an average blood alcohol level 2.5 times the legal limit for driving. On the roads over a quarter of all drivers killed have a blood alcohol level above the legal limit. Over half the adults killed in fires have alcohol in the blood[4]. Drowning and accidents at home are also associated with alcohol use.

Alcohol is not a new problem. In 1861 the manager of the Netham Alkai Works in Bristol wrote:

There was too much drinking and for want of the foreman checking their work, they often spoiled the materials by want of care and laziness, besides doing far less than a proper quantity for their day's work.

One night he found a 2 gallon stone bottle of beer in the Works. Nobody dared to own it and he had it removed[5].

1. British Medical Association, *Living with Risk*, Wiley, Chichester, 1987, p. 71
2. *The Problem Drinker at Work*, HMSO, London, 1981
3. *Alcohol Policies – A Guide to Action at Work*, The Industrial Society, London, 1986
4. British Medical Association, *Living with Risk*, Wiley, Chichester, 1987, pp. 60, 80 and 92
5. R Holland, *Chemistry and Industry*, 3 June 1985, p. 366

Alertness

Many an accident has been prevented because someone responded to an unusual observation.

For example, a plant used *nitrogen* in large cylinders. One day a cylinder of *oxygen*, intended for another plant on the same site, was delivered in error. The foreman noticed that the cylinder had an unusual colour and unusual fittings and he thought it strange that only one had been delivered as usually several cylinders were delivered at a time. Nevertheless he accepted the cylinder. He did not notice that the invoice said 'oxygen'. The invoice was sent as usual to the purchasing department for payment. The young clerk who dealt with it realized that oxygen had been delivered to a plant that had never received any before. She told her supervisor who telephoned the plant and the error came to light, fortunately before the cylinder had been used.

Similarly, an operator noticed that a pipe looked bigger than usual; this seemed unlikely but he reported his suspicions to his supervisor. The diameter had actually increased from 10 inches to 12 inches, due to creep (the result of exposure to higher temperatures than foreseen during design). If the creep had not been noticed, the pipe would ultimately have burst[1].

For other examples see *success*.

Many incidents have, of course, occurred because someone was not alert to an unusual observation. Before starting work several men who were constructing a ship lit cigarettes. They burnt rapidly, showing that oxygen was present. They assumed the cigarettes were faulty and did nothing about the unusual observation. When they started to weld, cables and cable ties caught fire and eight men were killed. Oxygen had leaked into the compartment in which they were working through hoses which had been left connected to *oxygen cylinders* overnight[2]. For another, similar incident see *oxygen*.

At *Flixborough* the temporary pipe which failed was in use for several months before it failed. There was no professionally qualified mechanical engineer on site at the time but there were many chemical engineers. Should they have noticed that the poor design and support of the temporary pipe? At the time some of my colleagues said that if they had been on the plant they would have questioned its design. Others thought that it would be unreasonable to expect a chemical engineer to do so and certainly those who were on the plant cannot be blamed for not noticing mechanical design errors.

If a chemical engineer had felt that the temporary pipe did not look right he might have felt that it was none of his business; perhaps the mechanical engineers would resent his interference. Presumably the mechanical engineers know their job; better say nothing. Flixborough, and the other incidents I have described, show us that we should never hesitate to speak out if we suspect that something may be wrong. (See *welding*.)

At *Aberfan* it occurred to the local Member of Parliament that a tip of colliery waste 'might not only slide but in sliding reach the village' but he did nothing as he feared that drawing attention to the tip might result in the closure of the colliery. When the tip did slide 166 people, mostly children, were killed.

Britain's worst railway accident, in which 226 people were killed, occurred at Quintinshill, north of Carlisle, in 1915, when a signalman forgot that there was train standing outside his signalbox and lowered his signal so that another train could approach. Ten other railwaymen (the drivers, firemen and guards of three trains and another signalman), all trained to read signals, were on the spot at the time but not one noticed that a signal was lowered when it should not have been, or, if they did notice, they did nothing.

1. *Loss Prevention Bulletin*, No. 085, February 1989, p. 9
2. *The Fire on HMS Glasgow, 23 September 1976*, HMSO, London, 1977

Alternatives

When evaluating the risks of a substance or action we should compare it with the alternatives. After the explosion at *Flixborough* it was described as the price of nylon. True, but the price of wool and cotton is higher as agriculture is a high risk industry. Similarly, if we are concerned about the risks from *nuclear power*, we should compare them with the risks of making electricity from coal or oil.

Here are some other examples:

- A new form of transport should be no more hazardous, preferably less hazardous, than the old form.
- Instead of installing a relief valve we sometimes install a high pressure trip which isolates the source of pressure. It should have a failure rate ten times lower than the failure rate of a relief valve. This factor is suggested because a relief valve which fails to lift at the set pressure may lift at a higher pressure but a trip which fails to operate at the set pressure will probably stay failed at higher pressures.
- If equipment which might cause *ignition* is introduced into a Zone 2 area it should be no more likely to spark than the electrical equipment normally used. See *electrical area classification*.
- Before banning artificial sweeteners we should compare their risks with those of sugar.

See *comparisons* and reference 1.

1. T A Kletz, *Hazop and Hazan – Notes on the Identification and Assessment of Hazards*, Institution of Chemical Engineers, Rugby, 2nd edition, 1985, §3.4 and 3.4.7

Amalgamation

A university lecturer refused to clean the blackboard if the previous lecturer had left it dirty. Instead he wrote on top of the existing notes with chalk of a different colour[1].

This eccentric behaviour may have taught his students a useful lesson. In real life we should not, metaphorically speaking, try to wipe the board clean but should make the best of what is there already.

Suppose two factories are amalgamated as the result of a take-over or reorganization. Tidy-minded people may be tempted to impose the same procedures and forms on both, the same permit-to-work procedure, for example and the same permit forms; this is a mistake. People get used to procedures and forms; change makes errors more likely and may cause resentment if it is imposed by outsiders. Unless the existing procedures and forms are clearly unsatisfactory, leave them alone. If they are unsatisfactory, say why you think they are wrong and let the staff of the factory modify them.

Of course, if the two factories share a common *maintenance* organization then the permit-to-work procedures should be the same. If the maintenance crew have to follow two different procedures then they will make mistakes.

Finally, if, despite what I have said, you do decide to go for standard forms and procedures, instead of standardizing on the taker-over's procedures, why not standardize on those of the taken-over?

If you think that the whole *culture* of the organization you have taken over is wrong then you cannot change it by decree, only by exposing the staff to your culture and hoping that they are so impressed that they wish to copy it. *ICI* staff throughout the world tackle problems (of all sorts) in a similar way to ICI staff in the UK. Nobody has told them that they have to do so, but many of them have worked in the UK and people from the UK have worked in the overseas companies. The overseas staff have seen the way ICI does things and liked it. This (not rule by decree) is the secret of a successful empire. *Du Pont* is probably similar.

See *reorganization*.

1. J Garside, *The Chemical Engineer*, No. 448, May 1988, p. 69

Amateurism

It is commonplace to say that responsibility for safety starts at the top and most of those at the top do accept reponsibility for the actions of their staff and do give them the money, manpower and encouragement that they need. However, it rarely occurs to those at the top that their own actions or inactions can cause or prevent an accident. Let me describe a case in which this occurred.

A chemical company, part of a large group, made all its products by batch processes. They were acknowledged experts and had a high reputation for safety and efficiency. Sales of one product grew so much that a large continuous plant was necessary. No one in the company had much experience of such plants so they engaged a *contractor* and left the design to him. If they had consulted the other companies in the group they would have been told to watch the contractor closely and told some of the points to watch.

The contractor sold them a pup. The process was good but the *layout* was very congested, the *drains* were open channels and the plant was a mixture of series and parallel operation.

When some of the parallel sections of the plant were shut down for extensive overhaul and repair other sections were kept on line, isolated by slip-plates. One of the slip-plates was overlooked. Four tonnes of a hot liquid, similar to petrol, leaked out and was ignited by a *diesel engine* being used by the maintenance team. Two men were killed and the plant was damaged. The congested design increased the damage and the open drainage channels allowed the fire to spread rapidly.

The immediate cause of the fire was the missing slip-plate. The foreman who decided where to put the slip-plates should have followed a more thorough and systematic procedure, but the underlying cause was the amateurism of the senior management and their failure to consult the other companies in the group. If they had consulted them they would have been given the following advice:

- Watch the contractor closely.
- Do not let the plant get too congested. Divide it into blocks with breaks in between, like fire breaks in a forest. (See *domino effects*.)
- Put the *drains* underground.
- If possible, do not maintain half the plant while the other half is on line. If it is essential to do so then build the two halves well apart and plan well in advance, at the design stage, where slip-plates will be inserted to isolate the two sections; mark these places clearly. Do not leave it for a foreman to sort out a few days before the shutdown.

I doubt if it ever occurred to the senior managers that their decisions had led to the accident, though they fully accepted that as the managers they were responsible for everything that went on. However, a few years afterwards the group was reorganized and responsibility for the continuous plant was transferred to another company in the group.

People can be expert in one field but children in another into which their technology has strayed. For more details of the above incident see *LFA*, Chapter 5.

Amateurism can be responsible for more than a poor safety record. According to the Chairman of the Manpower Services Commission, '. . . Britain's decline as a trading nation, and therefore its unemployment figures, are not the result of Government policies, intransigent unions, lazy workers or any of the other modern theories. They have their origins in history, and they are the culmination of persistent amateurism and lack of enterprise[1]'.

See *accident investigation, (are things as) black as they seem?, decisions* and *insularity*.

1. B Nicholson, *Journal of the Royal Society of Arts*, Vol. 135, No. 5375, October 1987, p. 804

Ammonia

The acute toxicity of ammonia is well-known with very young, old or unhealthy people being most vulnerable. The data are reviewed in reference 1 which concludes that exposure to 11 500 p.p.m. for 30 minutes will kill half those exposed. For other exposure times and concentrations:

$$ct^{0.5} = 6.3 \times 10^4$$

where: c = concentration for 50% lethality in p.p.m.
t = exposure time in minutes.

For other lethality levels between 20% and 80% a *probit* equation can be used:

Probit = $1.85 \ln[c^2 t] - 35.9$.

Ammonia is about 300 times less toxic than *chlorine*.

If an ammonia leak occurs, it is better to stay indoors with the windows shut than to try to escape. As ammonia is soluble in water, put wet towels around door edges and if necessary wrap one round your head. Go into the bathroom and run the shower.

About 225 people are believed to have been killed by ammonia leaks during the past 30 years, over a third of them in the US. The worst incident in the UK occurred in 1941 when 16 people were killed[1].

The chemicals most often involved in incidents attended by the London fire Brigade are *asbestos*, ammonia and hydrochloric acid[2].

The flammability of ammonia is less well-known than its toxicity and people are continually being surprised by the discovery that it can explode. Thus, after a *leak* from a refrigeration system the chief of the Houston, Texas Fire Department wrote, 'The hazards, it was believed, were limited to health; never had much thought been given to the flammability of ammonia' and 'It is hard to find any of the old, experienced ammonia hands who believe it possible for ammonia to explode[3].' Ammonia explosions are not common as its lower explosive limit is unusually high, 16%; the upper limit is 25%. Typical limits for hydrocarbons are propane: 2–9.5% and cyclohexane: 1.3–8.3%. Nevertheless ammonia explosions have occurred from time to time and the explosive limits of ammonia have been known since 1914. A paper presented in Houston in 1979[4] said that several ammonia leaks had exploded, though not all reported explosions were actually due to ammonia. It also said that there had been 11 explosions in aqueous ammonia tanks and several in nitric acid plants when the ammonia/air ratio became too high.

So far as I am aware ammonia has never exploded in the open air and I doubt if a concentration as high as 16% could be attained out-of-doors. See *WWW*, §19.1.

Another hazard of ammonia is that stress corrosion cracking is liable to occur in anhydrous ammonia storage tanks in the presence of traces of oxygen (over 0.05%); it is inhibited by traces of water, the amount required depending on the oxygen concentration. The subject is reviewed by L. Lunde and R. Nyborg[5].

The hazards of ammonia production are discussed in the annual volumes of *Ammonia Plant Safety*, published by the American Institute of Chemical Engineers.

See *indices of woe*.

1. *Ammonia Toxicity Monograph*, Institution of Chemical Engineers, Rugby, 1988
2. *Chemistry in Britain*, Vol. 23, No. 9, September 1987, p. 829
3. M H McRae, *Plant/Operations Progress*, Vol. 6, No. 1, January 1987, p. 17
4. P J Baldock, *Loss Prevention*, Vol. 13, 1980, p. 35
5. L Lunde and R Nyborg, *Plant/Operations Progress*, Vol. 6, No. 1, January 1987, p. 11

Anaesthetics

An article on the history of anaesthetics[1] describes many incidents that have parallels elsewhere.

From about 1867 onwards anaesthetics were supplied using a rudimentary type of inhaler. If it was connected up the wrong way round, liquid chloroform was blown into the patients, sometimes killing them. It was not difficult to modify the design so that it could be connected up only the right way round, but persuading doctors to use the modified design was another matter. They were reluctant to admit that it was possible to make such a simple mistake and as late as 1928 patients were still being killed by errors with the simple apparatus.

Several patients have died because they were given other gases instead of *oxygen*. In 1940 two patients were killed in a service hospital because carbon dioxide *cylinders* were used instead of oxygen cylinders. The green carbon dioxide cylinders had been painted black (the usual colour for oxygen) to smarten them up for an inspection!

The early anaesthetics, chloroform and ether, both had disadvantages. Ether could catch fire or explode and did so on several occasions. Chloroform could decompose in in the open flame of a candle or gas lamp, forming phosgene. Electric lighting, and the development of new anaesthetics removed these risks.

My purpose in relating these incidents is not, of course, to disparage the achievements of the early anaesthetists but to point out some of their mistakes so that we can learn from them. The incidents described show that:

- We should try to remove opportunities for error rather than hope that people will not make mistakes. See *human failing* and *friendly plants*.
- We should not change anything, even the colour of a cylinder, without authorisation by a competent person. See *modifications*.
- We should use safer materials whenever possible, instead of trying to keep hazardous materials under control. See *substitution*.

These incidents also show that we can learn a lot about safety by studying incidents in *other industries* and other walks of life. Because we are not involved we see the causes and the action required more clearly. (See *cognitive dissonance*.) And it will be relaxation rather than work.

1. W S Sykes, *Essays on the First hundred Years of Anaesthesia*, Vol. II, Churchill Livingstone, Edinburgh, 1960, 1982

Analysis

The most widely used safety analysis *instrument* is probably the combustible gas detector. Permanently mounted instruments are used for detecting leaks of flammable gas or vapour and portable instruments are used for checking that none is present before work with fire is carried out or a confined space is entered. (See *entry*.) However, conditions can change and so it is better to use a portable gas detector *alarm* than rely on a single test before a job is started.

Figure 1 Analysis

The instruments are very sensitive and can detect a tenth of the lower flammable limit. However, they will not detect flammable gas unless air is present. They do not always fail safe and so they should be tested before use, every time. (See Figure 1.) A test liquid, such as 30% v/v isopropanol in water, should be kept in the control room. Instruments with a sample head which is placed at the point of test are safer than those in which the sample is drawn through a long tube as the tube may be obstructed or may absorb the flammable gas[1].

Another very valuable analysis instrument is the *oxygen* analyser. Fixed or portable oxygen analysers are used:

- To check that the oxygen content is normal before a confined space is entered.
- To check that the oxygen content, in *stacks, tanks, centrifuges* and other equipment blanketed with *nitrogen* to exclude air, is low. (Most flammable materials will not burn unless 10% oxygen is present and so it is usual to aim for less than 5% oxygen in nitrogen-blanketed equipment.)
- To check that there are no leaks of oxygen.
- To measure the oxygen content of flue gas. Too much oxygen shows that fuel is being wasted; too little shows that combustion may be incomplete and that the flue gas may explode when mixed with air.

A wide variety of instruments are now available for testing, continuously or when required, for most hazardous materials. Before a confined space is entered, the atmosphere should always be tested for traces of any hazardous material that has been present.

When new and more sensitive analysis instruments are introduced, people often become concerned about the presence of impurities which were present all along, but previously undetected. For example, a plant installed instruments for measuring the level of carbon monoxide, instead of using mice; several leaks were detected during the following weeks. Similarly, improved methods of crack detection in vessels have shown up cracks that at one time would have been undetected.

Unusual instrument readings or laboratory results are often treated with scepticism until the instrument has been checked or a repeat sample analysed. Unfortunately, the delay can be hazardous or expensive. On one occasion a supply of raw material became contaminated (as a meter-prover was not cleaned before use); the laboratory detected the impurity in the product but the result was so unusual that they spent 5 hours analysing check samples before they reported the result to the control room. The operators were so sceptical that they spent a further 5 hours looking for a source of *contamination* and having more samples analysed before they shut the plant down. Consequently, off-specification product was manufactured for 10 hours.

On another occasion a foreman said to me, 'I'm having terrible trouble with the laboratories. I had to send them six samples before they gave me the right result.'

See *hazard analysis*.

1. Health and Safety Executive, *Industrial Use of Flammable Gas Detectors*, Guidance Note CS1, HMSO, London, 1987

Anomalies

See *failures*.

Area classification

See *electrical area classification*.

Arsenic

Arsenic is the poison most often used by murderers, in stories if not in real life; 100 mg is fatal. It has been so well studied by forensic scientists that anyone contemplating murder should look elsewhere. Arsenic also has long-term effects, including *cancer*, and in the UK the maximum exposure limit (see *COSHH Regulations*) is $0.2\,\text{mg/m}^3$.

Very small doses of arsenic, on the other hand, can be beneficial. They are said to improve the complexion and physical stamina. It is not unusual for very small doses of something to be beneficial when larger doses are harmful. Vitamins, ultra-violet light and perhaps *radioactivity* are other examples[1].

Arsenic is used in the production of gold and at one time there were many complaints of arsenic contamination in the Canadian goldmining areas[2].

See also *'Controversial chemicals'*

1. J H Fremlin, *Atom*, No. 390, April 1989, p. 4
2. *Controversial Chemicals*, edited by P Kruus and I M Valeriote, Multiscience Publications, Montreal, Canada, 2nd edition 1984, p. 7

Arson

Arson is a growing problem. In the UK convictions increased nearly sixfold from 1964 to 1981. However, the chemical and allied industries suffer less than many others. Those most affected are schools, transport and the communication and the distribution trades.

The motives for arson are financial gain, concealment of crime, protest, often by disgruntled employees, terrorism and vandalism.

To prevent arson we should:

- Control access, keep out intruders and supervise casual labour. See *security*.
- Detect and fight fires as soon as possible, using automatic equipment and regular patrols.
- Deny fuel to the fireraiser.

Case histories in the oil and chemical industries (many in warehouses) and advice on the investigation and prevention of arson are given in reference 1.

1. P A Carson, C J Mumford and R B Ward, *Loss Prevention Bulletin*, No. 065, October 1985, p. 1 & No. 070, August 1986, p. 15

Asbestos

Except for coal dust, no substance causes more deaths by industrial disease than asbestos. About a 100 people die each year in the UK from asbestosis, lung *cancer* caused by asbestos, and mesothelioma, a form of cancer caused by asbestos; about 600 die from pneumoconiosis caused by coal dust and about 1000 from all prescribed industrial diseases, including those mentioned. In addition an unknown number of people, possibly several thousand, die from industrial disease which is not officially prescribed. (See *under-reporting*.) Since industrial disease takes several decades to develop, these figures reflect working conditions several decades ago rather than today's conditions.

In the UK the concentration of white asbestos (chrysolite) in the workplace atmosphere must not exceed 0.5 fibre/ml and the concentration of blue asbestos (crocidolite) must not exceed 0.2 fibre/ml[1]; concentrations must be reduced below these figures if it is *'reasonably practicable'* to do so. A company has been prosecuted and convicted for not doing all that was reasonably practicable to reduce the asbestos concentration although no evidence was submitted to show that the control limit had been exceeded[2].

The *fatal accident rate* (FAR), i.e. the number of fatal accidents in a group of 1000 men in a working lifetime, for someone exposed to the official limit of asbestos throughout the working day for 50 years is 25^3, compared with the UK average of four, for all industrial accidents, for all premises covered by the *Factories Act*. The risk from asbestos is thus high, compared with accidental risks, but very few people are likely to be exposed to such a concentration for so long.

Smokers are eight times more likely to get lung cancer from asbestos than non-smokers and the FAR would be greatly reduced if only non-smokers could be employed where asbestos dust is present. Employing non-smokers is not, of course, an alternative to reducing the concentration of asbestos in the atmosphere but however low the concentration, smokers will still be at greater risk than non-smokers. (I am assuming, as is usual, that the risk is proportional to the dose and that there is no threshhold.)

See *'Controversial Chemicals'*.

1. *Control of Asbestos at Work Regulations, 1987*, HMSO, London
2. *Health and Safety Monitor*, August 1984, p.1. Quoted in *Chemical Safety Summary*, Vol. 55, No. 219, 1984, p. 74
3. J Peto, *Lancet*, 1978, i, p. 484

Assessment

After we have identified the hazards on a new or existing plant, using one or more of the methods described under *identification*, we have to assess them, that is, decide whether they are so large that we ought to remove them (or protect people from their consequences) or so trivial or improbable, compared with all the other risks around us, that we can leave them alone, at least for the time being.

Sometimes the answer is obvious. Sometimes our experience or a *code of practice*, even *'gut feel'*, may tell us what to do. Sometimes the answer is not obvious, we have no experience, there is no code and the arguments seem finely balanced. It may then be useful to use quantitative *hazard analysis* – to work out the probability that there will be an accident and the size of the consequences and to compare the results with a target or *criterion*.

The biggest errors made in hazard assessment are not in the assessment itself but in the preliminary identification. People spend time and effort assessing hazards which have been brought to their attention, by an accident or in other ways, and fail to realize that there are greater hazards which ought to be dealt with first. Before assessing hazards we should ask, 'Are we sure that we have identified all the major hazards and all the ways in which they can occur?'

Assumptions

In any activity we take certain assumptions for granted and do not write them down. We should do so, as then we might realise that they are not true every time. For example, in *hazard analysis*, the application of numerical methods to the assessment of safety problems[1], the following are some of the assumptions normally taken for granted and the circumstances when they may not be true:

Assumption	May not be true when ...
Failure is random	equipment is new or old or machinery has been repaired.
Testing is perfect	testing interferes with production.
Repair time is negligible	spares are not stocked.
Flows are not limited by inventory	flows are high but inventories small, the same material going round and round.
Substances have no unusual properties	substances have unusually high or low melting or boiling points or are near their critical points.
The plant is designed, operated and maintained to good standards	in overseas, subsidiary or remote plants which get less attention than main plants. Rot starts at the edges.

1. T A Kletz, *Hazop and Hazan – Notes on the Identification and Assessment of Hazards*, Institution of Chemical Engineers, Rugby, 2nd edition, 1986, §4.5

Astonishment

There is a story that Webster, compiler of Webster's dictionary, who always used words precisely, arrived home to find his wife in bed with another man. 'You have surprised me', she said. 'You have astonished me', he replied.

Mrs Webster knew all along that she might be discovered but decided that the chance was small, that it was an *acceptable risk*. Her judgement may have been wrong, or an unlikely probability may have occurred. She was surprised.

Mr Webster, however, had never considered it possible that the event might occur; he was therefore astonished. He had to change his views on fundamental matters. Lanir[1] calls such events 'fundamental surprises', a term which he coined to describe the Yom Kippur War. Although Mr Webster was astonished, perhaps he should not have been. Infidelity is not uncommon but he had not believed it possible that it could happen to him; he had to make a fundamental reappraisal.

Aberfan, Chernobyl, Flixborough and *Bhopal* were also 'fundamental surprises' and caused major reappraisals. Yet similar events, though less serious, had happened before.

Whenever plants are not *inherently safe* but made safe by adding on *protective equipment*, we accept that simultaneous failure of components could cause an accident. Simultaneous failure is unlikely, but possible; if it occurred it would be a surprise. Sometimes we do not foresee that someone might deliberately switch off the safety equipment. When they do, we are astonished. Anyone who operates hazardous plant or equipment, made safe by added-on protective equipment, should be able to demonstrate that he has foreseen that this could occur, has taken action to prevent it and will not be 'astonished'. (See *coincidences* and *'Normal Accidents'*.)

1. Z Lanir, *Fundamental Surprises*, Decision Research, Eugene, Oregon, 1987

Asymptote

Safety is often approached asymptotically. We wish to prevent a vessel being overpressured so we install a relief valve, properly sized and maintained. It may fail, so we install a second in parallel. The chance of failure is now much less but not zero as both relief valves may fail at the same time. We could install a third and make the chance of failure even smaller, but still not zero; where do we stop? To answer this question we have to set a target; we have to decide that over-pressuring of the vessel is acceptable if it occurs less than once in a 100 years, or a 1000 years or a 1 000 000 years. We cannot set a target of 'never', unless we can remove the hazard by a change in design, making the vessel so strong that it will withstand any pressure to which it might be subjected.

See *hazard analysis, inherently safer design* and *redundancy*.

Asymptotes can be illustrated by the story of the engineer who wooed a reluctant lady mathematician. She suggested that he stood some distance away and with each step halved the distance between them. As a mathematician she knew that they would never meet but as an engineer he knew that he would soon get near enough for all practical purposes. How near is 'near enough'?

Attenuation

Attenuation is one of the methods of achieving *inherently safer design*. If we have to use or store large quantities of hazardous materials then we should use or store them, whenever possible, in the least hazardous form. For example, some dyestuffs, which form explosive dusts, can be handled as slurries and liquefied gases (such as chlorine or ammonia) can be stored refrigerated at low pressure rather than under pressure at ambient temperatures. When the pressure is low the *leak* rate through a hole of a given size will be small and because the temperature is low there will be little evaporation. However, the risk of a leak from the refrigeration and heating equipment should be considered as well as the risk of a leak from the storage vessel and there may be no increase in safety unless the quantity stored is large.

Attenuation should be adopted only when *intensification* (reducing the amount of hazardous material) and *substitution* are not practicable[1].

1. T A Kletz, *Cheaper, Safer Plants – Notes on Inherently Safer and Simpler Plants*, Institution of Chemical Engineers, Rugby, 2nd edition, 1985

Attitude

The wrong attitude is often given as the reason for a poor accident record, by an individual or an organization. We say, 'We need a change in attitude' or 'The British workman has the wrong attitude'. Courses are designed to change people's attitudes; F Herzberg questions their value. These courses, he says, 'rest on the assumption that to change people's attitude is to change their behaviour. In truth it's the other way about. A change in behaviour leads to a change in attitude[1]'.

I doubt if attitudes can be changed by a direct attempt to do so. Furthermore, we have no right to try to do so. People's attitudes are their private affairs; we should concern ourselves with whether or not they achieve their objectives.

If someone has too many fires, let us discuss the reasons for the fires and what can be done to prevent them, also, perhaps, what they cost and what their consequences might be. After a while he may start to do a few things differently, he may have fewer fires and everyone will say, 'He's changed his attitude'. In short, don't try to change people's attitudes. Just help them with their problems.

See *abstractions* and *policy*.

1. F Herzberg, *Sunday Times*, 31 January 1971

Audits

An audit is a systematic examination of an activity to see if it is being carried out correctly. Safety performance should be audited as well as accounts. We need safety audits because:

1. Those who work in a plant soon fail to notice the hazards they see every day, for example, vibrating pipes, but they may be noticed immediately by an outsider.
2. Auditors may have specialized knowledge and thus see hazards not apparent to others.
3. Auditors have more time for in-depth investigation than those who work regularly on a plant.

Thus, safety auditing is not, or should not be, a police activity, to catch out people who are not doing their job properly, but an activity designed to help the local management who may miss hazards through familiarity, ignorance or lack of time.

Who should audit?

It has to be admitted that '. . . the standard of performance of any given function varies directly with the degree of probability that someone higher up is going to look at what has been done and what has not been done.' (The quotation is taken from Safe and Sound, the report of an investigation into safety auditing in the US, published by the Chemical Industries Association in 1970, a report which did much to introduce safety auditing to the UK.) Hence managers at all levels should carry out a certain amount of auditing as part of their job. They may wish to set aside special periods of time or elect to keep their eyes open as they go round the plant on their normal visits. It does not matter, so long as they do it.

However, when we talk about safety auditing we are usually thinking of special audits carried out by a team or an individual from outside the plant who spends a few days or even a few weeks on the job. A typical team might include a professional auditor, someone from another plant, a *safety professional*, and an expert on the hazards involved. In addition *Factory Inspectors* and *insurance* surveyors may carry out their own audits. Asking people from other companies to take part in audits has been suggested but not welcomed[1].

Weaknesses in safety audits

Many auditors look only for the obvious mechanical hazards and ignore *methods of working*; software as they are often called. Auditors should look at:

- The quality of the *training* and *instructions*.
- The procedures for preparing equipment for *maintenance*, controlling *modifications*, testing *protective equipment* and so on, and whether or not these procedures are actually followed.
- Procedures for investigating accidents, passing on the lessons learned and ensuring they are not forgotten.
- Process hazards as well as mechanical ones.
- The quality of the *management* and the effectiveness of their commitment to safety; vague protestations of goodwill are not enough. See *platitudes*.
- Places which others do not look at, behind and underneath equipment.

Auditors should visit the plant at night and not just during the day. However, time is always limited. Instead of trying to have a quick look at everything it may be better to pick one topic at a time and look at it in depth; such selective audits are usually known as *surveys*.

Some things seen on audits

The following are some of the things I saw during half-day *inspections* – really too short to qualify as audits – of about 20 plants, considered by others to have a high standard of safety:

- In every control room some *alarm* lights were lit up. In almost every case the operators were able to explain why and say what they were doing about them.
- Preparation for maintenance was good on some plants but on many others methods of *isolation* were poor: valves were not locked and slip-plates were not used.
- On many plants there were good sets of operating and safety instructions but on others the instructions could not be found, were out of date, were so clean that they were obviously never read – like poetry books in public libraries – or the folders creaked when opened like the

doors of a seldom-visited church[2]. If they were never read, perhaps no one could understand them or they did not give the information needed.
- Most trips and alarms were tested regularly but record keeping was poor. It was difficult or impossible to find out what tests had been done, when, what faults had been found and whether they had been repaired.
- Very few plants had any system for learning the lessons of the past. Accidents were investigated and reports written up, circulated, filed – and forgotten. In contrast, a few plants had excellent *'black books'*, (folders containing past reports of continuing interest); they are compulsory reading for newcomers while others dip into them from time to time.
- There was a surprising variation between different plants belonging to the same company and often located near each other. Each had good practices, which others could follow with advantage and each had problems that others had solved. This variation persisted despite interchanges of staff. It seems that people decide after a transfer that when in Rome they should do as Rome does. Great tact is needed when introducing 'foreign' ways into any group. Nevertheless, on moving to Rome we should try to introduce the Romans to other ways, if we think they are safer.

At a more mundane level, the following is an extract from the report of an audit on the storage area attached to a large chemical plant[3]:
- There are holes in the bunds surrounding the *tanks*, where bricks have been removed to pass pipes through.
- One bund contains a pump-out pump started automatically by a float control. This is intended to remove rain water but any process spillage will be pumped out.
- Many tanks are fitted with *level glasses* which are never used, and other small, unnecessary branches.
- Some flammable gas detectors (see *analysis*) are installed but are too few in number.
- Before *entry* is allowed to the inside of a tank, the atmosphere is tested with a portable flammable gas detector. If the tank has contained a toxic gas or liquid, such as benzene, then its concentration should be measured.
- Several tank vents are fitted with gauzes but no one knows whether they are *flame traps* or protection against birds.
- The need for careful study of *modifications* is recognized and we saw some excellent examples of this.
- We are alarmed at the methods used for entry to vessels. Men are regularly entering to clean under an atmosphere of *nitrogen*. Some of the slip-plates used for *isolation* are very thin. (Slip-plates used for isolation for entry should be at least ¼ inch (7 mm) thick.)

Audit results are usually qualitative but they are more effective if they can be made quantitative. See *criteria*.
See *breaking the rules* and *tests*.

1. P Varey, *The Chemical Engineer*, No. 458, March 1989, p. 3
2. D Lodge, *Write On*, Penguin Books, 1988, p. 205
3. T A Kletz, *Health and Safety at Work*, Vol. 4, No. 3, November 1981, p. 20

Australia

All the Australian *airlines* have outstandingly good safety records. Since 1963, *Flight International* has ranked Australia first or second in each of its periodic surveys of world airlines. As is often the case when organizations or people have good records, in safety or anything else, no one is quite sure why (see *Du Pont*). The following reasons have been suggested[1]:

- A strong commitment to safety at all levels. 'Qantas' standards are high because we want them to be high, not because they have been imposed.'
- An atmosphere in which there is good communication at all levels. The crew are not reluctant to question the captain's decisions and 'We don't have the attitude that pilots don't talk to anyone in overalls'.
- A good system for reporting incidents and defects and exchanging information with other airlines. Pilots and others who report incidents are guaranteed immunity from punishment.
- Good *training* programmes and low staff turnover.

1. J M Ramsden, *Flight International*, 1 December 1984, p. 1449 and 26 January 1985, p. 29

Auto-ignition temperature

This is the temperature at which a substance will catch fire spontaneously in air, without a source of ignition. References 1 and 2 give data, reference 3 gives more information, while *Bretherick*[4] quotes those that are unusually low (below 225 °C).

In a hydrocarbon series the auto-ignition temperature (AIT) falls as the boiling point rises. The AIT of ethylene is about 480 °C and that of heavy fuel oil about 250 °C. Processing is often carried out at higher temperatures and any leaks that occur will ignite. Pumps handling liquids above their AIT should be provided with remotely operated *emergency isolation valves*. Great care is needed when opening up for *maintenance* any equipment which contains liquid above its AIT; it should be isolated by double block and bleed valves and slip-plates, the valves being necessary so that the slip-plates can be inserted safely. See *WWW*, §1.1.1. and *Safety in Chemical Operations*[5]. If oil is spilt on *insulation* on hot pipes, degradation occurs, the AIT falls and the oil may ignite at a temperature far below its normal AIT.

1. C J Hilado *et al.*, *Chemical Engineering*, Vol. 79, No. 19, 1972, p. 75
2. *Fire Hazard Properties of Flammable Liquids, Gases, Solids*, National Fire Protection Association, Boston, 1969
3. T J Snell, *Loss Prevention Bulletin*, No. 081, June 1988, p. 25
4. L Bretherick, *Handbook of Reactive Chemical Hazards*, 3rd edition, Butterworths, 1985
5. *Safety in Chemical Operations*, edited by H H Fawcett and W S Wood, 2nd edition, Wiley, 1982, Chapter 36

Automation

We install automatic equipment for various reasons; it may be cheaper than men or quicker-acting or more reliable. People often say, 'Let's put in an automatic system and remove the human element'.

However, when we make a plant or an operation automatic we do not remove our dependence on men, we merely transfer it from one man to another.

After an operator has allowed a *tank* to overflow we may install a high-level trip which will close the inlet valve when the tank is nearly full. We have removed our dependence on the operator but we are now dependent on the men who design, construct, test and maintain the trip, and they also make mistakes. It may be right to make the change as the designers, constructors, testers and maintenance workers are probably less exposed to stress than the operator and may therefore make fewer errors but we must not kid ourselves that we have removed our dependence on men.

Automatic equipment should be tested regularly or we cannot be sure that it will work when required. (See *tests*.) *Relief devices* are very reliable and usually need testing only every year or two but instrumented trips need testing about once every month. It is easy to buy equipment: all we need is money and if we make enough fuss we usually get it in the end. It is much more difficult to make sure that the equipment is tested and maintained, year after year, after the initial enthusiasm for the new toy has faded. Procedures are subject to a form of corrosion more rapid than that which affects the steelwork, and can vanish without trace once managers lose interest.

Designers should not specify automatic equipment unless they know that the client has the knowledge, resources and commitment necessary to test and maintain it properly. See *Bhopal*.

Aversion to risk

Are we too risk-averse? Do we go too far in expecting every risk to be removed, no matter how trivial or unlikely to come to pass? Sometimes it seems so. On the other hand when we look at our attitudes to smoking or motoring we seem to be a risk-loving society. In fact we are neither risk-averse nor risk-loving but risk-illiterate. We do not know which risks are big and which are small and it is hard for the ordinary citizen to find out. The press ignore most road accidents (about 5000 people killed every year in the UK) but make a great fuss about an occasional accident on a railway level crossing. As a result, money is spent on reducing the risk at level crossings when more lives might be saved if it was spent in other ways. The press reports every air crash but give little space to the hazards of smoking though in the UK it kills as many people as a jumbo crashing every day.

A British Medical Association book[1] lists the major causes of death and discusses them dispassionately. It is the best available source of information for anyone who wants to know their chances of surviving the attentions of doctors, employers, drivers, brewers and tobacconists.

I have previously presented some information on risk in various ways that may be easier to understand than 'one in 10 000 per year'. For example, I assumed that there was a tax on risks proportional to the size of the risk. If the rate is set at £1M per life then we will have to pay the following taxes:

Item	Tax
Cigarettes	70 p each (£14 on a packet of 20)
Wine	£2 per bottle
Beer	25 p per pint
Petrol	50 p per gallon if used in a car
	£3 per gallon if used in a motor cycle
Chest X-ray	£1
Working in the chemical industry	£1.60 per week
Working in the steel industry	£3.80 per week
Diet soft drink	2 p (due to the saccharin)
Living near nuclear power station	20 p per year (due to radiation)

1. British Medical Association, *Living with Risk,* Wiley, Chichester, 1987
2. T A Kletz, *Interdisciplinary Science Reviews*, Vol. 8, No. 2, June 1983, p. 114

See *cause and effect, false alarms, perception of risk, perspective* and *water*.

Avoidance of risk

See *inherently safer design*.

Back flow

See *reverse flow*.

Batch processes

From the loss prevention viewpoint the most significant feature of batch processes is that as many more operations have to be carried out and many more changes made in operating conditions (compared with continuous processes) there are far more opportunities for error and thus more errors.

Many years ago I worked in a factory that was split into two sections: continuous plants and batch plants, separated by a railway line that emphasized the difference between them. The foremen and operators on the batch plants had a poor reputation as a gang of incompetents who were always overfilling *tanks*, putting material in the wrong tank, letting the temperature get too high or too low, and so on. Some of the best men on the continuous plants were transferred to the batch plants but with little effect. Error rates on the batch plants were actually lower than on the continuous plants but there were more opportunities for error: tanks were filled, pumps started up etc. many more time per day. (See *human failing*.)

In recent years *computer* control of batch processes has increased. This has on the whole reduced error rates but batch processes are now susceptible to errors in the software, equipment failures and so on.

If 1000 tonnes per year of a product are required it is usually assumed that this is too little to justify a continuous plant, and a batch plant is designed. However, 1000 tonnes per year will pass through a pipeline 1 cm diameter, assuming a velocity of 1 m/s, so a continuous plant made from 1 cm pipe might be considered. Heat removal should be less of a problem than in a batch reactor, as the surface area is large, and runaway *reactions* are unlikely.

See *knowledge of what we don't know*.

Benzene

After ethylene, benzene (boiling point 80 °C) is the most produced organic chemical; it differs in two respects from most other flammable hydrocarbons:

1. It has an unusually high melting point, 5 °C, and is liable to solidify in sample, drain and vent lines, relief valves etc., so steam or electric trace heating may be necessary. The heating should be checked frequently, say daily, in cold weather to make sure that it is in operation and has not failed or been switched off.
2. It is poisonous; 7500 p.p.m. in the atmosphere can cause death in an hour and 100 p.p.m. can cause fatigue, weakness and confusion. There is ample evidence that concentrations above 10 p.p.m. produce long-term effects[1] and in the UK and Western Europe the maximum exposure limit limit (see *COSHH Regulations*) has been set at 5 p.p.m. In 1978/1979 the US Courts overturned a proposal by the Occupational Safety and Health Administration (OSHA) to reduce the maximum

concentration for 8 hours exposure from 10 p.p.m. to 1 p.p.m. as there was no evidence that 10 p.p.m. was hazardous and no evidence of benefits to offset the cost. However, the level was reduced in 1987.

According to Richard Wilson of Harvard more people will be killed manufacturing and installing the extra equipment required than will ever be saved by reducing the concentration.

In many applications other solvents such as cyclohexane can be used instead of benzene[2], an example of *substitution*.

1. R J Fielder, *Toxicity Review 4: Benzene*, HMSO, London, 1982
2. R I Pollard-Cavalli, *Plant/Operations Progress*, Vol. 6, No. 4, October 1987, p. 185

Bhopal

This town in central India was the scene of the worst disaster in the history of the chemical industry. In 1984 a *leak* of a toxic chemical, methyl isocyanate (MIC) (from a chemical plant where it was used as an intermediate in the manufacture of the insecticide carbaryl) spread beyond the plant boundary, killing about 2000 people and injuring about 200 000. Most of the dead and injured were living in a shanty town which had grown up next to the plant.

The immediate cause of the disaster was the contamination of an MIC storage *tank* by several tonnes of water, probably a deliberate act of sabotage[1]. A runaway *reaction* occurred and the temperature and pressure rose. The relief valve lifted and MIC vapour was discharged into the atmosphere. The *protective equipment* which should have prevented or minimized the release was out of order or not in full working order, i.e. the refrigeration system which should have cooled the storage tank was shut down, the scrubbing system which should have absorbed the vapor was not immediately available and the flare system which should have destroyed any vapour which got past the scrubbing system was out of use.

The main lessons to be learnt from the disaster are:

'What you don't have, can't leak'

The material which leaked was not a product or raw material but an intermediate. It was convenient to store it, but it was not essential to do so. Following Bhopal the company concerned, Union Carbide, and other companies decided to greatly reduce their stocks of MIC and other hazardous intermediates. See *intensification*.

Plant layout and location

If materials which are not there cannot leak, people who are not there cannot be killed. The death toll at Bhopal would have been lower if a shanty town had not been allowed to grow up near the plant. It is, of course, much more difficult to prevent the spread of shanty towns than of permanent dwellings but nevertheless companies should try to do so, buying and fencing land if necessary. See *layout and location*.

Hazard and operability studies

The MIC storage tank was contaminated by substantial quantities of water and chloroform, up to 1 tonne of water and 1.5 tonnes of chloroform, and this led to a complex series of runaway reactions. The precise route by which water entered the tank is unknown but several theories have been put forward and sabotage seems the most likely. If any of the suggested routes were possible then they should have been made impossible. Hazard and operability studies are a powerful tool for identifying ways in which contamination and other unwanted deviations can occur. Since water was known to react violently with MIC, it should not have been allowed anywhere near it. See *Hazard and Operability Studies*.

Keep protective equipment in working order – and size it correctly

As already stated, the refrigeration, flare and scrubbing systems were not in full working order when the leak occurred. In addition the high temperature and pressure on the MIC tank were at first ignored as the *instruments* were known to be unreliable. The high temperature *alarm* did not operate as the setpoint had been raised and was too high. One of the main lessons of Bhopal is therefore the need to keep protective equipment in working order.

It is easy to buy safety equipment. All we need is money and if we make enough fuss we get it in the end. It is much more difficult to make sure that the equipment is kept in full working order when the initial enthusiasm has faded. All procedures, including testing and *maintenance* procedures, are subject to a form of corrosion more rapid than that which affects the steelwork and can vanish without trace once managers lose interest. A continuous *management* effort is needed to make sure that procedures are maintained. See *audits*.

Joint ventures

The Bhopal plant was half-owned by a US company and half-owned locally. The local company was responsible for the operation of the plant, as required by Indian law. In such joint ventures it is important to be clear who is responsible for safety, in design and operation. The technically more sophisticated partner has a special responsibility and should not go ahead unless they are sure that the operating partner has the knowledge, experience, commitment and resources necessary for handling hazardous materials; it cannot shrug off responsiblity by saying that it is not in full control.

Training in loss prevention

Bhopal makes us ask if those who designed and operated the plant received sufficient training in loss prevention, as students and from their employers. In the UK all chemical engineering undergraduates get some training in loss prevention but this is not the case in most other countries, including the US.

At Bhopal there had been changes in staff and reductions in manning and the new recruits may not have been as experienced as the original team. However, the errors that were made, such as taking protective equipment out of commission, were basic ones that cannot be blamed on inexperience of a particular plant. See *Training*.

Public response

Bhopal showed the need for companies to collaborate with local authorities and emergency services in drawing up plans for handling emergencies.

Terrible though Bhopal was we should beware of *overreaction* such as suggestions that insecticides, or indeed the whole chemical industry, are unnecessary. Insecticides, by increasing food production have saved more lives than were lost at Bhopal. Furthermore, Bhopal was not an inevitable result of insecticide manufacture. By better design and operations and by learning from experience, further Bhopals can be prevented. Accidents are not due to lack of knowledge but by failure to use the knowledge we have.

See *LFA*, Chapter 10.

1. A Kalelkar, in *Preventing Major Chemical and Related Accidents*, Symposium Series No. 110, Institution of Chemical Engineers, Rugby, p. 553

Bible

Quotations to suit every occasion can be found in the Bible. Here are a few on safety (mainly from the *Good News Bible*). Do not follow the advice on *protective clothing*!

Compensation

> If a man takes the cover off a pit or if he digs one and does not cover it, and a bull or donkey falls into it, he must pay for the animal. He is to pay the money to the owner and may keep the dead animal. – Exodus 21: 33.
> See also *history of safety*.

Repeated accidents

> What has happened before will happened again. What has been done before will be done again. There is nothing new in the whole world.
> No one remembers what has happened in the past, and no one in days to come will remember what happens between now and then. – Ecclesiastes 1: 9, 11.
> Whatever happens or can happen has already happened before. – Ecclesiastes 3: 15.
> If you dig a pit you fall in it . . . If you work in a stone quarry, you get hurt by stones. If you split wood, you get hurt doing it. – Ecclesiastes 10: 8–9.
> See *repeated accidents*.

The need for clear instructions

Write down clearly on clay tablets what I reveal to you, so that it can be read at a glance. – Habukkuk 2: 2.
See *instructions*.

The need for safety professionals

Where no counsel is the people fall: but in the multitude of counsellors there is safety. – Proverbs (Authorised Version) 11: 14.
See *safety professionals*.

Protective clothing

He gave his own armour to David for him to wear: a bronze helmet, which he put on David's head, and a coat of armour. David strapped Saul's sword over the armour and tried to walk, but he couldn't, as he wasn't used to wearing them. 'I can't fight with all this,' he said to Saul. 'I'm not used to it.' So he took it all off. – 1 Samuel 17: 38–39.

Safety training

Then shall the eyes of the blind be opened and the ears of the deaf unstopped. – Isaiah 29: 18.
The command that I am giving you today is not too difficult or beyond your reach. It is not up in the sky. . . Nor is it on the other side of the ocean. – Deuteronomy 30: 11–13.
See *training*.

Fire

. . . and after the fire a still small voice. – 1 Kings, 19: 12.
Probably someone explaining why the sprinklers did not work, why the fire brigade was not called sooner and why combustible stock was in the wrong place. See *fire*.

See *talebearing*.

Big plants

See *large plants*.

(Are things as) Black as they seem?

Only bad news is news and so those outside the chemical industry get the impression that those inside it spend their time blowing things up, spilling toxic chemicals and so on. Even those inside the industry who have to read accident reports may get a similar impression. It is therefore worth reminding ourselves that many plants go for months without a dangerous occurrence or even a minor injury and that every week thousands of permits-to-work are filled in correctly and thousands of pieces of equipment are prepared for *maintenance* without incident.

While a plant was starting up after a shutdown for overhaul there was a *leak* of hot, flammable liquid. It did not ignite but nevertheless there was an inquiry. Afterwards the manager said, 'We broke and remade 2000 joints during the shutdown and got one wrong. That's the only one anyone has heard about.'

Unfortunately, when hazardous materials are handled one leaking joint in 2000 is one too many and after this event the fibre gaskets in half the joints on the plant, i.e. all those exposed to liquid (there were 5000 in all) were replaced by spiral-wound gaskets which, if they do leak, do so at a much lower rate. However, fibre gaskets were left in any joints exposed to vapour. The chemical industry has to operate to extremely high standards, far higher than its critics realize. See *old plants* and *modern standards*.

Why were spiral-wound gaskets not installed when the plant was built? The plant was the one described in the item on *amateurism*. The decision was left to the *contractor*.

See *friendly plants*.

Black books

A black book (or memory book) is a folder of reports on past accidents, kept in the control room or other place where it is readily available. It should be compulsory reading for newcomers and others should dip into it from time to time to refresh their memories.

Past accidents are soon forgotten and repeated. Minor accidents seem to be repeated every couple of years, more serious accidents after ten or more years. (See *repeated accidents* and *(managerial) responsibility*.) A black book can help to keep the memory alive.

The black book should not be cluttered up with reports on slips and bruises. Only accidents of technical interest should be included, from other companies as well as our own.

Black books, like all other safety procedures (see *methods of working*) will not last unless managers take an interest in them. If they are seen to look at the book, others will. If they do not, others will assume that it is not important and after a few months no one will know where it is.

Blame

After an accident, policemen, newspaper reporters, Australian aborigines and other unsophisticated people look for someone to blame, as the following quotations show:

> Rites were performed often at the grave or exposure platform of the dead to discover the person to be blamed for the "murder". Since death was not considered a natural event a cause for it was always sought in the evil intentions of someone else, usually a member of another local group[1].

Divers Strangers, Dutch and French, were during the Fire (the Great Fire of London), apprehended upon suspicion that they contributed mischievously to it, who are all imprisoned, and informations prepared to make a severe inquisition thereupon by my Lord Chief Justice Kreling, assisted by some of the Lords of the Privy Council, and some principal members of the City[2].

No-one has yet been imprisoned for 'murder in the workplace', in spite of the Inspectorate's repeated admissions that the blame for workplace deaths lies with management[3].

The Robens Report[4] – the report that led to the UK *Health and Safety at Work Act* – gives a different view:

The fact is – and we believe this to be widely recognised – the traditional concepts of the criminal law are not readily applicable to the majority of infringements which arise under this type of legislation. Relatively few offences are clear cut, few arise from reckless indifference to the possibility of causing injury, few can be laid without qualification at the door of a single individual. The typical infringement or combination of infringements arises rather through carelessness, oversight, lack of knowledge or means, inadequate supervision, or sheer inefficiency. In such circumstances the process of prosecution and punishment by the criminal courts is largely an irrelevancy. The real need is for a constructive means of ensuring that practical improvements are made and preventative measures adopted.

They might also have added that an accident is usually the result of a *chain* of events which can be broken at any point. A typical accident can be prevented by many people, designers, operators, managers and others, who share the responsibility to varying degrees.

Another reason for not blaming people is that if we do they will say as little as possible, we may not find out what happened and may be unable to prevent it happening again. A tolerant attitude towards errors of judgement or omission is a price worth paying to find out the facts. See *accident investigation, Australia* and *prosecution*.

1. *Australian Aboriginal Culture*, Australian National Commission for UNESCO, 1973, p. 44
2. *London Gazette*, 3–10 September 1666, reprinted by HMSO, London, 1965
3. *Death at Work*, Workers Educational Association, 1987
4. *Safety and Health at Work: Report of the Committee (Chairman: Lord Robens) 1970–1972*, HMSO, London, 1972, § 26.1

BLEVE

A BLEVE (Boiling Liquid Expanding Vapour Explosion) occurs when a vessel containing liquid under pressure, above its normal boiling point, bursts, releasing its contents with explosive violence. The use of the term is usually restricted to vessels containing flammable liquids which burst as a result of exposure to fire. The escaping contents ignite immediately producing an intense fireball which causes more damage than the pressure wave from the bursting vessel or the bits of the vessel, though they may travel hundreds of metres.

How to protect pressure vessels from fire

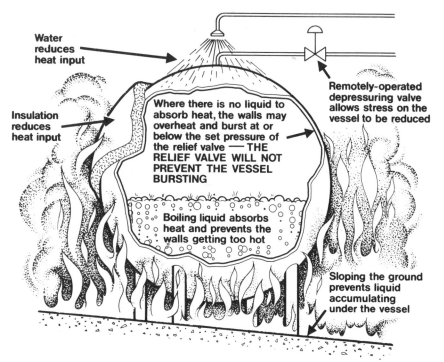

Figure 2 BLEVE

If a pressure vessel is exposed to fire the metal below the liquid level is kept cool by contact with the liquid, which cannot go above its boiling point. If, however, the flames are in contact with the upper unwetted portion of the metal, or all the liquid evaporates, then the metal becomes hot and loses its strength and the vessel will burst even though the pressure is not above the set point of the *relief device*. A relief valve or other relief device cannot prevent a vessel bursting if the metal gets too hot.

Gas filled vessels, or vessels containing liquids near their critical temperatures, where they have a low latent heat, can burst very quickly when exposed to fire. (See *useless equipment and procedures*.)

One of the best-known BLEVEs occurred at a refinery at Feyzin in France in 1966. While water was being drained from a $1200\,m^3$ pressure vessel containing propane, the drain valve stuck open and the escaping propane was ignited by a car on a motorway 150 m away. The refinery staff advised the fire brigade to use the available water supply for cooling the surrounding vessels, to stop the fire spreading. The relief valve, they said, would protect the vessel that was on fire. After 1½ hours the vessel burst, killing 15-18 men (reports differ) and injuring about 80, and the area was abandoned. The fire spread unchecked and several other pressure vessels containing liquefied petroleum gas also burst. See *WWW* §8.1.

To prevent a BLEVE occurring pressure vessels containing flammable liquids should be:

- Cooled with water if they are exposed to fire.
- Protected by *insulation*. Unlike water spray, insulation does not have to be commissioned and is immediately available as a barrier to heat input.
- Located on ground that is sloped so that leaks of liquid do not accumulate underneath but run off to one side. Catchment pits should be provided to collect spillages from large vessels.
- Protected by remotely-operated depressurising valves so that the pressure in the vessel can be reduced, thus decreasing the strain on the metal.

See Figure 2.

All these actions should be taken but there is some trade-off between them. If insulation is installed, less cooling water is needed and the relief valve and depressurising valve can be smaller.

Although most BLEVEs occur about 1 hour after the start of a fire they have occurred within 10 minutes so not much time is available for applying cooling water. The water cooling system should be fixed, especially if the vessel is not insulated[1].

Many BLEVEs have occurred, particularly in the US, when road or rail *tanker*s have been involved in an accident and have leaked. There have been none in the UK.

The worst BLEVE ever occurred in Mexico City in 1984 at a processing plant and distribution centre for liquefied petroleum gas (LPG). A pipe burst and the escaping LPG caught fire. The fire heated an LPG tank which burst, causing further fires and further BLEVEs. Altogether four spheres and 15 cylindrical tanks exploded during the next 1½ hours, some of the tanks landing 1200 m away. Over 500 people were killed and over 4000 injured. Most of them were living in a shanty town that had grown up near the plant, as at *Bhopal*.

Although the plant was only a few years old it was below the standards that many companies had adopted by about 1970. The water deluge system was inadequate, there was little or no insulation on the vessels (even their legs were not insulated) and bunds around the vessels allowed the liquid to accumulate where it could do most harm[2]. (See *WWW* §8.1.)

A small BLEVE, but big enough to kill someone, occurs when an *aerosol* can is put on a fire.

Some other BLEVES are described in reference 3.

1. T A Kletz, *Hydrocarbon Processing*, Vol. 56, No. 8, August 1977, p. 98
2. *BLEVE! The Tragedy of San Juanico*, Skandia International Insurance Company, Stockholm, 1985
3. J A Davenport, *Plant/Operations Progress*, Vol. 6, No. 4, October 1987, p. 207

Blind

A blind is another name for a slip-plate or spade. Inserted into a flange in a pipeline it provides a much more positive isolation than a valve.

The first step down the road to a serious accident occurs when a manager turns a *blind-eye* to a missing blind.

Blind-eyes

Managers and supervisers often turn a blind-eye to unsafe practice. One accident report was unusually frank. It said:

> There is, however, some doubt as to whether or not the method the injured man used was condoned by management. He gave the impression at the enquiry that he had done this sort of thing before.
>
> The foreman in his statement says that the procedure was incorrect but at the enquiry gave the impression that the practice was not uncommon, thereby indicating that he was aware of it.

The legal position in the UK is given in the *Health and Safety at Work* Act (1974), §37(1):

> Where an offence . . . committed by a body corporate is proved to have been committed with the consent or connivance of, or to have been attributable to any neglect on the part of, any director, manager, secretary or other similar officer of the body corporate, he, as well as the body corporate, shall be guilty of that offence and shall be liable to be proceeded against and punished accordingly.

Redgraves's Health and Safety in Factories[1] comments as follows on the meaning of 'connivance':

> It is submitted that it connotes a specific mental state not amounting to actual consent to the commission of the offence in question, concomitant to a failure to take any step to prevent or discourage the commission of that offence. Such a mental state has been termed 'wilful blindness'; that is to say, an intentional shutting of the eyes to something of which the percipient would, in his own interests, prefer to remain unaware. This construction accords both with the etymology of the word (Latin: *connivere*, to wink) and with the use made of it by the courts . . .

Managers and supervisors are not, of course, expected to stand over employees all the time but they should make occasional checks to see that rules are being observed (see *audits, breaking the rules* and *short cuts*) and they should keep their eyes open as they go round the plant.

I agree that a good manager should have less than perfect eyesight and sometimes be almost blind, have less than perfect hearing and sometimes be almost deaf – but not where safety is concerned.

In the UK prosecutions of individuals for offences against safety legislation are rare; a handful per year. The Health and Safety Executive prefers to prosecute the company unless there has been gross neglect by an employee. There are about 1200 prosecutions of companies per year in the UK but only a handful of individuals are prosecuted.

See *blame, contradictory instructions* and *prosecution*.

1. *Redgraves's Health and Safety in Factories*, edited by I Fife and E A Machin, Butterworths, 2nd edition, 1982, p. 16

Boilers

The history of boiler explosions is interesting in itself and as an example of public and private response to a new hazard created by a change in technology. The subject has been well documented[1,2,3]. (See *change – new problems*.)

Up to the year 1800 the steam pressure in boilers was low (up to 10 p.s.i.g.) and explosions were few. After 1800 higher pressures (30 p.s.i.g. and over) were used and there were many explosions, members of the public as well as employees being killed.

Following an explosion in a steamboat in 1817, which killed eight people, Parliament appointed a Select Committee. In the Committee's report the members noted their aversion to legislation but said that where the safety of the public (public note, not employees) was endangered by ignorance, avarice or inattention, it was the duty of Parliament to intervene. Precedents were quoted, including the regulation of stagecoaches, the qualification of physicians and party walls in buildings. The Committee recommended that passenger-carrying steam vessels should be registered, that boiler construction and testing should be supervised and that two safety valves should be used with severe penalties for tampering with the weights.

No action followed this report or similar ones in 1831, 1839 and 1843. Sir John Rennie, a steamboat manufacturer, told the 1843 enquiry that constant examination of boilers would cause serious inconvenience and would give no guarantee that the public safety would be assured. With so many varied engine designs it would be next to impossible to agree on methods of examination. Besides, there were really very few accidents.

Finally, in 1846 and 1851 the Board of Trade was given powers to inspect steamboats annually, to issue or deny certificates of adequacy and to investigate and report on accidents.

In the US it was steamboat explosions, more than anything else, which convinced people that the Government had to interfere in what was previously seen as people's private affairs. (The opposite view was put by Senator Stockton, in 1868, who said, '. . . what will be left of human liberty if we progress on this course much further? What will be, by and by, the difference between the citizens of this far-famed Republic and the serfs of Russia? Can a man's property be said to be his own, when you take it out of his control and put it into the hands of another, though he may be a Federal officer?')

In the UK fixed boilers had to wait until 1882 before they were covered by legislation but from 1859 onwards the *insurance* companies inspected boilers and did much to raise standards. However, until the first act was passed only about half the boilers in the UK were inspected regularly. Explosions were due mainly to lack of *inspection*, neglect of *maintenance* and to interference with the relief valves.

Locomotive boiler explosions, during the period 1853–90 were mainly due to *corrosion* and could have been prevented by regular inspection or by redesign to remove features which encouraged corrosion. Locomotive superintendants encouraged the belief that most explosions were due to drivers tampering with relief valves but less than ten occurred for this reason (and only four or five were due to low water levels). Beware of glib generalizations; we all like to blame someone else. If drivers did tamper with relief valves the answer should have been to make them tamper proof. As early as 1829 the Liverpool and Manchester Railway insisted that all locomotives must have two relief valves, one out of the reach of the driver.

Before 1853 reports were poor and no conclusions can be drawn. From 1890 to 1920 the main causes of explosions were poor maintenance or a deliberate decision to keep defective boilers in use. Many of the famous locomotive designers were abysmal maintenance engineers.

After 1920, standards of design and maintenance improved and the few explosions that did occur were due mainly to low water levels. Some occurred on imported US engines as the drivers were unfamiliar with the gauge glass design.

It is interesting to compare the legislative response to today's hazards, such as *Flixborough* and *Bhopal*. The response is quicker and there is little opposition but the successors to Sir David Rennie still suggest that industry can be left to put its own house in order. There is still little or no legislative action until after accidents have occurred.

For more information see *Brunner Mond, furnaces* and references 1–3.

1. J G Burke, *Bursting boilers and the Federal Power*, in *Technology and Culture*, edited by M Kranzberg and W H Davenport, Schocken Books and New American Library, 1972, p. 93
2. W H Chaloner, *Vulcan – The History of One Hundred Years of Engineering and Insurance*, 1959
3. C H Hewison, *Locomotive Boiler Explosions*, David and Charles, Newton Abbott, 1983

Boil-over

If an oil tank is on fire, hot unburnt residues may sink. When they reach the water layer, usually present at the bottom of the tank, the water may boil and the steam produced will shoot boiling oil out of the tank. There will be a sudden increase in the intensity of the fire. Boil-over, as it is called, is most likely to occur when the oil is a mixture of components with a wide boiling range. The oil is blown out with great force as the water is first superheated and then suddenly turns to steam. As the oil boils over the pressure on the water is reduced and boiling becomes more vigorous.

There have been some spectacular boil-overs[1]. In 1926 burning oil blown out of a tank covered an area 330 m diameter and the flames were 1800 m tall. In 1982 in Venezuela a boil-over killed 160 people.

See *foam-over*.

1. *Loss Prevention Bulletin*, No. 057, June 1984, pp. 26–30

Bolts

See *threads*.

Books on process safety and loss prevention

The standard work on process safety and loss prevention is:

- F P Lees, *Loss Prevention in the Process Industries*, Butterworths, 1980, 2 volumes, 1316 pages.

Shorter works are:

- V C Marshall, *Major Chemical Hazards*, Ellis Horwood, Chichester, 1987, 600 pages.
- G L Wells, *An Introduction to Loss Prevention*, Godwin, Harlow, 1980.

Shorter still:

- Chapter 9 of Coulson and Richardson's *Chemical Engineering*, Volume 6, edited by R K Sinnott, Pergamon, Oxford, 1983, 30 pages.
- *A First Guide to Loss Prevention*, Institution of Chemical Engineers, Rugby, 1981, 34 pages.

The Institution of Chemical Engineers publishes a number of books on particular aspects of loss prevention. Some of these are noted in the specific items; some other general books are listed in the Introduction.

The following is a comprehensive collection of articles of variable length and quality by many different authors, but some of the newer loss prevention methods are not discussed:

- *Safety and Accident Prevention in Chemical Operations*, edited by H H Fawcett & W S Wood, Wiley, New York, 2nd edition, 1982, 910 pages.

The Center for Chemical Process Safety of the American Institute of Chemical Engineers has published a series of guidelines on specific aspects of loss prevention such as *Hazard Evaluation Procedures, Safe Storage and Handling of High Toxic Hazard Materials, Vapor Cloud Dispersion Models* and *Vapor Release Mitigation*.

Older books still worth reading, if you can locate copies, are:

- G Armistead, *Safety in Petroleum Refining and Related Industries*, Security Publishing Corporation, St Michaels, Maryland, 2nd edition, 1964

and the series of nine booklets:

- *Hazard of Water, Hazard of Air, Hazard of Electricity*, etc. published by the American Oil Company, Chicago, Illinois between 1955 and 1966.

(safety) Books

Most plants have a safety book in which people write down details of jobs that need attention. Instead, or as well, why not have a book in which people write down details of jobs that they have done to prevent accidents? On a plant that tried this the jobs listed included:

- Steam hose rolled up and hung on wall.
- Start made on drawing up list of all steam traps.
- Absorbant spread over oil spillage.
- Tanker drivers told to wear protective clothing.
- Fitters asked to remove tools which they had left scattered around the compressor house.

Nothing world-shattering but several hundred small jobs which might otherwise have been left were put right and people were encouraged to do something themselves instead of just listing jobs that others should do. See *'they'*.

Boredom

Boredom, as well as *stress*, can make errors more likely. If we are bored we may 'switch off' and then not notice a change in an *instrument* reading or even an *alarm*. Night watchmen find it difficult to stay alert. Our error rate is at its lowest when we are under a little stress, but not too much.

Some people are concerned that *computer* control may leave operators with so little to do that they become bored and less alert. I question if this is true as even on highly automated plants there are many jobs for the operators: spare pumps to be changed over, equipment to be prepared for *maintenance*, samples to be taken, plant tours, instruments that are out-of-order to be watched and alarms to be followed up. However, if we feel that the operators need more to do, do not ask them to do jobs that computers can do better. Instead ask them to calculate and graph efficiencies, energy usage, catalyst life and so on or to study *training* material; this is all work that can be put aside if the plant needs their attention; it is the process equivalent of leaving the ironing for the babysitter. See *EVHE*, § 4.2.2 and 4.2.3.

What are the actions that computers can do better than men? They can monitor readings without getting bored and they can take rapid, accurate action in an emergency, if the action can be prescribed in advance. Men are better at diagnosis (though *expert systems* may change this) and at detecting meaning in a fog of confused instructions.

See *human failing*.

Breaking the rules

A report on an accident involving permits-to-work said that supervisors should be reminded in writing of their responsibilities every six months.

Unless there are special local circumstances I cannot support this recommendation. The best way of making sure that people do not break the rules is for managers to carry out regular checks. People doing routine jobs tend to take *short cuts* and if nothing is said it becomes habitual. A friendly word the first time a rule is broken is far more effective than disciplinary action after an accident. Managers will also find out if it is impossible to follow the rules or if people do not understand them or the reasons for them.

Regular checks need not take up a lot of time. Keep your eyes open as you go round the plant; follow a different route each time; if you are talking to the foreman and he is called to the telephone, use the opportunity to look at the permit-to-work book or one of the other safety records. Above all, do not turn a *blind-eye* when you see a rule being broken. (See *audits* and *magic charm*.)

A speaker at a safety conference said, 'We cannot expect the right *attitude* to be self-generating at the operational level'.

To quote from a judge's summing-up that was quoted in several newspapers in October 1968:

> The standard which the law requires is that (employers) should take reasonable care for the safety of their workmen. In order to discharge that duty properly an employer must make allowance for the imperfections of the human nature. When he asks his men to work with dangerous substances he must provide appliances to safeguard them; he must set in force a proper system by which they use the appliances and take the necessary precautions, and he must do his best to see that they adhere to it. He must remember that men doing a routine task are often heedless of their own safety and may become slack about taking precautions. He must, therefore, by his foreman, do his best to keep them up to the mark and not tolerate any slackness. He cannot throw all the blame on them if he has not shown a good example himself.

Foremen are human too and a manager cannot leave all the checking to his foreman. He should check himself, though less often. Senior managers should make occasional checks on junior managers.

In 1621 Robert Burton wrote, 'No rule is so general, which admits not some exception'. True, but exceptions should be authorized and recorded in an approved manner or everyone will be making them when it suits them to do so.

Bretherick

Bretherick's Handbook[1] is a 1852 page compendium of published information on hazardous chemical *reactions*. Most of the book, about 1400 pages, lists chemicals in formula order with details of hazardous reactions they can undergo, either alone or in combination with other substances. This is followed by a section in which classes of chemicals and general topics, such as *auto-ignition temperature* and *oxygen* enrichment, are discussed. There is an index of chemical names.

1. L. Bretherick, *Handbook of Reactive Chemical Hazards*, Butterworths, 3rd edition, 1985

Bridges

See *Tay Bridge* and *team working*.

Brunner Mond

Formed in 1873 by Ludwig Mond and John Brunner to manufacture soda ash at Winnington in Cheshire, Brunner Mond was one of the four companies which combined to form Imperial Chemical Industries (*ICI*) in 1926 and the one which more than the others was responsible for ICI's distinctive style, epecially in safety and employee relations. Both the founders showed a strong commitment to safety from the start. In 1874 a man was killed when a *boiler* burst:

> The death of the workman, however, haunted Ludwig all his life. He saw it as his personal responsibility and never again allowed high pressure steam to be used in his factories[1].

Many years later three employees of the Mond Nickel Company were killed by a leak of carbon monoxide:

> He was on holiday in the South of France when he learned of the accident and immediately his mind flashed back more than a quarter of a century to the day when a workman had been scalded to death by an explosion in a boiler plant at Winnington. The thought that he was responsible for the loss of human life was something he could no longer bear. His instinctive reaction was to order the closure of the nickel works and return the shareholders' capital from his own pocket. Alfred (his son) was more realistic[2]. . .
> Early in the war (the 1914–18 war) his firm had realised the imprudence of making TNT in a densely populated area (Silvertown in the dock area of London) and, as soon as their other works were in full production, had asked to be released from the Silvertown agreement[3].

Unfortunately, the Government refused his release and in 1917 an explosion killed 69 people and injured over 400.

In the Brunner Mond accident book for the 1890s many of the reports are stamped 'Copy sent to Sir John' (Brunner).

The Brunner Mond spirit still survives in ICI.

See *history of safety*.

1. J Goodman, *The Mond Legacy*, Wiedenfeld and Nicholson, London, 1982, p. 41
2. *Ibid.* p. 76
3. *Ibid.* p. 101

Buckets

See *containers*.

Buck-passing

A letter to The *Daily Telegraph*[1] gave a good example of buck-passing. An aeronautical engineer, travelling on a Jugoslav aircraft, noticed that the seats were so close to the four emergency exits that they could not be opened. He wrote:

- To the airline, who did not reply.
- To the UK Civil Aviation Authority (CAA), who forwarded the letter to their counterparts in Yugoslavia.
- To the tour company, who forwarded the letter to the airline.
- To the Airport Director, who said that it was a matter for the CAA and the airline.

In industry responsibilities are clearer – under the UK *Health and Safety at Work Act* (1974) and earlier legislation the employer or 'occupier' is responsible for the safety of employees and members of the public, and the Health and Safety Executive is responsible for enforcing the law – but departments in large companies have been known to pass the buck to other departments. Policy statements, required under the Health and Safety at Work Act, should make responsibilities clear.

Buck-passing of another sort occurs when we blame *abstractions* such as *attitude* or generic terms such as poor *layout* or gland failure for an accident and do not say who should do something about them. *Accident reports* should not *blame* people for past omissions but they should make it clear who should do what in future.

1. L Woodgate, *Daily Telegraph*, 25 September 1987

Buildings

See *layout and location* and *ventilation*.

Burns

See *BLEVE, corrosive chemicals* and *fire*.

Cancer

About a quarter of all deaths in the UK, 140 000 per year, are due to cancer. How many of these are due to exposure at work or to pollution of the atmosphere? According to Doll and Peto[1] about 4% of US cancer deaths have occupational causes and about 2% are due to pollution. Three-quarters of the occupational cancers are lung cancers and most of these are due to *asbestos*.

The position in the UK is probably similar, so about 5000 to 6000 cancer deaths per year have occupational causes. (However, a Royal Society Report suggests that the number is only about 1000.) Whatever the number, they are concentrated among relatively small groups of people, who are exposed to a relatively high risk, and so the detection and elimination of occupational hazards should have higher priority than the absolute number of deaths would suggest. A US report[3] claimed that 23–28% of cancers were occupationally caused but its methods were seriously flawed and it was probably written 'for political rather than scientific purposes'[1].

As cancer takes several decades to develop, the present number of deaths reflects conditions several decades ago rather than today.

Deaths due to pollution are due mainly to combustion products; tobbaco smoke and domestic fires are probably more important than industrial smoke. See *smoking*.

If occupation and pollution do not cause most cancers, what does? Doll and Peto suggest that about a third are due to smoking and that another third could be prevented by changes in diet. Possibly 10% are due to infection and 7% to sexual habits. See *cause and effect*.

See also *COSHH Regulations, fugitive emissions* and *under-reporting*.

1. R Doll and R Peto, *Journal of the National Cancer Institute*, Vol. 66, No. 6, June 1981, p. 1194
2. *Long-term Toxic Effects: A Study Group Report*, Royal Society, 1978
3. *Estimates of the Fraction of Cancer in the United States related to Occupational Factors*, National Cancer Institute, National Institute for Environmental Health Sciences and National Institute for Occupational Safety and Health, 1978.

Canvey Island

Canvey is an island in the Thames Estuary containing many oil refineries and chemical plants. In response to public concern the Health and Safety Executive carried out a detailed *hazard analysis* of the risks to the public[1]. The study can be criticized in detail but the principles it followed make it a landmark in the development of the official attitude to safety in the UK. The report concluded that if the risk to life exceeded a certain value, then it should be reduced. Once it was below this value there was no reason why the plants should not be extended. The value chosen was the subject of much discussion and many people thought that it was too high[2] but the significance of the report is the recognition that we cannot remove all risks completely and that we have to decide what risk is tolerable, taking into account the benefits of the activity and the other risks around us. See *acceptable risk* and *'reasonably practicable'*.

1. *Canvey: An Investigation of Potential Hazards in the Canvey Island/Thurrock Area*, HMSO, London, 1978, 2nd edition, 1982
2. T A Kletz, *Reliability Engineering*, Vol. 3, No. 4, July 1982, p. 325

Care

See *'reasonable care'*.

Carelessness

In a book published in 1932[1] Eric Farmer looked for the causes of accidents entirely within the men injured in each accident. He showed that some men are *accident-prone*, discussed the reasons and asked what could be done. He was too scientific to suggest that they were merely told to be more careful (in this he was ahead of his time) and he wrote:

> Accidents are said to be due to carelessness, but if we ask what carelessness is, we often get an answer that implies that carelessness is something that shows itself in having accidents. To give a name is not to give an explanation and if we are to study the human factor in accidents seriously we must do something more than name mental qualities that are assumed to be the cause of them. We must endeavour by experimental means to see to what extent any mental functions we are able to measure are involved in accident causation.

The author then went on to suggest tests to identify the accident-prone so that they can be placed where they can do least harm. Nowhere is there any hint or suggestion that it might be possible to reduce accidents by altering the plant *design* or method of operation so as to remove opportunities for error. People did not start to think in this way until several decades later.

See *friendly plants, human failing* and *(visit accident) sites*.

1. E. Farmer, *The Causes of Accidents*, Pitman, London 1932. (The author was an investigator for the Medical Research Council's Industrial Health Research Board).

Caring

Many *safety professionals* are motivated by a desire to help their fellow men but they are less effective than they might be because they lack some of the qualities the safety professional needs today, especially in the high technology industries. In particular, they are often 'firstest in asking for the mostest'; they do not realize that we cannot do everything possible to prevent every conceivable accident and therefore we have to allocate our resources, removing some risks but continuing to live with others. (See *hazard analysis*.) They spend lavishly to remove the risks that are brought to their attention, by an accident or in other ways, and ignore greater risks.

They do not realise that the most effective humanitarian is one who allocates his resources so as to maximize the benefit to his fellow men.

When appointing a safety professional, look for the technical abilities that are needed. Do not choose someone just because his heart is in the right place. A desire to do good is not enough. (See *old-timer*.)

Cause

We should use this word sparingly, and instead ask what action we should take to prevent further accidents, for three reasons:

1. Looking for a cause encourages us to list causes that we can do little or nothing to remove. For example, a source of *ignition* is often given as the cause of an oil or gas fire but sources of ignition are difficult to remove completely. If we wish to prevent further fires we should ask why a leak of oil or gas occurred and what can be done to prevent further leaks.

    Similarly *human failing* or human error are often given as the cause of an accident but human errors are difficult to prevent. If we ask ourselves what we should do to prevent further accidents we may think of ways of improving the design or method of operation so as to remove or reduce opportunities for human error. (See *friendly plants*.)
2. The word 'cause' has an air of finality about it that discourages further investigation. To say that a pipe failure was caused by *corrosion* is like saying a fall was caused by gravity. It may be true but it is not very helpful. To prevent further failures we need to know the answers to many more questions: Was the material of construction specified correctly? Was the specified material used? Were operating conditions those assumed by the designer? What corrosion monitoring was requested? Was it carried out? What was done with the results? and so on.
3. The word 'cause' implies *blame* and people become defensive. Instead of saying that an accident was caused by bad design, let us say that it could be prevented by better design. Most people are reluctant to admit they did something badly but are willing to admit that they could do it better.

See *LFA* and *simple causes*.

Cause and effect

In *accident investigation* we start with an effect, such as a *fire*, and look for the *cause* and for means of prevention. When dealing with long-term effects we usually work the other way round; we start with a possible cause such as a chemical and see if it has any effects. D E Broadbent[1] suggests that we treat long-term effects in the same way as we treat immediate ones, that is, start with effects and look for their causes.

It is difficult to open a safety magazine without reading about the 'carcinogen of the month'. Something – a drug or an industrial chemical, a radioactive discharge or an item of equipment such as a VDU – is under suspicion for causing *cancer* or some other disease and there are calls for it to be banned. Sometimes the evidence is so strong that no reasonable person would disagree; sometimes it is rather weak. There may be 'a sincere, almost frantic, effort to seek out the most remote conceivable hazards[2]'. If we feed massive doses of anything to rats it is liable to harm them. If we carry out 20 epidemiological studies to see, for example, if chips cause chilblains, one study is likely, by chance, to show a positive correlation with 95% confidence. More important, we can test or study only a small fraction of the materials we use and we usually test the newer ones. There are many substances in common use in industry or in the home, as foods or in other ways, that have never been tested with the thoroughness with which we test new ones.

Broadbent therefore suggests that instead of spending most of our resources studying the effects produced by various substances we should put more effort into studying the causes of various effects. For example, it is believed that at least 75% of cancers can be prevented, by giving up smoking, by changes in diet or sexual habits and so on, though the precise action we need to take is not always clear. If we put more effort into identifying and publicizing the major causes of cancer we could save more lives than we will ever save by stopping minute discharges of radiation or carrying extensive testing on drugs and pesticides.

It is now estimated that close to one million scientists, lawyers and bureaucrats are occupationally concerned with drugs and pesticides as potential carcinogens, in spite of the fact that with minor exceptions (mostly anti-cancer drugs) such chemicals have not been known to cause cancer in man. Perhaps we might by now have a cure for cancer, had we put half of these people onto cancer research[3].

Broadbent quotes *lead* as another example. Epidemiological studies show that children with high levels of lead in their blood have lower IQs than children with lower levels of lead and the decision has been made to remove lead from petrol, at great expense – £70M per year at 1983 prices – though petrol is only one source of blood lead. Instead of spending all this money removing one cause of low IQ perhaps we should spend some of it investigating and dealing with other causes, such as lead pipes and paint and neurotic symptoms in mothers. Yet there has been no concern in the newspapers or on TV with the damage caused to children in this last way and little or no research on the problem. It is a much more difficult problem than taking the lead out of petrol, but on the other hand the pay-off could be greater. Is it not worth a few million pounds per year? See *'Controversial chemicals'*.

1. D E Broadbent in *Risk: Man-made Hazards to Man*, edited by M G Cooper, Clarendon Press, Oxford, UK, 1985, p. 54
2. J Lederburg, quoted by C E G Smith, in *Risk: Man-made Hazards to Man*, edited by M G Cooper, Clarendon Press, Oxford, UK, 1985, p. 18
3. A Baklien, *Search*, Vol. 12, Nos. 1–2, January/February 1981, p. 30

Caution

Changes often have unforeseen and unwanted side-effects. See *modifications*. It is therefore right to be cautious about changes and check carefully, by experiment, if possible, or by techniques such as *hazard and operability studies*, before going ahead. But how far should we go? Following the thalidomide tragedy, when a new drug caused birth defects in children, stringent testing has to be carried out on all new drugs before they can be used. Many people now believe that the testing goes too far and that the loss to patients outweighs the risk[1]. (See *cause and effect*.) On the other hand changes to vehicle design, which could have as great an effect on health as a new drug, are introduced after no more testing than the manufacturer considers necessary[2]. Whenever governments are involved in safety, and indeed in many other activities, logic flies out of the window. See *cost of saving a life* and *iatrogenesis*.

1. W H W Inman in *Risk: Man-made Hazards to Man*, edited by M G Cooper, Clarendon Press, Oxford, UK, 1985, p. 35
2. M McKay, *Conference on Road Safety, 13 June 1978*, Department of Transport, London

Centrifuges

Centrifuges which handle flammable liquids should be blanketed with *nitrogen* or another inert gas and should be fitted with a low pressure or high *oxygen* alarm to give warning of blanketing failure. Several *explosions* have occurred because nitrogen blanketing was not installed or was ineffective. The source of *ignition* was probably friction between parts of the machine.

One explosion, which killed two men, occurred because a cover plate between a centrifuge and its drive housing was left off after *maintenance*. The nitrogen escaped and there was nothing to show that the oxygen content was high. A high oxygen or low pressure *alarm* is better than a low flow alarm as the latter would not have prevented this incident. We should always measure directly the property we need to know.

Another explosion occurred because the nitrogen supply failed and again there was no monitoring. A third explosion occurred because the nitrogen flow was too small. The nitrogen flowmeter had a full-scale deflection of 60 l/min but 150 l/min was needed.

It is not necessary to remove every trace of oxygen. 10% is needed for an explosion (unless hydrogen is present) so high oxygen alarms should be set at 5%.

Centrifuges may be damaged if they are connected to the electricity supply so that they turn the wrong way.

Other hazards of centrifuges have been documented by the Institution of Chemical Engineers[1].

1. *User Guide for the Safe Operation of Centrifuges*, Institution of Chemical Engineers, Rugby, 2nd edition, 1987

Chains

Most accidents are the result of a chain of events which could have been broken at any point. Many people could have prevented the accident and must therefore share the responsibility for it, morally if not legally.

For example, suppose that two men who were digging a hole in a factory went home and left it unfenced; several people passed by and saw the hole but did nothing and later that day a man fell in and was injured. The accident would not have occurred if:

- The men digging the hole had not left it unfenced.
- Their supervisor had reminded them to fence it before they left and had made sure that suitable materials were available.
- Those who passed by had done something.
- The injured man had looked where he was going.
- The managers had encouraged the development of a climate in which people did not leave jobs in an unsafe condition or ignore the hazards they saw. One suspects that other jobs had been left in an unsafe condition before and no manager had commented or done anything, so sloppy methods became a habit. Perhaps the managers had not made responsibilities clear. We do not live or work in a vacuum and the actions we take depend on the examples we have been set by those in authority. See *blind-eyes*.

A more technical (but imaginary) example is summarized in Figure 3. A bellows was incorrectly installed so that it was distorted. After some months it leaked and the escaping vapour was ignited by a passing vehicle. Damage was extensive as the surrounding equipment had not been fireprotected, to save cost.

The leak would not have occurred, or the damage would have been less, if:

1. Bellows were not allowed on lines carrying hazardous materials.
2. The use of bellows had been questioned during design. Was a *hazard and operability study* carried out?
3. The fitter who installed the bellows had done a better job. Did he know the correct way to install a bellows and the consequences of incorrect installation?
4. There had been *inspection* after *construction* and regular inspections of items of equipment whose failure could have serious consequences.
5. Gas detectors (see *analysis*) and *emergency isolation valves* had been installed.
6. The plant had been laid out so that vehicles delivering supplies did not have to pass close to operating equipment. (See *diesel engines*.)
7. There had been better control of vehicle movements.
8. The fire protection had been better.
9. An expert in process safety was involved during design, as he would have drawn attention to items 3, 5, 6, 7, and 8.

There were thus at least nine ways in which the chain of events leading to the damage could have been broken and many people who could have

60 Chains

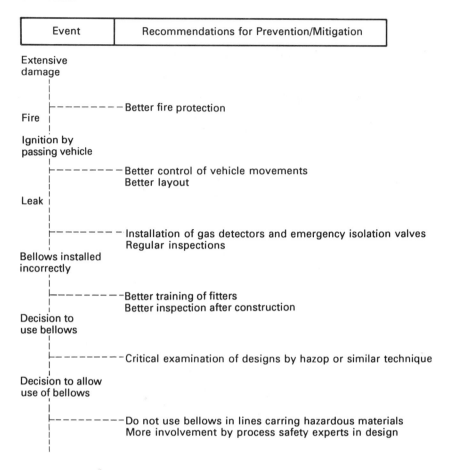

Figure 3 An example of an accident chain

broken it. They were all responsible to some extent and it would be wrong and unfair to pick on one or two of them to make into culprits. Unfortunately, after a serious accident the press thinks there must a 'guilty man' who is responsible and ask why there is no *prosecution*. There are also managers who lay a *smoke screen* over their own *responsibility* by blaming workmen who broke a rule or did a bad job, though they never enforced the rule or *audit*ed workmanship.

In *LFA*, I discuss a number of accidents in detail, illustrate them with similar diagrams, and show that *accident investigation* is like peeling an onion. Beneath one layer of causes and recommendations there are other layers. The outer layers are the immediate technical recommendations; the inner layers deal with ways of avoiding the hazard and with improvements to the management system. See *Challenger*.

Challenger

In January 1986 the US space shuttle Challenger was destroyed by fire soon after take-off and the crew of seven were killed. The US space programme was delayed by several years[1,2,3].

The immediate cause was the failure of a joint between two sections of a rocket motor which in turn was due to the failure of an O-ring. However, the underlying causes went much deeper and the accident provides a very good illustration of the concept of layered *accident investigation*. (See *chains*.)

The key errors are listed below. The failure of the O-ring was due to earlier errors by the engineering staff, which in turn were due to management failings, which were the result of political decisions.

Engineering errors

- Signs that the O-rings were unsatisfactory were not followed up.
- A segmented design was chosen for the rocket casing instead of a one-piece design.
- Inspection standards were allowed to slip.

Management errors

- Engineering judgements were overruled by a need to cut costs and meet deadlines. Senior managers were warned that the seal design was unsatisfactory, particularly at January temperatures. They were also warned that quality control was slipping.
- Key staff were overworked and overstressed.
- The feeling of being in a risky enterprise became dulled. A series of successful flights had made the operations seem almost routine and flights were allowed to continue despite known deficiencies in the design. See *complacency* and *familiarity*.

Political decisions

- President Nixon decided to build a shuttle, not because it was technically the best solution, but because he saw it as quick project from which he could reap some credit. Unlike President Kennedy he was not a space enthusiast and was less generous with funds. His successors were also unenthusiastic.
- Top men were changed whenever the president changed.

After the accident, the engineers who had tried to make the problems known were punished, but not the managers who ignored their advice.

1. J J Trento, *Prescription for Disaster*, Crown, New York, 1987
2. B Tolley, *European Safety and Reliability Newsletter*, Vol. 4, No. 2, August 1987, p 14
3. *The Rogers Report,* The official report on Challenger, US Government Printing Office, June 1986.

(absence of) Change

See *novelty*.

Change – new problems

> We have become amphibious in time. We are born into and spend our childhood in one world; the years of our maturity in another. This is the result of the accelerating rate of change.

This quotation from the novelist, B W Aldiss, might apply to industry. Anyone who has worked in industry for 20 years or more has moved, like frogs coming ashore for the first time, into a different world where there are new problems.

ICI's safety record worsened during the 1960s (See *ICI*, Figure 35.) A new generation of plants, larger than those built hitherto, produced new problems. Because the new plants were much bigger, and temperatures and pressures higher, breaking into a line that contained flammable material became more hazardous than it used to be. If we break into a 3 inch line and it is full of oil we may get away with it. If we break into a 12 inch line we probably will not. After a fitter broke into a line that was full of hot oil, new procedures for the preparation of equipment for maintenance were introduced.

It took the staff a few years to recognize that there had been a step change and that instead of piecemeal improvements after each accident a new approach to safety was needed, the *loss prevention* rather than the traditional safety approach. Figure 35 shows the success achieved when loss prevention was practised.

In the 1980s the growth of computer control and biotechnology produced new problems, many of which are not yet fully appreciated. It is easier to see change in retrospect than when it is occurring. It is hard to see the signal above the noise when we do not know what signal to look for and we rather hope there is no signal at all.

See *actions, boilers, failures* and *modifications*.

Change – old problems

There is a story about an economics professor who set the same examination questions every year. The students knew what the questions would be, but the answers the professor wanted changed every year!

In safety the answers to old problems also change, though not so quickly, for several reasons:

- New knowledge becomes available. For example, we now know far more than we did a few years ago about the *dispersion* of heavy gases in the atmosphere or the size of *relief device* required to vent a *runaway* reaction.
- Standards are rising. We now aim for lower concentrations of toxic gases in the atmosphere than we did a few years ago. See *COSHH Regulations*.
- New techniques are developed, such as *hazard and operability studies* for identifying problems and *hazard analysis* for assessing them.

- Most important of all, old ideas are seen to be unsatisfactory. The pathways of history 'are strewn with the wrecks' of once known and acknowledged ideas, discarded by later generations[1]. (See *culture*.) One of these 'truths' is the view that accidents are due to *human failing*.

Despite all the changes, some problems remain much the same. Accidents still occur, for example, because vessels are prepared for *entry* without sufficent thought. Either the factory rules are poor or are they are not followed. *WWW*, §11.3(e) describes an accident that occurred in 1910 and again in 1955 and which is similar to accidents that occur today.

1. F J Turner, quoted by B Tuchman, *Practicing History*, Ballentine Books, New York, 1982, p. 58

Check lists

Check lists are often used for identifying problems: tick those that apply, cross off those that do not. The disadvantage of a check list is that new problems, often the most important problems, are not on the list and are not brought forward for consideration. For this reason the process industries have come to prefer the open-ended technique of *hazard and operability studies*. Check lists are OK if we are repeating a design we have used many times before. All the problems have probably been identified and we just need reminding of them. But if we are innovating there will be some new problems which will not be on the check list.

Check lists are, of course, useful if we have a series of complex or unfamiliar tasks to perform, starting up a plant, for example, or carrying out an *audit*. If, however, the plant is started up every few days it is unrealistic to expect the operators to tick off a check list as they go along, even though they do make occasional mistakes. If we insist, they will simply complete the check list at the end of the shift. (I may occasionally forget to push in my choke after I start my car, but I would not be willing to complete a check list.)

Chemstar

See *small companies*.

Chernobyl

The world's worst nuclear accident occurred at Chernobyl in the Ukraine in 1986 when a water-cooled reactor overheated and radioactive material was discharged to atmosphere. Although only about 30 people were killed immediately several thousand more may die during the next 30 years, a one-millionth increase in the death rate from cancer in Europe.

There were two major errors in the design of the Chernobyl reactor, a design used only in Russia:

Figure 4 The Chernobyl nuclear reactor, at low outputs, was like a marble on a convex surface. If it started to move gravity made it move increasingly fast. Other reactors are like a marble on a concave surface

1. The reactor was unstable at outputs below 20%. Any rise in temperature increased the power output and the temperature rapidly rose further. In all other commercial nuclear reactor designs a rise in temperature causes a fall in heat output. The Chernobyl nuclear reactor, at low outputs, was like a marble balanced on a convex surface. If it started to move, gravity made it move increasingly fast. Other reactors are like a marble on a concave surface. See Figure 4.
2. The operators were told not to go below 20% output but there was nothing to prevent them doing so.

The accident happened during an experiment to see if the reactor developed enough power, while shutting down, to keep auxiliary equipment running during the minute or so that it took for the *diesel* generators to start up. There was nothing wrong with the experiment but there were two major errors of judgement in the way it was carried out: the plant was operated below 20% output and the automatic shutdown equipment was isolated so that the experiment could be repeated. When the temperature started to rise, it rose a hundred fold in one second.

The main lessons to be learnt from Chernobyl are therefore:

- Protective equipment should not be isolated and basic safety rules ignored. Perhaps the operators believed that the instructions to carry out the experiment overrode the normal safety instructions. See *contradictory instructions*.
- Plants should be designed so that crucial safety rules cannot be ignored.
- Complex plants cannot be controlled just by following the rules. Problems may arise which were not foreseen when the rules were written. Operators should be trained to diagnose faults and work out the action required, that is, to apply skill-based behaviour, not just rule-based behaviour. (See *human failing*.)
- Plants should be *audit*ed by independent outsiders. The Russians have no equivalent of the UK Nuclear Installations Inspectorate.
- Plants should be designed with negative power coefficients, i.e. so that a rise in temperature causes the heat output to fall.
- Plants should, whenever possible, be designed so that they are not dependent on added-on protective systems which may fail or may be neglected. See *nuclear power* and *Three Mile Island*.
- When planning experiments, list possible outcomes and decide what action to take. See *foresight*.

For more details see *LFA*, Chapter 12.

Chlorine

The acute toxicity of chlorine is well known. The data are reviewed in *Chlorine Toxicity Monograph*[1] which concludes that in the absence of subsequent medical treatment exposure to 400 p.p.m. for 30 minutes will kill half of those exposed. However, later work suggests that the risk may be lower. For other exposure times and concentrations:

$ct^{0.5} = 2200$

where: c = concentration for 50% lethality in p.p.m.
t = time in minutes.

Very young, old or unhealthy people are more vulnerable.

Concentrations for other lethality levels are not known with certainty but no deaths have been observed at 30 minutes exposure time at or below 50 p.p.m. for any species. Withers and Lees[2] suggest that other lethality levels can be estimated from the *probit* equation:

Probit $= 0.92 \ln[c^2 t] - 8.29$

If death does not occur, complete recovery from the effects of chlorine is possible.

If a chlorine leak occurs, it is better to stay indoors with the windows shut than try to escape.

Up to 1952 six major leaks of chlorine killed 115 people. There have been no major leaks since then. With one exception, all the deaths occurred within 400 m of the leak.

A method of deciding how far to go in preventing chlorine leaks from manufacturing plants is described by J G Sellers[3]. All leakage points are identified, the leak rate estimated and the concentration at places to which the public have access calculated, taking wind strength and direction into account. A leak which could cause a nuisance to the public is considered an *acceptable risk* if it occurs not more than once a year. For a leak which causes some distress but no permanent harm, the figure is not more than once in 10 years and for a leak which could cause injury or loss of life the figure is not more than once in 100 years. This is roughly equivalent to an average risk of death of 10^{-7} per person per year, amongst the exposed population.

In the manufacture of chlorine, leaks of mercury have caused long-term toxic effects, particularly in Japan. New plants now use membrane or diaphragm cells instead of the older mercury cells.

The use of chlorine for purifying water has saved countless lives from cholera and other diseases but now there is concern that the chlorine might react with traces of organic chemicals in water and produce carcinogenetic substances[4].

See *cylinders, dead-ends* and *toxicity*.

1. *Chlorine Toxicity Monograph*, Institution of Chemical Engineers, 1987
2. R M J Withers and F P Lees, *Journal of Hazardous Materials*, Vol. 12, 1985, pp. 231 & 283
3. J G Sellers, in *Process Industry Hazards – Accidental Release, Containment and Control*, Institution of Chemical Engineers, 1976, p. 27
4. *Controversial Chemicals*, edited by P Kruus and I M Valeriote, Multiscience Publications, Montreal, Canada, 2nd edition, 1984, p. 58

Choice of problems

There are always more *problems* awaiting our attention than we have time or money to solve. How do we decide which to investigate first? Sometimes we make curious choices. For example, since *Flixborough*, there has been a flood of papers on the probability of a *leak* and the behavior of the leaking material: how long it will take to disperse in various wind and weather conditions, what pressures will be developed if it explodes, what heat will be radiated if it ignites, and so on. Expensive experiments have been carried out. In contrast, little thought has been given to the reasons for the leaks and the action needed to prevent them. Yet this work would be much cheaper to carry out. See *LFA*, Chapter 16.

Investigating the problems that are easy or interesting to investigate rather than those that most need investigation is often considered an academic failing but in this case the experiments have been paid for by industry.

Just as there is never enough time and money to investigate all the problems on our agenda so there is never enough money to remove all the hazards. We have to decide on our priorities and the methods of *hazard analysis* may help us.

See *cause and effect, lead, overreaction, perception of risk, perspective* and *problems*.

Chokes

The main causes of chokes in pipelines are:

- Deposition of solids.
- Rubbish left after construction or maintenance.
- Failure of the heating or inadequate heating on lines carrying materials (including *water* and *benzene*) which may be solid at ambient temperature. The lines are unlikely to freeze when the liquid is flowing but may freeze as soon as flow stops.
- Unforeseen reactions.

An example of the last item is the solid hydrates formed between light hydrocarbons and water. Propane hydrate caused the choke at Feyzin (see *BLEVE*) which cleared suddenly, causing a serious leak and fire. On another plant hydrates caused a choke in an 18 inch blowdown line. Wet condensate was blown down for a much longer period than was foreseen during design. The choke was cleared by external steaming. When it cleared the sudden movement of the plug caused the blowdown line to fracture at a T-joint where it joined a larger line.

A *blind* was made concave by the impact of a plug of solid pushed along by gas pressure. If a 0.5 kg (1 lb) plug in a 2 inch line is moved 15 m (50 ft) by gas at a gauge pressure of 3.5 bar (50 p.s.i.) its exit velocity will be nearly 200 km/h (300 m.p.h.). These incidents show that chokes should not be cleared by gas pressure. If external steaming is used the upstream pressure should be blown off. If lines are dismantled, beware of *trapped pressure*.

The results of an attempt to free a stuck pig show the power of compressed gas. Air at a gauge pressure of 7 bar (100 p.s.i.) was used, the pig chamber door, which was not secured properly, was blown off and the pig travelled 230 m, hitting various objects on the way[1].

Many incidents have occurred because vents were choked. They should be adequately heated and regularly inspected and, in addition, if vents are dependent on steam heating or are liable to choke for any other reason, an emergency vent such as a blow-off panel should be provided. See *LFA*, Chapter 7.

Trace heating, like all *protective equipment*, should be inspected frequently to check that it is in operation.

Stacks carrying cold gas have been choked with ice from steam that was passed up the stack to disperse the gas or even from spray from an external steam ring at the top of the stack. See *WWW*, §2.5, 6.2.

1. *Petroleum Review*, July 1983, p. 27

CIMAH Regulations

After the explosion at *Flixborough* in 1974 the UK Government set up two committees, one to enquire into the immediate causes of the disaster and a second, the Advisory Committee on Major Hazards, to consider the wider issues. This Committee issued three reports (1976, 1979 and 1984) which resulted in 1984 in the Control of Industrial Major Accident Hazard Regulations (CIMAH)[1]. These require companies which have more than defined quantities of certain hazardous chemicals in storage or process to prepare a 'safety case' which lists the hazards, shows how they have been identified and describes the action, including emergency plans, taken to control them. The public have to be told about the risks. The Regulations apply when there is, for example, more than 25 tonnes of *chlorine* in a plant or 75 tonnes in storage.

Following the toxic release at *Seveso* in 1976 the European Community issued the so-called Seveso Directive[2]. The CIMAH Regulations apply this directive to the UK.

The CIMAH Regulations provide a good illustration of the UK methods of legislating for safety. Extensive consultations take many years but result in regulations that, by and large, are accepted by those to whom they apply.

In another country with similar regulations:

> A difficulty has arisen in some cases where the studies appear to have been aimed at proving that the plant is satisfactory, rather than open-mindedly investigating whether it is. This can be a problem . . . and requires expertise and determination . . . to penetrate the glossy facade presented by the resulting reports.

See *COSHH Regulations* and *indices of woe*.

1. *A Guide to the Control of Industrial Major Accident Hazard Regulations 1984*, Booklet HS(R)21, HMSO, London, 1985
2. European Community, Council Directive of 24 June 1982 on the Major Accident Hazards of Certain Industrial Activities, *Official Journal of the European Communities*, 1989, No. L230, 5 August 1982, pp. 1–18

Cleanliness

... industrial health has become a more popular subject, both in fact and fiction, and is increasingly the concern of press, radio and television. For this reason, it is perhaps important to state that over large areas of industry any risk to health can be eliminated by the vigorous application of soap and water, i.e., by reasonable cleanliness.[1]

According to Lord Sieff, former chairman of Marks & Spencer:

You can tell a great deal about the general standards of a firm by looking at the standards maintained in their washrooms and lavatories. If they are poor, standards in the factory as a whole will often be poor. When visiting suppliers, I nearly always ask to see the toilet facilities[2].

A clean plant is safer than a dirty one as cleanliness encourages tidy and systematic methods of working. See *tidiness*.

If *solvents* are used for cleaning floors or equipment, use non-flammable, non-toxic ones, not a process liquid that happens to have good solvent properties. Some painters ran out of the detergent they used for cleaning their equipment so they helped themselves to some kerosene from the plant. They took it to a point where it was above its *auto-ignition temperature* and it caught fire.

1. *Annual Report of HM Chief Inspector of Factories for 1971*, HMSO, London, 1972, p. xiii
2. M Sieff, *Don't Ask the Price*, Collins, London, 1988, p. 224

Codes of practice

These distillations of accumulated experience play a large part in safety and loss prevention, particularly in the UK where most regulations do not give detailed instructions on what should be done but instead define an objective to be achieved. (See *Health and Safety at Work Act*.) The regulations are often accompanied by an approved code of practice, in which case, failure to follow it is *prima facie* evidence of failure to comply with the regulations. However, the employer can argue that the code is inapplicable to his circumstances and/or that he is doing instead something that is as safe or safer. If there is no approved code then failure to follow any generally accepted industry code is evidence of failure to provide a safe plant or system of work but the evidence is not as strong as with an approved code.

Many companies have their own codes of practice. *Factory Inspectors* are in a strong position when they point out to a company that it is not following its own codes.

Do not relax if you are following all the codes; they may be out-of-date. See *rules and regulations*.

Lees, Appendix 4, lists UK and US standards and codes.

Cognitive dissonance

> We received the impression that many delegates were present not to share knowledge but merely to observe rivals' progress; as a result attitudes to novel processes were somewhat entrenched. Many industrial delegates seemed sceptical to the point of myopia regarding the viability or worthinesss of new ideas[1].

This report is an example of cognitive dissonance (an unpleasant noise in the mind), a term used by psychologists to describe our reactions to information which upsets our beliefs and, if correct, means that we have to change our ways. So at first we deny it, if we can, and then try to play down its importance.

Safety professionals often meet this reaction. Some new information, perhaps an explosion in another company, shows that that perhaps we ought to modify our designs. The design engineer is not so sure. Perhaps the facts have not been reported correctly. Their design is not quite the same as ours; in fact, it differs in critical features. Their operating team is less experienced. Their factory is more congested. Their legal system is different. The new equipment required is expensive and difficult to obtain. We knew all along that an explosion was possible if the plant was not operated correctly, so what's changed? These may be fair comments but they can easily become cognitive rigidity – 'Don't confuse me with the facts; my mind is made up'.

How should we react? With *patience*. We cannot expect people to discard in a few minutes or even a few weeks the beliefs they have held for years; seduce them by degrees. Answer their questions patiently and in time they may come round to a different view. However, they may not.

Many years ago I was a member of the team investigating an *explosion* in a compressor house, the one described under *insularity*. One of our recommendations was that the compressor houses on future plants should have no walls so that leaks could be dispersed by natural *ventilation*; many of the staff agreed. Other companies already had open compressor houses and found that working conditions were tolerable, even in the winter, and that maintenance quality was not affected. The chief engineer, who had been designing closed compressor houses for years, was not convinced and no open compressor house was built until he retired. See *LFA*, Chapter 4.

The opposite of cognitive dissonance is *overreaction* and after an accident it is just as common. Before we blame our colleagues for rigidity and unwillingness to change their views let us be sure we are not overreacting to an incident.

See *action, innovation, myths* and *underestimated hazards*.

1. A Crerar and R Low, *The Chemical Engineer Diary and News*, November 1987, p. 4

Coincidences

We are often told, after an accident, that it occurred because several safety devices or procedures failed simultaneously, an unlikely coincidence that could not reasonably have been foreseen and so no one can be blamed. In fact, what usually happens is that all the safety measures are neglected and left for long periods of time in a failed state. When the final triggering event occurs the accident is inevitable.

For example, a *tank* was overfilled and some of the contents were lost to drain. Three precautions had been taken to prevent this occurring:

- The operator was supposed to watch the level but he never did so. He waited until the high level *alarm* sounded.
- The tank was surrounded by a bund but the bund drain valve had been left open; it was never checked.
- When the high level *alarm* failed – it was bound to fail after a year or two – a spillage was inevitable.

Another example: when an oil tanker was struck by lightning the vapour coming out of the tank vent was set alight and the flame travelled back along the vent line into the tanks where an *explosion* occurred, killing two men. Three independent safety systems would have prevented the explosion:

- *Nitrogen* blanketing: it was not being used.
- A *flame trap* in the vent line: it was fitted incorrectly, with a gap round the edge.
- A pressure/vacuum valve: this opens only momentarily to discharge vapour and the flow rate is too high for the flame to travel back but a by-pass round the valve had been left open[1].

In both these cases there was no coincidence of instantaneous events. There were two or three ongoing unrevealed faults, which stayed unrevealed for long periods of time because there were no regular *tests* or *inspections* of *protective equipment* or *methods of working*. When the alarm failed in the first case, and lightning occurred in the second case, a spillage or explosion was inevitable. (See *astonishment*.)

S J Gould puts forward a different view of coincidences[2]:

> Many events, although they move forward with accelerating inevitability after their inception, begin as a concatenation of staggering improbabilities[2].

However, he is writing about evolution where time is unlimited and, if we wait long enough, unlikely events can occur.

1. *Hazardous Cargo Bulletin*, Vol. 4, No. 7, July 1983, p. 8
2. S J Gould, *Hen's Teeth and Horses's Toes*, Penguin Books, 1984, p. 263

Cold readings

A cold reading is a statement that is always true and always safe to say. Thus a doctor, if he cannot diagnose, might say, 'It would be a good idea to give up smoking and lose a little weight'. Similarly, managers asked to comment on safety tend to produce *platitudes* such as, '. . . matter of major importance . . .' or '. . . responsibility of all employees. . .'; trade union officials ask for more *Factory Inspectors*, bigger fines (see *prosecution*) and greater commitment by *management*. *Safety professionals*, when asked for their opinions, tend to ask for more resources and support whoever is asking for the most.

From time to time we should look at our own statements and see if we are addressing specific problems or just producing cold readings.

Combustible gas detectors

See *analysis*.

Combustion

See *oxidation*.

Common mode failures

See *redundancy*.

Communication

In a small organization everyone knows what everyone else is doing without making any special effort. In a large organization this is not the case; systems have to be set up so that people are told at least what they *need to know*. These systems do not always work very well, so far as as safety is concerned. An accident in one factory or department has lessons for others. Often the right people are not told, or do not get the message or soon forget.

If they do not get the message, perhaps it is not clear. Often the essential message is lost inside a mass of detail, of no interest to other factories or departments. The *safety professional* should not just circulate an *accident report* as it stands but should pull out the essential messages and rewrite them. His readers are busy people who will look at his reports in odd moments so he should write clearly, so that 'He who runs may read', not so that he who reads will run away. We should give our safety bulletins as much care, in preparation and presentation, as we give our sales literature. (See *language*.)

If the right people are not told, perhaps the report is sent to the factory manager or head of department. A busy man, he may not have time to read everything. Aloof from the detail he may not see that there is a message for some of his staff. Reports should be circulated directly to those who are able to act on them. During my time as a safety adviser in industry I circulated a monthly newsletter directly to all levels of management and many copies were seen by foremen and operators.

Many people are not great readers. Reports get put aside to be read when they have time and are subsequently forgotten. So staff should be brought together from time to time to discuss accidents and safety problems; discussion is more productive than lecturing. See *training*.

There should be some system for reminding people of the lessons of past accidents. Old accidents as well as new ones should be described in safety bulletins and discussed at safety meetings.

If communication inside a company is bad, communication between it and other companies is worse. Many companies are reluctant to publish their accident reports, or send them to other companies. Nevertheless they should do so so that we can all learn from each other's mistakes. See *publication*.

At a different level, accidents have occurred because people misheard each other or failed to understand each other. For example:

- An operator asked to shut down pump JA1001 thought that the foreman said J1001.
- A tanker driver who collected a load of 'slops' from a refinery thought they were dirty water; he did not realise that they were flammable.
- Ice and 'dry ice' (solid carbon dioxide) have been confused.
- Over the telephone an analyst reported that a sample was OK. Unfortunately he had not carried out the analysis requested. Safety analyses should be accepted only in writing.

See *WWW*, Chapter 4 and §17.5.

Comparisons

Comparing the safety record of one company, or part of a company, with another is full of pitfalls. (See *accident statistics*.) For example:

- A higher *lost-time accident rate* may be due to a greater readiness of injured men to lose time, rather than a higher accident rate.
- The minor accident rate may be higher because the ambulance room is nearer the plant, or the nurse is prettier. (See *over-reporting*.)
- The dangerous incident rate can be reduced by not counting borderline incidents.
- Marks may be given for *audit* results but different plants and companies have different problems.

On the other hand, comparisons of the same company or plant over periods of time are much more valid. There is no doubt that the trend in *ICI*'s fatal accident rate indicated a real worsening followed by a real improvement (see Figure 35).

Some companies spend less on safety than others. If you are told that you spend more than another company, ask how their fire and accident records compare?

If we are assessing a risk we should compare it with the alternatives. For example, the risks of one form of transport with the risks of others and of manufacture on the spot, the risks of saccharin with those of sugar, the risks from nuclear waste with those from coal tips (see *Aberfan*), the risks of making nylon (see *Flixborough*) with those of cotton and wool production. The latter are higher as agriculture is a high-risk industry. (See *alternatives* and *operations research*.)

Complacency

When there have been no accidents for some time people become complacent, relax precautions and an accident often occurs. During the 1939–45 war a factory produced a very toxic chemical. When accidents were steady at a moderate level, the managers were happy. When the accidents were very few and far between they got worried. Experience had shown that people became careless and then there would be a sudden jump to one or more serious accidents[1]. See *Challenger* and *familiarity*.

Today we would not, of course, try to steady out accidents at a moderate rate in order to keep people alert. We would try to make the accidents as few as possible and would try to keep people alert in other ways. For example, by reminding them of accidents that had happened in the past – this can be done by discussing them regularly and describing them in safety bulletins (see *communication*) – and by involving people in *audits* of the workplace and of operating procedures. At least that is what we should do; do we always do it? Left to themselves people have short memories, but there is a lot we can do to remind them of the hazards around them and keep them alert. See *training*.

1. A H Little, *Chemistry in Britain*, Vol. 23, No. 4, April 1987, p. 323

Compressed air

See *air*.

Compromise

In technology (but not politics) compromises are sometimes worse than either extreme. For example, suppose it is possible to reduce the size of a liquid phase reactor by increasing the temperature and pressure. A very small reactor, operating at high temperature and pressure may be safe as it contains so little hazardous material. (See *intensification*.) A very large reactor, operating at low temperature and pressure, may be safe for a different reason: because of the low pressure, leaks are unlikely and any that do occur will be small; because of the low temperature evaporation will be low. (See *attenuation*.) A compromise solution, a moderate size reactor, operating at moderate temperature and pressure may combine the disadvantages of the extremes. If a leak occurs the pressure may be big enough for a large leak rate, the temperature may result in considerable evaporation and the inventory may be big enough for a serious explosion or toxic incident.

Another example: strong buildings can survive earthquakes. So can very weak ones such as traditional Japanese wood and paper houses. Intermediate strength buildings, such as heavy mud block or stone and timber walls with heavy roofs – typical housing in the earthquake zones of the Mediterranean and Latin America – are about the worst possible for an earthquake area[1].

For another example see Hazop and Hazan[2].

1. A Wijkman and L Timberlake, *Natural Disasters – Acts of God or Acts of Man?*, International Institute for Research and Development, London, 1984, pp. 87, 88
2. T A Kletz, *Hazop and Hazan – Notes on the Identification and Assessment of Hazards*, Institution of Chemical Engineers, Rugby, UK, 2nd edition, 1986, §3.6.5

Compressor houses

See *ventilation*.

Computers

Computers are in increasing use for plant control. A survey of incidents on computer-controlled plants[1] (see *WWW*, Chapter 20) shows that:

- Hardware failures on the computer itself are rare, though not unknown[2]. Most hardware failures occur on the measuring devices and valves connected to the computer; the failure rate of a computer-controlled system is therefore similar to that of a conventional system.
- Thorough testing is needed to detect software errors. It may take twice as long as design[3].
- The commonest cause of incidents is using the computer as a black-box which will do what we want it to do without the need to understand what goes on inside it. It is not necessary to understand the electronics but it is essential to understand the logic, and *hazard and operability studies* should be carried out on the instructions to the computer as well as the line diagrams.
- Another common cause of incidents is failure to assess the way people will react to the computer if, for example, it gives them more information than they can handle in the time available.
- Other possible causes of incidents are incorrect entry of data and failure to tell those concerned that changes have been made.

The computer does not introduce new causes of error but it does provide new opportunities for old causes. Accidents can occur on manually controlled plants if operators are given instructions that have not been clearly thought through. Operators, however, may do what you want them to do even if your *instructions* are not entirely clear – people are good at detecting meaning in a fog of imprecise verbiage – but a computer can do only what it is told to do.

Computers can also be used for:

- The storage and retrieval of information.
- The analysis of *alarms*, to identify the origin of the upset.
- *Expert systems*.

See *timebombs* and *viruses*.

1. T A Kletz, *Plant/Operations Progress*, Vol. 1, No. 4, Octber 1982, p. 209
2. I Nimmo *et al*, *Safety and Reliability Society Symposium*, Altrincham, November 1987
3. J Love, *The Chemical Engineer*, No. 443, December 1987, p. 36

Confidence limits

If someone is estimating the probability of an accident (see *hazard analysis*) he may quote confidence limits. However, the meaning of these limits is not always clear. They show the errors that may be present in the estimate because sample size produces uncertainties in the data used. They do not allow for errors in the logic or failures to identify all the ways in which the accident can occur. These are usually much more important than errors in the data[1].

Estimates of the consequences of an accident often differ greatly, especially when gas dispersion is involved (see *Lees*, Chapter 15), but as methods of calculation improve, the estimates are tending to converge.

In estimating the consequences of a *leak* of flammable or toxic gas or liquid, the biggest uncertainty is usually the size of the hole and the consequent rate of release. Time is often spent estimating the *dispersion*, or the *explosion* over-pressure, with increasing accuracy, though the quantity released is no more than a guess.

In one of the lesser-known works by the author of *Alice in Wonderland*[2] a boy says that there are about a thousand and four sheep in a field. Someone corrects him: 'You mean about a thousand. You can't be sure about the four'. The boy says he is sure about the four and he points out a group he has counted; it is the thousand he isn't sure about.

1. T A Kletz, *Hazop and Hazan – Notes on the Identification and Assessment of Hazards*, Institution of Chemical Engineers, Rugby, 2nd edition, 1986, Chapter 4
2. L Carroll, *Sylvie and Bruno Concluded*, 1894

Confidentiality

See *publication*.

Construction

Many accidents occur during construction; others occur after construction because the construction team did not follow the *design* or *good* engineering *practice*.

Accidents during construction

The construction industry has a poor accident record. This is due to poor *management*, not to the inherent hazards of the task, as some in-house construction teams, employed by large chemical companies, have good accident records, typical of the chemical industry rather than the construction industry.

In the process industries many accidents occur during the construction of extensions or modifications to existing plants because the client does not give adequate warning of the hazards or set up and enforce a proper *method of working*. Giving a rule book to the construction foreman is not enough. The rules must be explained to those who will have to follow them. See *contractors*.

Some construction hazards could be avoided by giving more thought to the hazards of construction during design. The designer should ask, 'Is it safe to build?'

For example, on a chemical plant which was being extended, new equipment had to be lifted over operating plant. If it fell there would be a leak of flammable gas, a fire, and perhaps an explosion. The construction team were prepared to lift it but the plant management worked out the probability that the *crane* would drop its load, decided it was too high and shut down the plant for a few days. If the designer had been aware of the hazard he might have located the new equipment elsewhere.

While building a box girder bridge over the Yarra river in Melbourne, Australia in 1970 the construction workers forced together components which did not fit; a common practice. The designers had not told them that with this type of bridge, then new, components which do not fit should be modified, not forced together. The bridge collapsed during construction. The design took little account of construction problems. See *team working*.

Insulation had to be stuck on to some pipes which were in a trench, close to the ground. Men had to lie in the trench breathing the heavy, toxic, flammable vapour given off by the adhesive. Whoever decided to use this method of insulation gave no thought to the hazards of application. On one occasion some mastic dropped into someone's eye. Although he should have worn goggles, the accident was due to poor design as well as *human failing*.

Accidents after construction

Many accidents have occurred because construction teams did not follow the design or did not construct well (in accordance with that rather elusive quality, good engineering practice) details that were not specified in the design but left to the discretion of the construction team.

For example, a compressor house was designed so that the walls would blow off at a low pressure, if an explosion occurred inside, and thus minimize damage to the equipment. (See *explosion venting*.) The walls were made from plastic sheets secured by weak fasteners. The construction engineer did not understand the design philosophy and substituted stronger fasteners. When an explosion occurred the walls did not blow off until the pressure was higher than intended and the resulting damage was greater than it should have been. (See *instructions*.)

The most effective action we can take to prevent pipe failures, the cause of most large *leaks*, is to specify designs in detail and then inspect thoroughly, during and after construction, to make sure that the design has been followed and that details not specified in the design are in accord with good engineering practice. The construction inspector has to look out for malpractices that no one would ever dream of prohibiting specifically such as:

- Drain pipes or relief valve tail pipes so close to the ground that they could be blocked by dirt or frozen puddles.
- Pipes touching the ground.
- Temporary supports not removed.

- Leaks from flanges dripping onto cables.
- Support springs fully compressed or extended.

In *LFA*, Chapter 16 many more similar items are listed.

Describing the pyramids, Flinders Petrie wrote, '. . . on the one hand the most brilliant workmanship was disclosed, while on the other hand it was intermingled with some astonishing carelessness and clumsiness[1]'.

As Rudyard Kipling said (in *A Truthful Song*),

How very little, since things were made,
Things have altered in the building trade.

See *extravagance* and *timebombs*.

1. F Petrie, *Ten Years Digging in Egypt*, Religious Tract Society, London, 1893

Containers

Hazardous chemicals (flammable, toxic or corrosive) should always be handled in closed containers, never in open containers such as buckets.

Some drips of petrol were collected in a bucket and while a man was carrying it away for disposal it caught fire. The fire spread to his clothing and he died from his burns. The source of *ignition* was never found.

In plants that handle flammable liquids a bucket is more dangerous than matches. Matches are dangerous only if they are struck when there is a leak and in a well-run plant leaks are few and short-lived. If a bucket is allowed on the plant sooner or later someone will use it to carry a flammable liquid and then there will be a flammble atmosphere above the bucket. If a source of ignition is present, a fire is inevitable.

A man was carrying phenol in a bucket when he slipped and fell and some of the phenol was splashed onto his legs. Despite first aid treatment he was dead within an hour.

Contamination

Contamination is a regular cause of accidents but one that is often ignored. People say that the materials they handle are not flammable, toxic, corrosive or reactive but they forget that they may be contaminated with materials that are. Here are some accidents caused by contamination:

1. One of the UK's worst oil *fires*, at Avonmouth in 1951, occurred because a tank containing gas oil, a relatively safe oil with a high flash-point, was contaminated with petrol. The gas oil had arrived in a ship with a leaking partition between the gas oil and petrol tanks. The oil was handled in a way that was unsuitable for petrol, though safe for gas oil, and a *static electricity* spark ignited the petrol vapour, blew the roof off the *tank* and set fire to the oil.
2. A tank was filled with water for pressure testing using a pipeline that had previously been used for petrol. Some petrol that had been left in the line was flushed into the tank and ignited by a welding spark. The tank exploded.

3. Several fires have occurred in oil tankers because the bunker oil, used for the ship's boilers, was contaminated with lighter oil from the ship's cargo tanks. Sometimes this has been done deliberately[1].
4. A wooden pallet caught fire when it was dragged over ground that was contaminated by sodium chlorate.
5. A laundry explosion is believed to be due to ignition of the fumes from industrial rags[2].
6. A chemical, a brown liquid, was spilt in a van. Some sacks of flour were later carried in the van and became contaminated. The contamination was not noticed as the flour was brown. A number of people became ill after eating bread made from the flour. Finding the cause of the illness required some skilful detective work and makes a fascinating story[3].
7. There is a story, perhaps apocryphal, that a welder constructing pipework for a nuclear reactor sat on a pipe to eat his lunch and sprinkled some salt on a hard-boiled egg. Some of the salt fell on the stainless steel pipework and the chloride ions caused stress *corrosion cracking*[4].
8. *Service lines* containing steam, water, compressed air or nitrogen have often been contaminated.

Hazard and operability studies are a powerful technique for identifying ways in which contamination, and other unwanted deviations, can occur.

See *analysis, COSHH Regulations, nothing, oxygen* and, the opposite of contamination, *(accidental) purification*.

1. *Hazardous Cargo Bulletin*, July 1983, p. 36
2. *Safety Management (South Africa)*, Vol. 11, No. 5, May 1985, p. 42
3. M Howell and P Ford, *The Ghost Disease*, Penguin, 1986, p. 85
4. H Petrowski, *To Engineer is Human*, St Martin's Press, New York, 1982, p. 118

Context

See *culture*.

Contractors

Contractors, especially those hired for a short time, are involved in many more accidents than regular employees as they are usually less well-trained, less familiar with the hazards and less well-motivated. The following are typical of accidents in which they have been involved. For others see *construction* and *timebombs*.

- A contractor's supervisor entered a *tank* to estimate the cost of cleaning it; entry had not been authorised. He had a copy of the works rules but had not read them.
- Two contractor's men repaired a lift without permission and without immobilizing it. One of them fell down the shaft.
- A contractor started to acid-wash some pipelines without waiting for a permit-to-work. Result: *explosion*.
- A contractor used welding gas to inflate tyres.

- A contractor connected new equipment to live process lines without authority. Result: explosion. (See *WWW*, §5.4.2.)
- Contractors left rubbish inside the skirt of a distillation column. When a flange leaked, the rubbish caught fire.
- A contractor filled a tank for pressure test with water using a line that contained some petrol. Result: explosion. He had strengthened the roof-to-wall weld so the tank failed at the bottom instead of the top. (See *domino effects*.)

To prevent incidents such as these we should:

- Tell contractors in detail what they can and cannot do. The emphasis is on tell. It is not sufficient to give them a book of rules; the rules must be explained.
- Explain the reasons for the rules. Many contractors do not understand the reasons for our rules and do not realise the results that follow if the rules are not followed. Slides of incidents can be quite effective. See *training*.
- Keep an eye on contractors; do not turn a *blind-eye*.

Companies should not hire contractors who operate to lower safety standards than they do themselves. It is hypocritical to proclaim high standards and then farm out jobs to contractors with lower standards. Some companies publish the *accident statistics* of their contractors, for comparison with their own, and all should do so.

Contradictory instructions

After an accident, a manager has been known to say, 'I didn't know that was going on. If I had known, I would have stopped it'.

If he really did not know what was going on, then he was not doing his job. Managers are not expected to stand over people all the time but they are expected to keep their eyes open and to carry out occasional checks. (See *audits*.)

Sometimes a manager may turn a *blind-eye*, because if he saw what was going on he would have to stop it. His staff may have been given contradictory instructions. They may have been told that it is important to achieve a certain output, or complete a repair, by a certain time. It may be difficult to do this without disregarding one of the normal safety instructions. What do they do? Perhaps the manager prefers not to know.

At *Chernobyl* the operators were instructed to carry out a series of experiments. They seem to have assumed that these instructions overrode the normal safety instructions which they disregarded. Probably no one actually told them that the normal instructions were suspended but they may have got that impression from a lot of talk about the experiments without any mention of the need to follow the normal safety instructions.

A runaway reaction occurred in a *batch* reactor because material was added at the wrong temperature. The operators believed that if they added it at the temperature specified in the instructions they could not add it in the specified time.

Would the crew of the Herald of Free Enterprise (see *Zeebrugge*) have found it difficult to keep to the timetable if the doors were always closed before the ship departed?

In cases like these the unfortunate operators are in a 'can't win' situation. If there is an accident, they are in trouble for breaking the safety rules. If they stick to the rules, and the programme is not completed in time they are also in trouble. The manager can say that he was let down by his staff, who either broke the safety rules or followed them to the letter when it was inappropriate to do so. ('Heads I win, tales you lose.')

A responsible manager should never put his foremen or operators in this position. If he really feels that a relaxation of the usual safety procedures is justified – sometimes it is – then he should say so clearly, preferably in writing. If he feels that the usual safety procedures should be followed, then he should remind people, when asking for experiments or urgent repairs or extra output, that they are not to be obtained at the cost of violating the safety procedures. What you don't say is as important as what you do say. If you talk a lot about output or repairs and never mention safety then people assume that output or repairs are what you want and all that you want, and they try to give you what you want.

In 1987 a US Congressional Committee investigated allegations that arms were sold to Iran and the proceeds used, illegally, to buy arms for Nicaraguan rebels. The report said that President Reagan bore 'the ultimate responsibilty' for wrongdoing by his aides. 'If the president did not know what his national security advisers were doing, he should have'. 'The President created, or at least tolerated, an environment where those that did know of the diversion believed with certainty that they were carrying out the President's policies[1]'.

1. *New York Times*, 19 November 1987

Control

See *instruments* and *protective systems*.

'Controversial chemicals'

A book with this title[1] surveys 30 chemicals, from *alcohol* to vinyl chloride, via *benzene, dioxin, lead* and mercury, which have been the subject of heated controversy leading, in many cases, to banning or restrictions on use.

In many cases the chemicals are harmful, alternatives are available (for all or most applications) and today no one will dispute the need for the restrictions; *asbestos* is an example. In other cases the materials are not particularly harmful but controversy has surrounded particular uses, for example, fluorides, sometimes added to water supplies. In other cases the chemical has suffered from association with a particular incident; *cyclohexane*, not discussed in the book, is the best example.

The book devotes about 4–8 pages to each chemical and gives a fair summary of the cases for the prosecution and defence. Though biased toward the Canadian situation it is nevertheless thoroughly recommended. See *cause and effect, perception of risk* and *perspective*.

1. *Controversial Chemicals*, edited by P Kruus and I M Valeriote, Multiscience Publications, Montreal, Canada, 2nd edition, 1984

Corner cutting

See *breaking the rules* and *short cuts*.

Corrosion

The use of the word corrosion in accident reports illustrates very clearly the superficiality of much *accident investigation*. To say, as we often do, that corrosion was the *cause* of a leak or other failure is rather like saying that gravity was the cause of a fall. It may be true but is not very helpful. Before we can prevent another leak or failure we need to know the answers to many more questions, such as:

- Was corrosion foreseen?
- Was a suitable material of construction specified?; was it actually used?
- Were operating conditions the same as those foreseen by the designer?; if not, who changed them and after what consideration?
- Was corrosion monitoring or inspection specified?; was it actually carried out as often and as thoroughly as specified?
- Was any attention paid to the results?
- What was the rate of corrosion?

When we have the answers to these and other questions we can decide whether or not we should change the material of construction or the operating conditions, monitor regularly for corrosion or just accept further corrosion.

Examples could be given of major incidents caused by almost all forms of corrosion and by all the factors implied by these questions. See *LFA*, Chapters 8 and 16 and *WWW*, Chapter 16. For example, at *Flixborough* in 1974 cooling water was poured over a reactor to condense the vapour leaking through a stirrer gland. Nitrate in the cooling water caused stress corrosion cracking of the mild steel and a crack appeared. The reactor was removed for repair and replaced by a temporary pipe. The pipe failed in service and about 50 tonnes of hot *cyclohexane* escaped and exploded, killing 28 people and destroying the plant. The fact that nitrate causes stress corrosion cracking of mild steel was known at the time to materials experts but not to most chemical engineers. The main lesson to be learnt is therefore a general one: no change should be made in operating conditions, outside the accepted range, unless the hazards have been systematically considered, specialist advice taken and the change authorised at managerial level. See *modifications*.

Most accidents involving corrosion are similar: the lessons to be learnt are managerial rather than technical. It is unusual for the materials expert to be surprised by unexpected corrosion. More often, he was not consulted, the wrong grade of steel was used, operating conditions were changed or corrosion monitoring or inspection was neglected.

See *(better) late than never, inspection* and *penny-pinching*.

Corrosive chemicals

Acids, alkalis, phenols and many other chemicals are harmful to the skin and cause chemical burns. People who handle them should be trained in the hazards and in the precautions to be taken, namely:

- Keep the stock to a minimum. Use alternative materials if possible. See *intensification* and *substitution*.
- Provide suitable *protective clothing*. For operators of closed plants, who are exposed only if a *leak* occurs, gloves and goggles are normally adequate but *maintenance* workers and *drum* or *tanker* fillers should be fully protected. Even though the plant is cleaned before maintenance, experience shows that pockets of corrosive material may remain, trapped in *dead-ends* or behind *chokes*, sometimes under pressure. (See *trapped pressure*.)
- Provide remotely-operated or enclosed *sampling* points.
- Avoid the use of *hoses* as far as possible and make sure that those that are used are suitable for the materials handled and are in good condition.

The cause of every spillage and every chemical burn should be investigated, however slight. See *LFA*, Chapter 15, especially the Appendix.

Most chemical burns seem to occur during maintenance and sampling or as a result of leaks from *pump* and *valve* glands and joints.

See *containers*.

COSHH Regulations

Under the UK Control of Substances Hazardous to Health (COSHH) Regulations (1988) employers must ensure that the exposure of employees to substances hazardous to health, by inhalation, ingestion or skin absorption, is prevented or, if this is not *'reasonably practicable'*, adequately controlled. The Regulations apply to listed substances and to others of equivalent toxicity and these may include tobacco smoke. *Contamination* should be considered when deciding if a substance is hazardous. *Asbestos, lead* and radioactive materials are covered by other regulations and flammable and explosive materials are covered only if they are also toxic.

'Maximum Exposure Limits' (MELs) and 'Occupational Exposure Standards' (OESs) are published by the Health and Safety Executive for many hazardous substances. The MELs, formerly 'control limits', have greater force and are set when there is sufficient evidence to justify them. They have been set for only a few substances. OESs, formerly 'recommended limits' have been set for more substances, and set so that all workers may be repeatedly exposed day after day without adverse effects.

Control and recommended limits replaced Threshold Limit Values (TLVs); concentrations which, it was believed, could safely be breathed for 8 hours per day, year in, year out. However, there is an important difference between TLVs and MELs. If a TLV was achieved there was no need go below it. Now there is a requirement to go below the limit if it is reasonably practicable to do so, since, in setting MELs, cost and feasability have been taken into account as well as toxic effects. (See *asbestos* and *fugitive emissions*.)

If there is no MEL or OES for a substance then employers must decide on a suitable figure, for example, by analogy with other chemicals. While they should err on the safe side, they are not expected to treat every unknown as highly toxic. If full *protective clothing* is requested, for example, for substances which are probably harmless, less is likely to be worn than if a more practical solution is found.

In the US, short-term exposures limits are published as well as the limits for prolonged exposure.

The COSHH Regulations require employers to determine the nature and degree of risk, define control measures (which should include both equipment and *methods of working*), ensure that they are properly used and monitor exposure. Health surveillance is required for significant exposure to defined substances and to other substances which are likely to produce ill effects. Control by the use of protective clothing should be used only as a last resort. Information, *instruction* and *training* must be given to those who are exposed.

Employers are not expected to carry out a large programme of tests but to make full use of the information that is available. Suppliers have to pass on the information they have. The Regulations are expected to have little effect in the chemical manufacturing industry, where standards are already high, but big effects in *small companies* which use chemicals. Do not overlook solvents used in offices, print rooms, etc.

When many chemicals are handled, for example, by laboratory workers or samplers, a generic assessment is sufficient. Records of exposure and medical surveillance have to be kept for 30 years after an employee leaves; if he had 40 years' service then some records will have to be kept for 70 years.

See *cause and effect, CIMAH regulations, under-reporting* and various leaflets available from the Health and Safety Executive.

Cost estimates

See *new processes*.

Cost of saving a life

We cannot do everything possible to prevent every accident that might occur. Sufficient resources are never available. How then do we allocate our resources and decide our priorities? When the public and politicians are involved it often seems that those who shout the loudest get the most. A more defensible method is target setting: reduce first those risks that exceed a certain level. See *criteria*. An alternative might be to reduce first those risks that are cheapest to reduce, that is, spend our money where the cost of saving a life is lowest so that we save most lives per million pounds spent.

The actual sums spent to save a life vary over an enormous range. Doctors can save lives for a few thousands or tens of thousands of pounds, road engineers for hundreds of thousands. In the chemical industry millions of pounds are spent to save a life, in the nuclear industry tens or hundreds of millions. However, in the Third World, famine relief and immunization can save lives for trivial sums (only hundreds or thousands of pounds)[1]. Many lives would be saved if governments spent our money more wisely.

Nevertheless, within industry I suggest we should use target setting rather than the cost of saving a life as our main criterion, for three reasons:

- Moral: We should not tolerate risks that are high but expensive to reduce; it may be better for society, but not for the unlucky person exposed to the risk.
- Pragmatic: If we say that we will remove risks that are cheap to remove but will accept those that are expensive to remove, then many design engineers and managers may say that their risks are expensive to remove. On the other hand, if we say that all risks that exceed our target or criterion must be reduced, we can usually find a *'reasonably practicable'* way of reducing them.
- Custom: The normal practice in industry has been to use targets. Thus we say that handrails must be a certain height (see *history of safety*) or that the concentration of a toxic chemical in the atmosphere must not exceed a certain figure (see *COSHH Regulations*).

The cost of saving a life is useful, however, as a secondary criterion. If it is high, we should not tolerate the risk, but we should look for a cheaper solution. In practice a solution can usually be found; there are usually several solutions to every technical problem.

Note that in the chemical and nuclear industries the cost of saving a life is usually notional. If we spend the money, it is unlikely that any lives will be saved. All we will do is to make the low risk of a fatal accident even lower. In contrast, real lives will be saved if doctors and road engineers spend extra money. For this reason they make more use of the cost of saving a life than we do in the process industries.

1. T A Kletz, *Setting Priorities in Safety* in *Engineering Risk and Hazard Assessment*, Vol. 1, edited by A Kandel and E Avni, CRC Press, Boca Raton, Florida, 1988

Costs

What does industry spend on safety? Obviously it varies from one industry to another. (See *cost per life saved*.) Here are some figures from the oil and chemical industries.

A survey of ten *ICI* chemical plants showed that on average about 10–15% of capital cost was spent on safety measures over and above the irreducible minimum necessary for a workable plant. Relief valves are an example of equipment not included as they were considered 'basic engineering'. Another 5% was spent on pollution control. The amounts spent varied a great deal from one plant to another; on one plant 50% of the cost was for safety equipment[1].

A US estimate of expenditure in the hydrocarbon processing industry is even higher though this may be due to different definitions of safety equipment. It states that 18% of capital is spent on fire protection, 12% on pollution prevention and a sum equal to 9% on health and safety, though much of this is spent on revenue items such as protective clothing and training[2].

Safety equipment has to be tested and maintained and this also costs money; for *instruments* these costs roughly double the installed capital cost; instruments cost twice what you think.

Within an industry some companies spend more on safety than others, money that their competitors may not spend. In the end the spenders may get this money back in greater freedom from fires, explosions and other accidents. However, the money has to be spent today; the saving is in the future. A short-sighted company which looked only at this year's profits might spend less. See *penny-pinching*.

A company found that the 55% of their plants that achieved above average ratings in their 2-yearly *audits* had 40% of the *dangerous occurrences*, 25% of the major ones and 15% of the consequent financial losses.

Of course, money can be spent with *extravagance* on safety as on everything else. See *persistence*.

All these costs have to be paid in the end by the consumer and amount to perhaps 5-10 pence in every pound spent on plastic goods, man-made fibres, paint, drugs and other products of the chemical industry. This figure is derived as follows: product costs are made up of capital charges, labour and raw materials. 15-30% of capital is spent on safety and pollution prevention, as shown above. A similar proportion of labour costs will be spent operating and maintaining the safety and pollution control equipment. Raw materials costs are other people's capital and labour, apart from taxes and royalties. Allowing for these and for the fact that the products of the oil and chemical industries are further processed by other industries (including transport and retailing) which spend less on safety, we get a figure of perhaps 5-10% of final cost spent on safety and pollution prevention.

In a democracy, in the end the people choose how much is spent on safety and pollution prevention. If the public wishes, industry can make itself safer and cause less pollution – and send the public the bill.

Accidents cost money; they also cost lives and injuries. Having had an accident we should see that we learn as much as possible from it and do not waste the experience. See *accident investigation*. A high price was paid for the information in this and other books on safety. You get this information for a few pounds. It is the best bargain you will ever have.

See *time and money*.

1. D O Hagon, *Fourth Environmental Seminar*, Harwell, April 1982
2. *Hydrocarbon Processing*, November 1983, p. 34-P

Cranes

Cranes are involved in more accidents, mainly falling over or dropping their loads, than might reasonably be expected, perhaps because expensive equipment is often entrusted to unskilled people. The following accidents, picked at random from those I have collected without making a particular effort to do so, are typical of those that occur:

1. An oil refinery hired a crane for *construction* work. On arrival the driver was told the route to follow to the construction site. On the way he stopped to assemble his jib. A sudden gust of wind blew it onto a storage tank[1].
2. On several occasions cranes have been moved with their jibs raised and have bumped into overhead pipelines. Some companies now protect pipelines on busy roads with 'goalposts' – steel beams across the road a few metres away on each side.
3. A cranedriver tried to lift a pipe before it was unbolted and in doing so stretched and weakened the lifting lug. When the pipe was unbolted and lifted, it fell.
4. A crane was left overnight with its jib raised and was blown over by a strong wind.
5. A temporary hook stretched and a 15 tonne steel beam fell to the ground; a man lost both legs[2].
6. The driver of a telescopic jib crane lowered his jib and then fully extended it. He attempted to lift a light load but the crane fell on to the plant. The driver did not understand the need to keep within the maximum jib radius for a given load. There was no alarm to warn the driver. Telescopic cranes have an extra degree of freedom compared with mechanical cranes and therefore need an extra alarm but this was not generally realised at the time. See *EVHE*, §3.3.5 and *Lees* Appendix A3.2, item B59. For details see *rules and regulations*. You cannot relax just because you are following all the *codes of practice*; they may be out of date.

See *instructions*.

1. *Petroleum Review*, November 1973, p. 445
2. *Safety Management (South Africa)*, August 1987, p. 59

Creativity

See *'need to know'*.

Creep

See *alertness* and *furnaces*.

Criminals

The chapter house at Batalha, Portugal – one of the widest stone vaults in medieval architecture – collapsed twice during construction and was finally built by prisoners condemned to death[1].

See *prosecuiton*.

1. S J Gould, *Ever Since Darwin*, Norton, New York, 1979, p. 177

Criteria

We need a method of measuring safety and a criterion or target if:

1. We wish to compare the riskiness of different tasks, occupations or factories, either with each other or the same one at different times.
2. We wish to decide whether a particular risk is so high that it should be reduced or is so low, compared with all the other risks around us, that we can ignore it, at least for the time being. (See *hazard analysis*.)

The *Fatal Accident Rate* (FAR) is often used as a method of measuring safety, rather than the *lost-time accident rate*, as injuries differ so much in severity and are hard to define. (See *accident statistics*.) The loss of *expectation of life* can also be used. Other criteria have been used do define *acceptable risks* from leaks of *chlorine*.

If we are comparing equipment, then the failure rate may be used. To compare plants, we can use the probability of *fire* or *explosion* (preferably obtained from the historical record rather than estimated), *insurance* claims or *indices of woe*.

More subjective criteria are based on *audits*, marks being awarded under various standard headings. One of the best-known schemes of this sort is the Five Star Grading System devised by the National Occupational Safety Association of South Africa[1,2]. For example, Element 2.45 covers respiratory equipment. Up to two marks are awarded for correct identification of areas and situations where the equipment is needed, six marks for provision of the correct type and for adequate checking and control and two marks for training in its use. Altogether there are 126 elements covering housekeeping, safeguarding, fire prevention and protection, accident recording and investigation and safety organization. Over 90% of the total marks available are required for five stars and 40–50% for one star. The system is applicable to any organization and does not cover specific technical hazards but could be expanded to include them.

1. Leaflets available from NOSA, P.O. Box 26434, Arcadia 007, South Africa
2. J Bond, *Loss Prevention Bulletin*, No. 080, April 1988, p. 23

Culture

Companies and nations both have cultures (beliefs, values and forms of behaviour) that have an influence on safety.

A company's (or factory's) culture is more important than its official statements of *policy*, which may or may not reflect the culture. Safe action, at any level, is consistent and effective when it is part of the company culture, when everyone takes for granted that it is the right thing to do. In contrast, a book of rules imposed by a government, head office or other outside body may be followed unenthusiastically, and to the letter rather than the in spirit.

One night a young manager took a chance and allowed someone to carry out an unsafe act in order to keep a plant on line. He was not thinking or working in a vacuum. He was influenced by his assessment of what would be said the next day, if the plant was shut down. He was affected by the factory culture. For details see *LFA*, Chapter 3.

Company cultures are not unchangeable. See *Du Pont*. National cultures can also change and sometimes we blame those who went before us for actions which seemed right at the time. In the 1940s and 1950s UK *Factory Inspectors* did not enforce the *asbestos* regulations to the letter, probably because they thought that, if they did so, factories might close and they would be blamed for causing unemployment. At that time the public culture put employment above safety. By 1976 the culture had changed and the Inspectorate was blamed for following the public's wishes of 25–35 years earlier[1-3]. (See *change – old problems*.)

National cultures are, however, more deep-rooted and harder to change than company cultures. In the East 'face' is much more important than in the West. Juniors are reluctant to question the views of seniors and may be less willing to suggest that their superior's proposals are unsafe. Safety officers are reluctant to tell managers, senior to them, what is wrong. In some cultures it is impolite to say 'no'. Therefore, ambiguous answers are given and Westerners are left thinking that the answer is 'maybe'.

In many eastern countries there is a more fatalistic attitude to death than in the West. One dies when one's time comes, not before and not after, so why bother to take precautions? One cannot postpone the inevitable. It is difficult for outsiders to counter such views. It is better tackled by those who have grown up in the culture and understand it but realise that a change is necessary.

See *amalgamation* and *fate*.

1. *Occupational Safety and Health*, Vol. 6, No. 5, May 1976, p. 13
2. *Sunday Times*, 28 March 1979
3. Lord Hale, *Hansard*, 11 May 1976

Custom and practice

This term is used to justify practices which cannot be justified in any other way. It is most often met within industrial relations but sometimes also in safety. When a rule is regularly broken or an obviously unsafe practice is followed, then we may be told that it is 'custom and practice' to do it that way.

As with many other safety problems (see, for example, *modifications*) a two-pronged approach is necessary:

- We should explain the hazards of the customary method and, if possible, describe accidents which have occurred when it was used.
- We should monitor to make sure that the unsafe method is not being followed in the future. Habits die hard and once we relax the old method may return. It may have started in the first place because a manager turned a *blind-eye*.

See *'reasonable care'*.

Cyanates

See *Bhopal*.

Cyanides

Cyanides, including sodium cyanide, hydrogen cyanide (in liquid or gaseous form) and related substances such as acrylonitrile are rapid and deadly poisons[1]. People who handle them should be trained in the hazards and in the precautions to be taken, namely:

- Keep the stock to a minimum. Use alternative materials if possible. See *intensification* and *substitution*.
- Keep cyanides and acids apart as they react together to form hydrogen cyanide.
- Use only labelled, closed *containers* in a place accessible only to trained personnel.
- Provide suitable *protective clothing*: gloves for people who have to handle solids or liquids, breathing apparatus for people who may be exposed to vapour.
- Have trained first aid workers available.
- Wash thoroughly after handling.

A *hazard and operability study* on a laboratory unit disclosed that:

- Hydrogen cyanide was to be used.
- The laboratory was to be located on the top floor of a building.
- The hydrogen cyanide cylinders would be taken up in the lift!

1. *What you should know about Cyanide Poisoning*, Health and Safety Executive.

Cyclohexane

In 1974 at *Flixborough*, a leak of about 50 tonnes of hot cyclohexane exploded, destroying the plant and killing 28 people. Cyclohexane became a *'Controversial Chemical'* and it was difficult to persuade the public that its physical properties were much the same as those of petrol and that it was no more hazardous. Even within the industries that handled cyclohexane it was treated with more respect than petrol.

Cyclohexane has been involved in a number of other fires and explosions but this is because large quantities are used in the manufacture of nylon. It is oxidized hot, in the liquid phase, in large reactors and conversion is low so that most of the feedstock is recycled many times. The process, rather than the raw material, is unusually hostile. See *oxidation, friendly plants* and *LFA*, Chapter 8.

Cylinders

According to one report[1] about 60 000 one-tonne *chlorine* cylinders are in use in the US and of these about one per year fails catastrophically, usually as a result of falling off a lorry or fork lift truck and striking a sharp object. In addition about 100 other cylinders per year lose their contents at a lower rate as a result of failure of the fusible plugs.

Most compressed gases are handled in smaller, stronger cylinders, without fusible plugs; these presumably have even lower failure rates though, so far as I am aware, no figures are available.

Cylinders have failed catastrophically when exposed to fire. See *BLEVE*.

Cylinders should be stored in well-ventilated locations. If they are used regularly in a laboratory they should be stored elsewhere and connected by pipelines. An operator was overcome by *nitrogen* from a leaking cylinder stored in a basement underneath the room he was working in. Had he not been found in time he would have died[2].

In 1944 a leak developed on a 50 kg chlorine cylinder as it was being transported through New York. The truck driver stopped, unfortunately within a metre of the air intake of a subway ventilation system. The chlorine was sucked into the station and passengers on the station and in the trains were affected. Consequently, 418 people received hospital treatment; 40–50 were affected in the street and the rest in the station and trains[3].

See *anaesthetics, oxygen* and *threads*.

1. R W Johnson, *Air Pollution Control Association 79th Annual Meeting*, Minneapolis, Minnesota, June 1986
2. *Safety Management* (South Africa), Vol. 16, No. 1, January 1989, p. 27
3. *Loss Prevention Bulletin*, No. 086, April 1989, p. 1

Damage control

Damage control (or total loss control) is a systematic attempt to record, cost and investigate all forms of loss: damage to plant and materials, consequential losses and plant upsets as well as injuries. Often only the injuries are followed up and we fail to realise the size of the company's financial loss or the number of accidents that nearly happened. Directors might be more effective in preventing accidents if they kept their eyes on the profits and practised damage control as well as taking a humanitarian interest in injuries.

Pioneered in the US steel industry by Frank Bird[1], damage control has been practised mainly in industries in which damage to equipment or materials is easy to recognize, if we want to do so[2,3]. It is harder to practise damage control in the process industries where many losses are due to *instrument* faults and operator errors and often go unrecognized. The attempt to practise it is rarely made.

1. F Bird, *Management Guide to Total Loss Control*, Institute Press, Atlanta, Georgia, 1974
2. J A Fletcher and H M Douglas, *Total Loss Control*, Associated Business Publications, London, 1971
3. F E Bird and H E O'Shell in *Selected Readings in Safety*, edited by J T Widner, Academy Press, Macon, Georgia, 1973, p. 15

Danger

See *hazards*.

Dangerous occurrence

Accidents which do not cause death or injury, only damage to plant or loss of production, are often called dangerous occurrences, dangerous incidents or near-misses and in some companies the *safety professionals* take little interest in them; this is a mistake. The next time the dangerous incident occurs someone may be killed or injured and, in any case, damage and loss of profit should be avoided. In *loss prevention* and *damage control* attention is paid to all accidents.

If those concerned had learnt from dangerous occurrences the disaster at *Aberfan* would not have occurred.

It is sometimes said that that dangerous occurrences will never be reported, unless the damage is major, and that it is therefore impossible to analyse them and learn from them in the same way as we analyse and learn from accidents that cause injury.

Dangerous ocurrences will be reported if managers make it clear that they want them reported and make a fuss when they find out that an incident has not been reported. Stories picked up from the grapevine can be followed up; fires can be detected from the fire service reports, spillages from the cleaning gang's worksheets; claims for damaged clothing can be probed to see if the incident concerned has been reported. In these and other ways managers can bring dangerous occurrences out into the light of day.

Many years ago I was showing a visitor a summary report on all the dangerous ocurrences that had occurred in a works in a month. One was: 'Mr X, process operator, found a sample point running full bore.' My visitor was surprised. 'How', he said, 'do you get an atmosphere in which people are willing to report such incidents?'

To illustrate a point: in the Yorkshire Museum in York there is (or was) a map showing the places in Yorkshire where bats have been sighted; they cluster around Helmsley. Perhaps this district is popular with bats but it is more likely that several bat-watchers live (or lived) in the area. Similarly, a survey in one company showed that 70% of the reported dangerous occurrences occurred in one of their five factories. It was not the most dangerous or unsafe works but its managers made more effort to follow up dangerous occurrences.

See *accident investigation, non-events* and *triangles*.

Data

See *reliability data*.

Dead-ends

Water, present in traces in most oil streams, collects in dead-ends in pipework and may freeze, breaking the pipe. When the plant warms up the water may turn to steam with explosive violence, or corrosive materials may dissolve in the water and corrode the pipe. For example, water and corrosive impurities collected in a dead-end branch, 3 m long and 12 inches in diameter, in a natural gas line operating at a gauge pressure of 38 bar (550 p.s.i.). The pipe failed, the escaping gas ignited and five men were killed[1]. For other examples see *WWW*, § 9.1.1 and *LFA*, Chapter 16.

Dead-ends are often the result of *modifications* but are sometimes designed into new plants to make future extension easy; they should point upwards, not downwards. However, dead-ends can arise because equipment such as a control valve is by-passed or a spare pump is rarely used.

In olefine plants explosive NO_x gums and ammonium nitrate can accumulate in dead-ends and can decompose catastrophically during draining or cleaning[2].

If a process liquid is denser than water, any water present will collect in dead-ends that point upwards; dead-ends should therefore point downwards. If solids are liable to collect in the dead-ends then it may be necessary to install a filter. In one incident, water collected in an upward-pointing expansion pot in a *chlorine* line. Corrosion occurred and the pot burst. Fortunately there was an *emergency isolation valve* in the line and the leak was soon stopped[3].

1. US National Transportation Safety Board, *Safety Recommendations P-75-14 and 15*, 1975
2. F C Politz, *Plant Operations/Progress*, Vol. 6, No. 4, October 1987, p. 09
3. *Loss Prevention Bulletin*, No. 086, April 1989, p. 27

Debts

A father helped his son when he was in difficulty. The son offered to repay his father, by helping him when the need arose. The father said, 'I don't want repayment. Just do the same for your son'.

Similarly, we learn from many people, who tell us about the accidents they have had and the action they have taken to prevent them happening again. We may not be able to repay these people directly but instead we can tell others about our accidents and the action we took. Debts need not always be repaid to the person who helped us. (See *publication*.)

Decisions

A Board of Directors once asked an economist and a psychologist to join their meetings to predict the decisions they would make about the issues and problems raised. These decisions were mainly in the realm of economics and finance yet the psychologist was more often right in his predictions than the economist. The psychologist understood how the emotional side of a man's nature often rules his reason and he had been observing this committee for signs of emotional 'music', for the 'hidden agenda' of personality clashes, power struggles, friendship patterns, hopes and fears and temperamental influences. This Board of Directors, whose apparent common purpose was the running of a large company, existed also for two other powerful reasons – to maintain itself and to meet the emotional needs of its individual members. These two aspects had, at times, become more important that the acknowledged task[1].

The decisions we take on safety matters may also be influenced by other factors besides the merits of the case. What are these other factors?

- The foremost is the nearness of the accident site. (See *action*.) For example, drums are sometimes transported on a flat lorry in the vertical position with a horizontal layer on top. At one depot a drum rolled off the top and killed a man. Afterwards this depot would not accept lorries with a top horizontal layer. There may be a case for prohibiting the top horizontal layer in all cases. There may be a case for allowing it to continue despite the accident. It is illogical to prohibit it at the place where the accident occurred but not elsewhere. (However, sometimes, after an accident, we have to do more than is necessary on technical grounds out of respect for the feelings of the people involved.)
- Sometimes, for reasons that are hard to define but include technical novelty, a particular accident catches the imagination and everyone rushes to take precautions. This occurred when a *diesel engine* ignited a *leak* of hydrocarbon. More attention was paid to the unusual source of ignition than to the reason for the leak. See *amateurism* and *LFA*, Chapter 5.
- We are sometimes reluctant to make a change if this could be construed as a reflection on our previous decisions – we do not like to admit we were wrong. If, however, the accident was clearly the fault of someone else, an equipment supplier for example, we are more willing to make a change and, incidentally, be more open about it.

Courses on decision-making stress the need to compare the effects of each possible decision on profit, output, quality, labour relations, etc. The effects on safety should be included[1].

1. M Brown, *The Manager's Guide to the Behavioural Sciences*, The Industrial Society, 1972
2. W B Howard, *Chemical Engineering Progress*, Vol. 84, No. 9, September 1988, p. 25

Defence in depth

The control of hazards is based on defence in depth: if one line of defence fails there are others in reserve.

A common failing is to ignore or neglect the outer lines of defence because an inner one is considered impregnable; but if it fails there is nothing to fall back on. For example, companies have been known to ignore leaks of flammable gas because they had, they believed, eliminated all sources of *ignition* and so *leaks* could not ignite. When an ignition occurred, a *fire* or *explosion* was inevitable. (See *insularity*.) At *King's Cross* no one worried about small fires on escalators as they believed they could be extinguished without difficulty. A small fire which was not extinguished developed into a major conflagration. Effective loss prevention lies far from the ultimate result, the top event of the fault tree. (Confusingly, Americans sometimes call this the 'bottom line'.) See *LFA*, Chapter 4.

The nine commonest lines of defence are summarized below. All apply to flammable liquids and gases but only the first six apply to toxic ones.

1. The first line of defence is to avoid, when possible, large inventories of hazardous materials in storage or process. 'What you don't have, can't leak.' This can be done by:
   (a) *Intensification*: using so little hazardous material that it does not matter if it all leaks out.
   (b) *Substitution*: using safer materials in place of hazardous ones.
   (c) *Attenuation*: using hazardous materials in a safer form.
2. Most of the materials handled in the oil and chemical industries are not flammable or explosive in themselves but only when mixed with air in certain proportions. The second line of defence is therefore to keep the air out of the plant and to keep the fuel in the plant. The first is easy as most plants operate at pressure. *Nitrogen* blanketing is widely used to keep air out of equipment such as *tanks, stacks* and *centrifuges* which operate at low pressure, and equipment which is opened up for repair.

    Keeping the fuel in the plant is a little more difficult. The commonest reason for large leaks of hazardous materials is *pipe failures* and the commonest cause of pipe failures is the failure of *construction* teams to follow the design or to follow *good* engineering *practice* when details have not been specified in the design. The most effective action we can take to prevent leaks of hazardous materials is therefore to specify the design in detail and then inspect thoroughly during and after construction – more thoroughly than has been customary – to see that the design has been followed and that details not specified in the design have been constructed in accordance with good engineering practice.

3. Despite our efforts some leaks will still occur and therefore we should be able to detect them early. Gas detectors (see *analysis*) are a cheap, valuable and worthwhile investment but nevertheless, even on plants where they have been installed, many leaks are still detected by men. There is no substitute for a walk round by the operators.
4. We should be able to warn everyone on the plant that a leak has occurred so that they take appropriate action. Those who are not required to deal with the leak should leave the area. (If there is a leak of toxic gas it may be safer to stay indoors or go indoors. See *ammonia* and *ventilation*.)
5. We should be able to isolate our leaks by using remotely-operated *emergency isolation valves*. We cannot install then on the lines leading to all equipment which might leak but we can install them when experience shows that the probability of a leak is higher than usual, for example, we should install them on the lines leading to very hot or cold *pumps*. The emergency valves must, of course, be tested regularly or they may not work when required.
6. We should be able to disperse leaks by open construction – whenever possible, equipment containing flammable materials should be located in the open, not in closed buildings – assisted if necessary by steam or water curtains. (See *ventilation*.) If, however, a closed building is considered essential then the walls should be of light construction so that they blow off when an explosion occurs and minimize damage to the equipment inside them. This is known as *explosion venting*.
7. We should do what we can to remove known sources of *ignition*. Though this seems to be one of the strongest lines of defence it is actually one of the weakest. Even though we do all we can to remove known sources of ignition some may still exist.
8. Despite our efforts some leaks will ignite and so we should protect our equipment against the effects of fire and explosion. Protection against fire is usually achieved by *insulation* or water spray, and protection against explosion by strengthening buildings or by locating them so far away that they will not be affected.
9. Finally, we should provide fire-fighting facilities.

See *redundancy*.

Definitions

See *Nomenclature for Hazard and Risk Assessment in the Process Industries*, Institution of Chemical Engineers, Rugby, 1985.

Demolition

The hazards of dismantling and demolition are similar to those discussed under *construction* and *contractors*. For example, to make room for new equipment, contractors had to dismantle some old pipelines. They used a powered shovel, broke one of the lines on the wrong side of an isolation valve (see Figure 5) and spilt over 300 tonnes of oil. It is easy to blame the

Figure 5 A contractor used a powered shovel to remove an old pipeline and broke it on the wrong side of an isolation valve

contractors but they are not noted for their gentleness of touch and should have been supervised more closely.

Other accidents have occurred during demolition because equipment was not freed from hazardous chemicals. Equipment that is to be demolished should be prepared with the same care as equipment which is to be maintained. It should be freed from hazardous substances to an agreed standard and isolated from other equipment. (See *isolation*.) A programme should be agreed with the contractors who should be told exactly what they may and may not do and they should be closely supervised.

If equipment is taken out of use but is not being demolished until later, clean it immediately; do not leave it until demolition is about to start.

Equipment which has to be demolished should be clearly marked, not just pointed out. If pipelines have to be cut they should be marked at the precise point. If these rules are not followed it is only a matter of time before before the wrong equipment is removed or the wrong lines cut. (See *identification of equipment*.)

See *old equipment* and *traps*.

Design

Safety by design should always be our aim but sometimes it is impossible or too expensive and we have to achieve safety by procedures, that is, by *training, instructions, methods of working, audits, inspections* and so on. I often use the phrase 'work situation' to cover both design (or hardware) and procedures (or software).

Unfortunately changes in design can be troublesome and expensive. It is cheaper and easier to tell someone to be more careful and some managers are too ready to take this easy way out. However, telling people to be more careful will not prevent accidents. (See *human failing*.) Instead we should change the work situation, if possible the design, so as to reduce or remove opportunities for error.

This applies not only in industry but elsewhere. If we wish to prevent road accidents, changing the design of vehicles or roads, and software changes such as enforcing drink-driving laws are more effective than telling drivers to be more careful. (However, see *risk compensation*.)

Various safety studies are carried out during design. In the chemical industry these include *hazard and operability studies*, relief and blowdown reviews, *electrical area classifications* and model reviews. However, these all come late in design when all we can do is to add on protective equipment to control the hazards. If we could carry out critical reviews of the design much earlier, at the conceptual and flowsheet stages, then we might be able to avoid many hazards by a change in design. See *early involvement in design, feedback to design, friendly plants* and *inherently safer design*.

Safety should not be something added on to design like a coat of paint but an integral part of it. The designer may leave the choice of paint to a paint expert but he cannot leave the safety to a *safety professional*. He should be able to understand and apply the principles of technical safety. The safety professional is there to provide information, stimulate, question, suggest changes (forcibly at times) but the primary responsibility remains with the designer.

Finally, before operating staff criticize a design they should remember that the designer has to compromise between conflicting requirements. A design is never 'the logical outcome of the requirements' simply because, the requirements being in conflict, they do not have a logical outcome[1].

1. D Pye, quoted by H Petrowski, *To Engineer is Human*, St Martins Press, New York, 1986, p. 218

Deterministic accidents

See *will and might*.

Detail

According to David Lodge[1] the Englishman does not worry about big things such as *policies*, religion and death but is 'more empirical; he worries whether he is on the right train, about how much to tip the taxidriver; he gets up in the middle of the night to see if he turned off the living room light.'

This sort of worry (Have the *protective systems* been tested? Are there any gaps in the fire *insulation*? Are the foreman's *instructions* clear?), rather than worry about policies, produces success in safety and loss prevention.

1. D Lodge, *Write On*, Penguin Books, 1988, p. 5

Diesel engines

At one time it was thought that diesel engines, unlike spark-ignition engines, could not ignite *leaks* of flammable gas or vapour and could be used with safety in areas where leaks might occur. However, diesel engines can cause ignition in various ways:

- Gas can be sucked into the engine through the air inlet so that the engine continues to run when the fuel supply is isolated. The engine may

race until valve bounce occurs and the gas outside the engine ignites. Devices for isolating the air supply are available. (The leak described in the item on *amateurism* was ignited in this way.)
- Sparks or flames from the exhaust, a hot exhaust system or auxiliary electrical equipment may cause ignition.
- Use of a decompression control may cause ignition.

Diesel engines, like petrol engines, should not be allowed to operate in areas where leaks of flammable gas may occur except under controlled conditions, the degree of control depending on the length of time that the engine is allowed to operate. Engines that operate permanently, or for a large part of the time, should be fully protected[1]. Vehicles used during occasional maintenance may be protected to a slightly lower standard. Vehicles which are just passing through do not require any special protection but should not be allowed to enter unless the plant is steady and there are no leaks.

For further information see *emergency equipment*, *Myths*, §14, and *LFA*, Chapter 5.

1. Oil Companies Materials Association, *Recommendations for the Protection of Diesel Engines Operating in Hazardous Areas*, Wiley, Chichester, 1977

Difference of opinion

See *Factory Inspectors*

Dioxin

The dioxins are a family of chemicals but the name is usually used to describe the best-known member of the family, 2,3,7,8-tetrachlorodibenzo-para-dioxin (TCDD) See Figure 6:

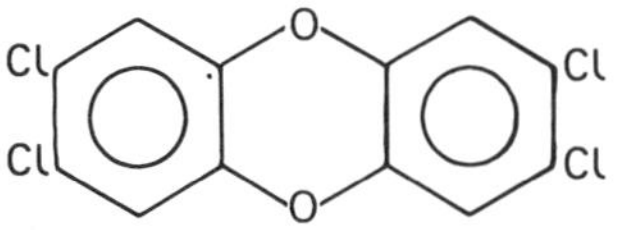

Figure 6 Dioxin

It is formed in minute amounts in the manufacture of the weedkiller 2,4,5-T (2,4,5-trichlorophenoxyacetic acid) but if the reactor gets too hot larger amounts are formed. This occurred at *Seveso* in 1976 when the TCDD was discharged through a relief valve and contaminated the surrounding area and also in Ludwigshaven in 1953 when 55 men were affected and the building had to be destroyed[1]. See also *Lees*, Appendix 2.

TCDD is very poisonous – 500 times more poisonous than strychnine and 10 000 times more poisonous than *cyanide* (in guinea pigs) – and causes an unpleasant skin disease, chloracne. However, it has caused no deaths in industrial accidents though some may have occurred during the war in Vietnam when 2,4,5-T contaminated with TCDD was used as a defoliant.

2,4,5-T has been banned in some countries though there is no proof that the pure material is harmful[2].

Traces of dioxin are present in car exhausts and incinerator effluents[3].

1. L Bretherick, *Handbook of Reactive Chemical Hazards*, 3rd edition, Butterworths, 1985, p. 541
2. A Hay, *Nature*, No. 269, 1978, p. 749
3. G H Eduljee, *Chemistry in Britain*, Vol. 24, No. 12, December 1988, p. 1233

Disbelief

See *instruments*.

Discretion

The responsibilities of many of those who work in industry have increased enormously. At one time many people had to carry out only simple tasks which left little room for discretion or judgement – they just had to do what they were told. Now many more people are free to exercise discretion and judgement, but have they been given the skill and knowledge they need?

Scrap plastic had to be dissolved in methanol, a highly flammable solvent, so a method was devised to prevent methanol and *air* coming into contact. The plastic was put in a vessel which was boxed up and then swept out with *nitrogen*. The methanol was then added. When the plastic had dissolved the solution was pumped into a *tank* and the dissolving vessel swept out with nitrogen before it was opened up to add more plastic.

The operators got into the habit of adding methanol as soon as the plastic was in the dissolving vessel, without bothering to box it up or sweep it out with nitrogen. One day a *fire* occurred and a man was injured.

It is easy to say that the fire was the fault of the operators who did not follow *instructions*, but why did they not follow them? Had the manager and foremen made them aware of the hazard, that mixtures of flammable vapour and air may catch fire even though we have tried to remove all known sources of *ignition*? See *human failing*.

A driver got out of his van at a country filling station just in time to prevent the attendant putting the nozzle of the petrol pump in the ventilator on the roof. Another attendant used his cigarette lighter to check the petrol level in a road *tanker*. What training, if any, had the attendants been given?

On many occasions, after a *dangerous occurrence*, it has been found that a temperature or pressure has been dangerously high for many hours. The operators wrote down the temperature or pressure on the record sheet but did nothing about it; they did not even tell the foreman. Had the operators been told what temperatures or pressures were dangerous and why and that their job was not just to take readings but to do something about them?

People need to be given the skill and knowledge necessary to replace rule-based behaviour by skill-based. In addition regular *audits* are needed to check that they are acting correctly and that rules are being followed.

See *unexpected hazards*.

Dismantling

See *demolition*.

Dispersion

If, despite our efforts at prevention (see *defence in depth*), there is a *leak* of flammable gas then we want to disperse it quickly, if possible before it reaches a flammable concentration. Plants handling flammable liquids or vapours should therefore, whenever possible, be built in the open air. Natural *ventilation*, even on a still day, is more effective (and cheaper) than forced ventilation. A roof over equipment is acceptable but not walls.

Although many explosions have occurred in closed buildings such as compressor houses (see *LFA*, Chapter 4) companies often take a short-sighted view and enclose equipment for the comfort of operators and maintenance workers or to reduce noise levels outside.

If calculation shows that likely leaks will not disperse before they reach a known source of ignition, such as a furnace or roadway, then the effects of natural ventilation can be supplemented by steam or water curtains. These prevent leaks spreading beyond the curtain and also help to disperse them.

There is extensive literature on methods of calculating the extent to which gas will disperse and many experiments have been carried out[1]. See *Lees*, Chapter 15.

1. *Guidelines for Use of Vapor Cloud Dispersion Models*, American Institute of Chemical Engineers, 1987

Distillation

Many distillation columns contain large inventories of hazardous materials, on the trays or packing and in the base. The hold-up per theoretical plate varies from 20 mm to 100 mm for various trays and packings. Whenever possible, designers should choose a tray or packing with a low hold-up. (See *intensification*.)

If the bottoms product is liable to degrade, its inventory is often reduced by narrowing the base, so that the column appears to balance on the point of a needle (see Figure 7). This can be done just to reduce the inventory, though it rarely is.

Much bigger reductions in inventory, up to a 1000 times, can be achieved by the use of ICI's Higee distillation process (see Figure 8). It is based upon the observation that distillation (or any liquid-vapour contacting process) will take place more efficiently if gravity can be increased, or simulated by centrifugal force. The rotating packed bed is about 1 m thick and about 1 m diameter. The liquid travels outwards and the vapour inwards. The diameter corresponds to the height of a normal column and the thickness to the diameter. Two units are required, one for the stripping section and one for the rectifying section. (See *innovation*.)

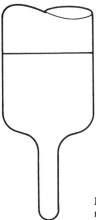

Figure 7 The inventory in a distillation column can be reduced by narrowing the base

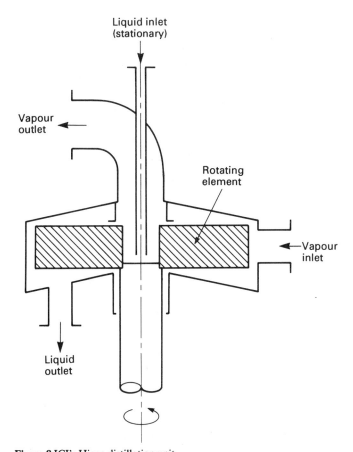

Figure 8 ICI's Higee distillation unit

If distillation columns contain large inventories of hazardous materials or the materials are particularly liable to leak from pump glands, then the reflux and bottoms pumps should be fitted with *emergency isolation valves*.

If flammable materials are handled in vacuum stills, the vacuum should be broken with *nitrogen*, not air.

Distraction

Human errors increase when *stress* and distraction (and *boredom*) are high. (See *human failing*.) The following description of life on a farm is intended to be a picture of life in the State Department in Washington but could apply equally well to a busy day on the plant[1]:

> Here a bridge is collapsing. No sooner do you start to repair it than a neighbor comes to complain about a hedgerow which you haven't kept up - a half-mile away on the other side of the farm. At that very moment your daughter arrives to tell you that someone left the gate to the hog pasture open and the hogs are out. On the way to the hog pasture you discover that the beagle hound is happily liquidating one of the children's pet kittens. In burying the kitten you look up and notice that a whole section of the barn roof has blown off, and needs instant repair. Someone shouts pitifully from the bathroom window that the pump must have busted – there's no water in the house. At that moment a truck arrives with five tons of stone for the lane. As you stand helplessly there, wondering which of these crises to attend to first, you notice the farmer's little boy standing silently before you with that maddening smile that is halfway a leer, and when you ask him what's up, he says triumphantly, 'The bull's busted out and he's eating the strawberry bed'.

No wonder operators sometimes forget to close a valve.

1. G Kennan, *Memoirs 1925–1950*, Atlantic, Little and Brown, New York, 1967

Diversity

See *redundancy* and *variety*.

Dolls

See *meccano or dolls?*

Domino effects

After the *explosion* at *Flixborough* in 1974 there was concern that a similar incident on a large site might produce a domino or knock-on effect, that is, an explosion on one plant might damage another so severely that a large leak occurred, followed by another explosion, and so on across the site.

Fortunately, many large sites are so spaciously laid out that this is unlikely to occur.

Domino effects should always be taken into account when deciding on plant *layout and location*. Plants containing hazardous chemicals should be located so that, if an explosion occurs on one, the peak incident overpressure on the others will not exceed 0.3 bar gauge (5 p.s.i.g.)[1].

To avoid domino effects with *fire*, large plants should be divided into blocks, with a space of at least 20 m between them, like fire breaks in a forest.

Other ways of avoiding domino effects are:

- Handling flammable gases in the open and not in a closed building. See *ventilation*.
- Constructing low pressure storage *tanks* with a weak seam roof, so that, if the tank is over-pressured, the roof lifts and the contents are retained in the tank.

Domino effects are at their most extreme on off-shore oil platforms. People and equipment are concentrated into a small space, ventilation is poor and escape is difficult so that a leak which, on a normal shore-based plant, might result at the worst in a few casualties can kill hundreds, as occurred on *Piper Alpha*.

1. T A Kletz, *Loss Prevention*, Vol. 13, 1980, p. 147

Drains

Drains, like all *service lines*, are often taken for granted, receive little management attention and are involved in more than their fair share of incidents. The main hazards are *explosion* and *reverse flow*.

As long as there is an oil/water separator, many people do not worry if oil gets into the factory drains. 'There is no source of *ignition* down there', they say, 'so it cannot explode.' However, when some light oil got into a factory drain it exploded. Some manhole covers were blown off, the contents of the drain blown out and the roadway 4 m above was cracked. As so often happens, the source of *ignition* was never found, showing once again that if air and fuel are mixed in the right proportions they are liable to ignite even though we do what we can to remove known sources of ignition.

In order to reduce the chance of fire or explosion in the drains:

- Drains should be underground (see *amateurism*) and should be fully-flooded (surcharged) so that there is no vapour space.
- All inlets and manholes should be sealed.
- Any oil that gets into the drains should be flushed out as soon as possible.
- Combustible gas detectors should be installed at the drain outlet so that oil is detected promptly. (See *reorganization*.)

The worst drain explosion in history occurred in Cleveland, Ohio in 1944 when a liquefied natural gas tank ruptured and the entire contents were spilt; some of the liquefied gas entered the town drains and exploded, killing 128 people. See *Lees*, Appendix 3, Item A2.

A similar but less serious incident occurred in the UK. A spillage of 20 tonnes of petrol soaked into the ground, percolated into a porous sewer and ignited. Two miles away two men were injured when the vapour exploded.

Reverse flow in drains, sometimes caused by excessive flow into another section, has caused hazardous materials to appear in some unexpected places. When some vinyl chloride was spilt and entered the drains its vapour pressure was sufficient to overcome the level of water in the inlet U-bends and the vapour came out in a laboratory and exploded. *Hazard and operability studies* should be carried out on all service lines including drain lines.

Drains should, of course, be covered when welding is taking place nearby. See *LFA*, Chapter 1.

Drugs

I know of no study of the extent of drug taking in industry or its effects on safety. The general view is that it is less of a problem than *alcohol*, but growing.

K Hayton[1], gives some advice on recognizing signs of drug taking and responding to them: We should avoid moral judgements; a decision to intervene should be based solely on deteriorating performance in the job; if we intervene early, chances of recovery are better.

1. F Hayton, *Alcohol and Drug Dependency Study and its Application to the Fire Service*, Clwyd Fire Service, 1987

Drums

It might be thought that drums are safer than *tanks* for the storage of flammable liquids, as each one contains so little, but this is not so. It would be difficult to find a better way of burning liquid quickly than putting it in thin-walled containers which are then stacked with spaces between them. If a leak ignites the drums soon burst and some of them may *BLEVE*.

There are, however, far fewer restrictions on the storage of liquids in drums than in a tank and stacks of drums are often allowed to grow uncontrolled on any piece of spare ground. Put there as a *temporary job* they soon become part of the scene. I suggest that drums of flammable liquids should be segregated from other materials and stacked on sloping, impermeable ground, at least 15 m from buildings or plant. There should be no more than 1500 drums in a stack, stacked not more than five drums high, and at least 5 m should be left between stacks (9 m for particularly hazardous liquids such as ether or carbon disulphide). These recommendations are based on those for storing timber at docks.

See *small companies* and *Myths*, §19.

In Scotland about four million tonnes of whisky are stored in barrels, on racks up to 10 m high. This probably exceeds by far the amount of any other liquid in storage in drums or similar containers and the Health and Safety Executive has published a Guidance Note on the subject. It will interest anyone who stores drums on racks.

Du Pont

The *United States* chemical company Du Pont have an outstandingly good safety record. Their fire and explosion record and their process safety standards are much the same as those of the best UK chemical companies but in everyday safety they are streets ahead. Their *lost-time accident rate* is about a twentieth of that of the best UK chemical companies. Why?

A senior Du Pont engineer once told me that when he was a young manager one of his men lost the tip of a finger in an accident. Twenty years later he still looked upon the incident as the most unpleasant in his career as so many senior managers descended on the plant and made him feel he had let the side down. If a similar incident occurred in the best UK companies the manager concerned would be genuinely distressed but it would not be the traumatic experience it is in Du Pont.

Success in any *management* function depends on the effort put in. If we are told that we must improve product quality or go out of business we usually manage to improve the quality with our existing plant before the research department have finished designing the new equipment needed. Similarly, if senior managers make it clear by their actions, rather than their words (see *platitudes*), that safety really is important, then the safety record will improve. Many years ago, after visiting Du Pont, an *ICI* manager wrote, '. . . accidents constitute a costly interference with production and therefore must be eliminated as far as possible. Having taken this decision, they have set about implementing it in the same thorough manner as they would were the interference caused by faulty machinery, bad material, or failure in organization'.

This Du Pont attitude seems to date back to about 1912. In that year their *lost-time accident rate*, the number of lost-time accidents in a million worked hours, was 43; in 1922 it was 10, in 1931 2. A plant acquired by Du Pont in 1929 had a rate of 28 that year; in the first half of 1931 it was 6, in the second half, zero.

See *culture, history of safety* and *LFA*, Chapter 3.

Dusts and powders

Many dusts are explosive. To prevent explosions we can:

- Use a non-explosive or less explosive dust instead. (See *substitution*.)
- Use the dust in a non-explosive or less explosive form. For example, some powdered dyestuffs can be handled as slurries, an example of *attenuation*.
- Eliminate all sources of *ignition*. Dusts have higher ignition energies than gases so this policy, impracticable for gases, is sometimes followed. However, it is difficult to be certain that all sources of ignition have been

eliminated, especially if there are any moving parts. Dusts, likes gases, can be ignited by *static electricity*.
- Use *explosion venting*; it is widely used.
- Inert with *nitrogen* or other inert gas.

Dust explosions can occur in equipment and in buildings. A feature of explosions in buildings is that very often a small explosion disturbs dust which has settled and this dust then explodes causing far more damage than the original explosion. Buildings in which dusts are handled should therefore be designed so that surfaces on which dust can settle are as few as possible and access should be provided so that those surfaces that cannot be eliminated can be kept clean.

For more information see the references given under *explosion venting* and *Industrial Dust Explosions*[1].

Dusts, of course, can also be harmful to inhale and if the concentration in the workplace atmosphere cannot be reduced to a safe level, dust masks may have to be worn.

The storage of dusts and powders presents some problems which do not arise with liquids, namely:

- Arching.
- Eccentric filling and discharge.
- Piping or rat-holing; flow stops when the central pipe is empty.
- Consolidation.

These, especially the first two, set up stresses in storage silos which are not allowed for in most codes and which have caused many silos to collapse[2].

Parabolic silos (the walls and roof are the shape of a parabola) have been used for many years for storing fertilizers. The contents exert no pressure on the walls, which can be made very weak. The contents have to be removed mechanically and cannot flow out.

See *unexpected hazards*.

1. K L Cashdollar and M Hertzberg, *Industrial Dust Explosions*, ASTM, Philadelphia, Pennsylvania and Hitchin, UK, 1987
2. A Boniface, *The Chemical Engineer*, No. 458, March 1989, p. 21

Early involvement in design

Many *safety professionals* waste their company's money on the accepted and conventional ways of controlling hazards when they could be avoided during the early stages of design. See *friendly plants* and *inherently safer design*. This is not due to any fault of the safety professional but to the fact that they are not usually involved in the early stages of design.

For example, relief valves on plants which handle hazardous materials should, and usually do, discharge to a flare or recovery system. The safety professional usually and rightly asks for these systems to be installed and sees that the relief valves are tested regularly. How often, however, does he ask if it is possible to avoid the need for relief valves and the associated flare or recovery systems by constructing equipment able to withstand a higher pressure? If the question is asked, it is often found that the equipment is already on order and it is too late to change the design.

Another example: On some plants the biggest fire risk comes from flammable heat transfer liquids, including refrigerants. The safety adviser asks for the necessary fire prevention, detection and fighting equipment but how often does he ask if a non-flammable or less flammable liquid or water can be used instead? It is no use asking this question when the plant is designed or built; it should be asked in the early stages of design.

If safety professionals are to give their employers better value for money, and achieve a higher standard of safety, they must equip themselves to contribute to the early stages of design and make sure they are involved.

Most safety professionals spend a good deal of time on the plant and so they should; this is where accidents happen. Many safety professionals get involved in detailed design and so they should; alterations then are cheaper and easier than when the plant is built. Few safety professionals get involved in the early stages of design when it is possible to avoid hazards rather than tinker with them. The safety professional's motto should be 'Get in early'.

Effects

See *cause and effect*.

Electrical area classification

Plants handling flammable gases or liquids are divided into Zones (called Divisions in the US and in old UK papers) according to the likelihood that a flammable mixture will be present. (See British Standard 5345.):

Zone 0:	Continually present or present for long periods.
Zone 1:	Likely to occur.
Zone 2:	Not likely to occur under normal operation and if it does occur will exist for only a short time.
Non-hazardous:	Not expected to be present in quantities such as to require special precautions.

These official definitions are *qualitative* and the following are often used:

Zone 0: Present for more than 1000 hours per year.

Zone 1:   Present for 10–1000 hours per year.
Zone 2:   Present for less than 10 hours per year.
(Non-hazardous: Present for less than 0.1 hour per year.)

The last item is in brackets as it is not usually quoted but I have added it to complete the list.

The purpose of the area classification is to guide the electrical engineer in the choice of electrical equipment. In Zone 2 areas equipment may be used which does not spark in normal operation but may spark if a fault develops, say once in 100 years. The chance that this will coincide with the presence of a flammable mixture and that a person will also be present is so low that it can be accepted. The Zone 2 concept is thus one of the earliest applications of *hazard analysis* in the oil and chemical industries[1]. (See *Fermi estimates*.)

Although the area classification provides information essential to the electrical design engineer the classification of the plant into Zones is a matter for the process engineer who has to assess the likelihood that various items of equipment will leak, the duration of the leak and the extent of the flammable cloud. There are various codes to assist him, notably the ICI code[2] and the NFPA guide[3]. At the time of writing the UK engineering institutions are jointly preparing a new code.

In modern plants most of the plant area is Zone 2 and only small areas are Zone 1. Zone 0 is restricted to a few places such as the insides of storage tanks where there is normally no need for electrical equipment.

If it is difficult or expensive to provide electrical equipment to suit the area classification the electrical engineer should ask if the classification is certain or if the equipment could be moved. *LFA*, Chapter 2 describes a complex protective system in a Zone 2 area which was neglected and produced a bigger hazard than the one it was designed to remove. If the equipment it was protecting had been moved a few metres it would have been in a safe area and there would have been no need for the protective system.

A wide variety of equipment is now available for use in Zone 1 and 2 areas. It includes:

- Type p – pressurised: air or inert gas prevents flammable gas entering the housing.
- Type d – flameproof: able to withstand an explosion without igniting surrounding gas.
- Type i – intrinsic safety: the energy available is too small to cause ignition.
- Type N – 'non-sparking', that is, does not spark in normal use, only when a fault develops. It is suitable for Zone 2 only.

Intrinsically safe equipment has a long history. In the eighteenth century attempts were made to illuminate a gassy coalmine by the phosphorescent glow from putrefying fish skins.
See *Lees*, §16.6.

1. J M Benjaminsen and R H Wiechen, *Hydrocarbon Processing*, Vol. 47, 1968, p. 121
2. ICI, *Electrical Equipment in Flammable Atmospheres*, Royal Society for the Prevention of Accidents, Birmingham, 1972
3. P J Schramm and M W Early, *Electrical Installations in Hazardous Locations*, National Fire Protection Association, Boston, Massachusetts, 1988

Electricity

See *static electricity*.

Emergencies

Handling emergencies and planning for them is too big a subject to be discussed here. See *Lees*, Chapter 24. All I can do is to draw attention to a few points that should be obvious but are often overlooked.

The first is the the need for adequate warning systems, so that those who have to deal with the emergency can do so and those who are not required can leave the area (or take refuge if the emergency is a release of toxic gas). If there are different alarms for different sorts of emergency (for example, fires and toxic releases) their sounds should be easy to tell apart and should be made known to all concerned by regular *tests*.

The routes to be followed when leaving the plant should be made clear and there should be planty of exits from fenced areas. When a leak occurred in a fenced area, several men were seriously injured because they took the normal route to the assembly point; it led them close to the leak, which ignited as they were passing. See *LFA*, Chapter 5.

The second point I wish to make is that spectators should be kept out of the way. They endanger themselves and get in the way of those who have to deal with the emergency. Nine men were injured when a fire which had been extinguished started again without warning. (See Figure 9.) Keeping

Figure 9 Emergencies

people out of the way is particularly important if an emergency, such as a tanker accident, occurs outside the factory and the public are tempted to gather. A boy was killed and two others injured when they were watching a van fire; a propane cylinder exploded and was projected 17 m across the street[1].

Following the accident at Moorgate underground railway station in London in 1975 there were complaints that visiting VIPs got in the way of the rescuers. The Chief Fire Officer was reported as saying that 'parties of VIPs should be conducted at certain times only, i.e. during operational pauses to the incident face and not allowed to stay on the scene indefinitely[2]'.

Plans for handling emergencies should be prepared and rehearsed in advance. They should include, as well as plans for calling and liaising with the fire, police and ambulance services, plans for briefing the press and handling public inquiries. After the *Zeebrugge* disaster the holding company referred all inquiries to its public relations office which had just two telephone lines! M Langford[3] advises those who speak to the press that they should, 'Never try to blame someone else even if you think it's their fault. If a manager shows that his only interest is the public safety, they will forgive him almost anything.'

See *alarms*.

1. *Health and Safety at Work*, Vol. 10, No. 2, February 1988, p. 19
2. S Holloway, *Moorgate – Anatomy of a Disaster*, David and Charles, Newton Abbot, 1988, p. 160
3. M Langford, quoted in *Telegraph Weekend Magazine*, 18 March 1989, p. 5

Emergency equipment

Many factories, offices and hospitals have installed *diesel engines* for use in emergencies, to supply electricity or fire water. An article by an *insurance surveyor*[1] shows that these engines often fail to start or operate satisfactorily when they are needed, not because their design or manufacture is poor but because the engines are not looked after properly or tried out regularly.

A quarter of the failures were due to overheating of the engines, because cooling water pumps were not working, cooling water lines were clogged or valves in cooling water were closed. A tenth of the failures were due to freezing of the cooling water, easily prevented with anti-freeze. Some failures were due to overspeeding, because governors were never tested or properly maintained. Other were due to lack of lubrication or water in the lubricating oil.

One failure occurred on an engine which was fitted with a low oil pressure switch. It failed to operate because the operator had connected a wire across the terminals. The switch was a nuisance, he said, liable to shut the engine down.

A few failures were due to poor *maintenance* rather than lack of maintenance, for example, over-tightened bolts, misaligned valves, incorrect clearances or poorly reconditioned crankshafts.

All emergency equipment should be tested and maintained regularly or it cannot be expected to work when required. The paper includes a *check list* for the testing and inspection of diesel engines.

There is no point spending money on emergency equipment if we are not prepared to give it the attention it needs. Once again we see accidents caused not by lack of money but by a failure to look after the equipment we have. See *extravagance* and *Summerland*.

1. R Stevens, *Plant/Operations Progress*, Vol. 2, No. 4, October 1983, p. 203

Emergency isolation valves

Remotely-operated emergency isolation valves have successfully isolated many leaks and paid for themselves many times over[1]. They cannot be installed in the lines leading to and from all equipment which might leak but they can be installed when experience shows that equipment is particularly liable to leak, for example, very hot or cold pumps, or when, if a leak occurs, the quantity of hazardous material that leaks out will be large, for example, the bottoms pumps on large *distillation* columns. The valves can be operated electrically or pneumatically and the operating buttons should be located so that they can be reached when a leak occurs. (See Figure 10.)

On plants handling flammable materials the valves and their impulse lines should be fire-protected, unless safe by location.

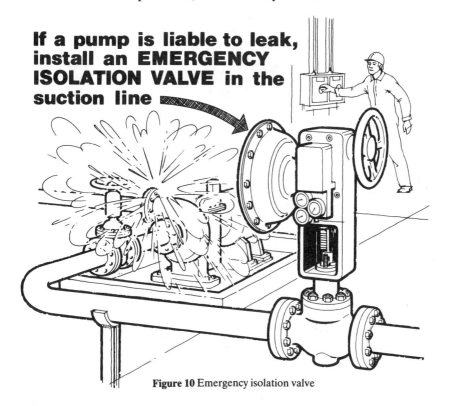

Figure 10 Emergency isolation valve

Figure 11 Emergency isolation valves

Emergency valves, whether remotely-operated or hand-operated, should be lubricated and tested regularly, say once a month. (See Figure 11.) If testing interferes with plant operation the valves should be moved halfway and tested fully at shutdowns.

See *dead-ends, defence in depth, expert systems, furnaces*, and *WWW*, §7.2.1

1. *Chemical Engineering Progress*, Vol. 71, No. 9, September 1975, p. 63

Employer's responsibilities

An employer's responsibilities are numerous but can be summarized by the following:

The standard that the law requires is that (employers) should take reasonable care for the safety of their workmen. In order to discharge that duty properly an employer must make allowance for the imperfections of the human nature. When he ask his men to work with dangerous substances he must provide appliances to safeguard them: he must set in force a proper system by which they use the appliances and take the necessary precautions, and he must do his best to see that they adhere to it. He must remember that men doing a routine task are often heedless of their own safety and may become slack about taking precautions. He must, therefore, by his foreman, do his best to keep them up to the mark and not tolerate any slackness. He cannot throw all the blame on them if he has not shown a good example himself[1].

The Factories Act is there not merely to protect the careful, the vigilant, and the conscientious workman, but, human nature being what it is, also the careless, the indolent, the inadvertent, the weary and even perhaps in some cases the disobedient[2].

(A person) is not, of course, bound to anticipate folly in all its forms, but he is not entitled to put out of consideration the teachings of experience as to the form those follies commonly take[3].

See *(legal, managerial* and *personal) responsibility*.

1. A quotation from a judge's summing up which appeared in several newspapers in October 1968
2. A quotation from a judge's summing up, origin unknown
3. A House of Lords' judgement quoted by M Whincup, *Guardian*, 7 February 1966

Empty vessels

See *nothing*.

Enforcement

See *breaking the rules*.

Entry

The insides of vessels and other confined spaces are common sites of serious, often fatal, accidents. Sometimes it seems that vessels used for storing hazardous chemicals are more hazardous empty than full!

Some incidents have occurred because vessels were not cleaned before they were entered, others because they were not thoroughly isolated and hazardous materials leaked in, some because hazardous materials were deliberately introduced for testing or painting and others because people entered before preparation was complete. Sometimes a man has been overcome inside a vessel, others have gone in to rescue him and have been overcome themselves.

Before anyone is allowed to enter a vessel or other confined space we should take the following precautions:

- Isolate it from all sources of danger by disconnecting or slip-plating all pipelines, including the lines leading to internal coils, and isolate any electrical supplies.
- Free it from hazardous materials. The method used will depend on the materials present. Traces of oil can be removed by steaming; water-soluble materials by washing out with water; flammable vapours by sweeping out with nitrogen followed by air.
- Test to make sure that hazardous materials have been removed. Test well inside the vessel, not near the manhole, and in several places if the vessel is large or has a complex shape. Check continuously or at frequent intervals to make sure conditions have not changed.
- If it is impossible to reduce the concentration of hazardous materials to a safe level then see that breathing apparatus and/or protective clothing are worn. No one should be allowed to enter a vessel if the concentration of flammable vapour exceeds 20% of the lower flammable limit, as the concentration may be higher in other parts of the vessel. Remember that any sludge present may give off fumes when it is disturbed.
- Do not introduce hazardous materials except under controlled conditions. For example, if the inside of a tank has to be painted, adjust the ventilation and rate of application so that the concentration of vapour never exceeds the Maximum Exposure Limit (or Occupational Exposure Standard) (see *COSHH Regulations*) or 20% of the lower flammable limit, whichever is the lower. Check that these limits are not exceeded. A carpet fitter was killed by trichlorethylene vapour while glueing tiles in a basement toilet[1].
- If you cannot see the whole of a vessel, for example, the space behind a baffle, assume that the unseen part is dirty and contains some of the material previously present in the vessel, particularly sludges and other materials which are hard to remove. (See Figure 12.)
- Do not allow entry until a responsible person has checked that all these precautions have been taken and has issued an entry permit detailing any additional measures considered necessary such as presence of stand-by men. It should be physically impossible to enter the vessel until the permit is issued. If the manhole has been removed then a temporary barrier should be fixed across it. It is not necessary to enter a vessel to be overcome; just looking in is enough. Men have been killed because they took a quick look inside a vessel to see if it was clean. (See Figure 13.)

One report[2] describes an extraordinary incident. A shaft had to be fitted into a bearing; it was a tight fit, so the men on the job decided to cool the shaft and heat the bearing. They cooled the shaft by pouring liquefied petroleum gas on it while they heated the bearing with a welding torch.

In the UK most of these precautions are required by the Factories Act, § 30.

For more information and accounts of accidents involving entry see reference 2, *WWW*, Chapter 11, *Lees*, § 21.2.5 and *unexpected hazards*.

1. *Health and Safety at Work*, Vol. 11, No. 4, April 1979, p. 6.
2. W W Cloe, *Selected Occupational Fatalities related to Fire and/or Explosion in Confined Workplaces*, Report No. OSHA/RP-82/002, US Dept of Labor, April 1982

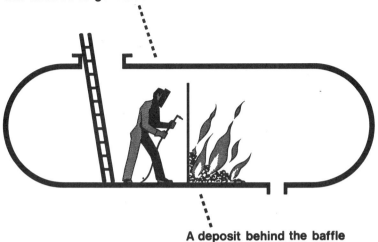

The vessel looked clean so the welder was allowed to go inside

A deposit behind the baffle caught fire

If you cannot see the whole of a vessel, assume it contains hazardous materials

Figure 12 Entry

Joe put his head inside a vessel to see if it was clean

Figure 13 Entry

Equipment

See *emergency, identification of, old, poor* and *protective equipment* and *removing equipment or procedures*; also *boilers, centrifuges, cranes, cylinders, distillation, fin-fans, furnaces, level glasses, pipe failures, plugs, pumps, reactors, relief devices, stacks, tanks, valves* and *vessels*.

E shift jobs

Continuous plants are usually operated by four shifts known as A, B, C and D. When something has been done wrongly, it is often impossible to find out which shift did it, when or why. No one will admit to doing it.

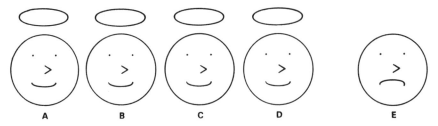

Figure 14 Who did the jobs that A, B, C and D shifts say they never did?

These jobs are sometimes said to have been done by a mythical E shift who work only nights and weekends so that the day staff never see them (see Figure 14). Here are some examples of E shift jobs.

- A *leak* on a liquefied petroleum gas pipeline was traced to a sub-standard drain valve, made from brass and stocked for use on domestic water lines. No one knew who had fitted it or when.
- When a plant was being brought up to pressure after a shutdown a loud bang was heard and it was found that a low pressure hose had burst. No-one knew who fitted it or when but it may have been fitted before the shutdown in readiness for a catalyst change which was later cancelled. It was not on the list of jobs to be done and no permit-to-work could be found.
- A small tank was used for storing, as liquid, a substance that melts at 100°C; it was kept hot by a steam coil. The inlet line to the tank was being blown with compressed air to prove that it was clear, the usual procedure before filling the tank. The end of the tank was blown off, killing two men who were working nearby. It was then found that the vent on the tank – a hole 3 inches in diameter – was choked. Sometime in the past the original 6 inch vent had been blanked and a 3 inch hole used instead. It was not known when, by whom or why. Possibly it was done to prevent dirt entering the tank. See *LFA*, Chapter 7.

Who really does E shift jobs? The other shifts are not being dishonest when they say that they cannot remember doing the jobs. The jobs have been forgotten because there was no proper system for authorizing, controlling and recording what was done and no one kept their eyes open for unauthorized jobs. No change should be made to plant or process until it has been authorized at an appropriate level and specified in detail; it should be inspected on completion and a record kept. Maintenance organizations should refuse to carry out *modifications* which have not been authorized.

If E shift has been busy on your plant, do not blame the other shifts but look at your *management* systems.

Evangelicalism, evangelism

Many *safety professionals* are evangelists for their ideas, that is, they try enthusiastically to convert others to their ways. But evangelicalism in the narrower sense, the nineteenth century religious movement of that name, had a big influence on industrial safety, an influence rarely mentioned in histories of factory legislation. Barbara Tuchman writes[1]:

> A lot of ridicule has stuck to the reputation of Lord Shaftesbury, the archetype as well as the acknowledged lay leader of the Evangelical party. It hurts the economic historians, the Marxians and Fabians to have to admit that the Ten Hours Bill, the basic piece of nineteenth century labor legislation, came down from the top, out of a private nobleman's private feelings about the Gospel, or that abolition of the slave trade was achieved, not through the operation of some 'law' of profit and loss, but purely as the result of the new humanitarianism of the Evangelicals. . . Granted that they were not thinkers, not reasonable or graceful or elegant; granted that, including Lord Shaftesbury, they were in some ways rather silly. Yet they were the mainspring of early Victorian England. . .
>
> In our day it has become almost impossible to appreciate justly the role of religion in past political, social and economic history. We cannot do it because we have not got it. Religion is not part of our lives; not, that is, comparable to its part in pre-twentieth century lives.

See *fate*.

1. Barbara Tuchman, *Bible and Sword*, Redman, New York, 1956, Macmillan, London, 1982, p. 180

Exchanging one problem for another

There is an old story about a factory inspector who complained about the dust in a factory. The manager installed equipment to collect the dust and blow it outside. Another inspector then complained about the pollution. (See Figure 15.)

As this story shows, when we solve one problem we often acquire another. Here are some examples.

- In parts of the US in the 1970s, the authorities insisted that petrol filling stations should collect the fumes emitted when petrol is put into a car.

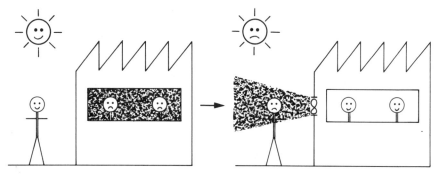

Figure 15 Installing ventilation equipment to improve the atmosphere inside a factory may produce a problem outside

The mixture of petrol vapour and air was collected by a fan and blown into a recovery system. Twenty fires occurred in four months[1].
- The amount of air used in a furnace is sometimes decreased to save fuel and to reduce the formation of oxides of nitrogen. This can result in the production of fuel-rich mixtures and a risk of *explosion* when the unburnt gas is mixed with air[2].
- Bag filters and precipitators, installed to avoid discharge of dust to atmosphere, have produced *dust* explosions[2].
- The burning of waste products in plant *furnaces*, to save fuel and reduce pollution, has caused serious corrosion and tube failure.
- Shortening of pipe-runs to reduce heat loses and save pipe has resulted in congested plants; if a fire occurs, damage is increased. Stacking of equipment above pipe runs, to save space, is particularly short-sighted. (See *domino effects*.)
- *Nitrogen*, widely used to prevent explosions, has caused many deaths by asphyxiation.

For other examples see *aerosol cans, iatrogenesis, modifications* and *small companies*.

There is no easy way of foreseeing the results of change. We are so relieved that we have found a solution to our problem that a mental block makes it difficult for us to see the snags. *Hazard and operability studies* may help.

1. W H Doyle, *Loss Prevention*, Vol. 10, 1976, p. 79
2. F Bodurtha, *Loss Prevention*, Vol. 10, 1976, p. 88

(loss of) Expectation of life

How can we compare risks? (See *criteria*.) We often compare the probability of death. (See *fatal accident rate*.) An alternative is the loss of expectation of life. Thus a man who becomes a deep sea fisherman at age 20 and plans to retire at age 65 loses on average over 3½ years of life. If he entered the paper, publishing and printing industry instead he would lose only 12 days on average[1].

These figures can bring out very forcibly the relative risks of different occupations but they can be misleading. A few people lose a lot of life and most lose none. For the man who is unemployed the unemployment rate is 100%.

Suppose the probability of being killed quickly by an accident or slowly by a toxic chemical is the same; the accident produces a much bigger loss of expectation of life. The publication *Living With Risk*[1] is highly recommended to anyone who would like to know his chances of surviving the attentions of surgeons, employers, drivers, brewers and tobacconists and how these chances have changed.

See *smoking*.

1. British Medical Association, *Living with Risk*, Wiley, Chichester, 1987, p. 69

Expenditure proposals

See *persistence*.

Experience

Experience is a good teacher but the fees are heavy; it has two more drawbacks. First, an accident has to occur before we can learn from it. When plants were small and the size of accidents limited this might have been acceptable. We built our plant or piece of equipment; when an accident occurred we said, 'Sorry. I didn't know that could happen, we'll make sure it doesn't happen again'. This 'every dog is allowed one bite' philosophy is not acceptable now that we keep dogs as big as *Bhopal* (over 2000 people killed in one bite) or even *Flixborough* (28 killed). We have to find ways of preventing accidents before they occur. See *hazard and operability studies*.

The second drawback of experience is that it is not readily transferable. People who have blown up a plant rarely blow it up again, at least, not in the same way, but after a few years they leave and their experience is lost. Organizations have no memory.

What can we do to see that the lessons of the past are not forgotten? See *learning from others' experience* and *lost knowledge*.

(learning from others') Experience

Legally and morally we are expected to learn the lessons of accidents that have occurred elsewhere, if the information is readily available. Every dog is allowed one bite but in industry we are expected to muzzle our dog if a similar dog has bitten someone elsewhere.

What is meant by 'readily available'? Nobody would expect us to know about an accident described in the Journal of the Outer Mongolian Chemical Society but information published in well-known British and American Journals is certainly readily available. *Safety professionals* should peruse these journals and tell their colleagues about incidents that should concern them.

Sometimes information is readily available but only in journals not normally read by members of the profession concerned. See *Abbeystead*. Engineers should read widely.

The following quotation from a judge's summing up describes the legal position in the UK:

> (He) pointed out that a reasonable and prudent employer should take positive thought for the safety of his workers in the light of what he knows or ought to know. Where there is a recognized and general practice which has been followed for a substantial period without mishap in similar circumstances he would be entitled to follow it unless in the light of commonsense or newer knowledge it is clearly bad. Where knowledge is developing he must keep reasonably abreast of it and not be slow to apply it. Where he has, in fact, greater knowledge of the risks he may therefore be obliged to take more than the average or standard precautions.

Although we should learn from the past we often do not. See *Aberfan* and *lost knowledge*.

(limitations of) Experience

When an unsafe practice is pointed out someone often says, 'It must be safe. We have done it this way for 20 years and never had an accident.'

Next time someone talks like this, ask them if an accident in the 21st year is acceptable. If not, their experience is too short to prove that the practice is safe. It does not even prove that the average accident rate is less than once in 20 years. All it proves is that we are 86% confident that the average rate is less than once in 10 years. See *Myths*, § 33.

Here are some examples of accidents that occurred after the same operation had been carried out without incident for many years:

- Road *tankers* were splash-filled with gas oil. The filling arm should have reached to the bottom of the tankers to prevent splashing but it did not. The oil had a high flash-point (over 50 °C) and was much harder to ignite than petrol but the splashing produced a mist which was easy to ignite, and a *static electricity* spark set it alight. The tanker and filling arm were earthed but the spark passed between the gas oil in the tanker and the filling arm (or the top of the tanker). Thousands of tankers had been filled without incident before but this time conditions were right for an ignition.
- At Oppau, Germany in 1921 explosives were used to break up storage piles of a 50/50 mixture of ammonium nitrate and ammonium sulphate. Two explosions occurred, killing 430 people. The operation had been carried out without mishap about 16 000 times. See *Lees*, Appendix 3.1, §A1.
- While starting up a plant for making coke an employee forget to open a valve and this led to an explosion in which a man was killed. The plant, and a similar one, had been shut down and started up every few days and had had 6000 successful start-ups[1].

- From 1940 to 1958 nitromethane was considered stable and safe to transport in rail tankers. In 1958 two tankers exploded in separate incidents in the US; both were triggered by shunting. Two people were killed and damage was extensive[2].

See *extrapolation, King's Cross* and *will and might*.

1. O A Pipkin in *Fire Protection Manual for Hydrocarbon Processing Plants*, edited by C H Vervalin, Gulf Publishing Company Houston, Texas, Vol. 1, 3rd edition, 1985, p. 95
3. D J Lewis, *Hazardous Cargo Bulletin*, Vol. 4, No. 9, Sept 1983, p. 36

Experts

See *knowledge of what we don't know* and *welding*.

Expert systems

Much loss prevention knowledge is suitable for storage and retrieval on expert systems. At the time of writing much development work is going on but little use is being made of expert systems by those seeking advice on loss prevention. In time their use will grow though I suspect that the systems will be used more by loss prevention professionals who wish to be reminded of information they have forgotten or overlooked, than by their clients.

Bunn and Lees[1] have considered the feasibility of using expert systems to decide when *emergency isolation valves* should be installed and in the design of flare *stacks*. They identified several problems. 'One is that the abstraction of rules from the literature is liable to give rules which are contradictory or out-of-date. Another is that some rules are stronger than others and means need to be found to incorporate this. Similarly some rules are subject to exceptions and this also needs to be taken into account'. They also point out that 'in applying classification methods in order to induce rules it is important to take care in the definition of the attributes and in the provision of a set of examples which is not only sufficiently large but specifically designed to cover the domain.'

Other areas in which expert systems may be found useful are fault diagnosis and specification of equipment, including control equipment, during design.

Andow points out[2] that expert systems are best at applying 'narrow and deep' knowledge but that often we need to apply broad knowledge, a combination of many different types of knowledge and commonsense. For this reason an expert system is unlikely ever to take over a *hazard and operability study*[2,3].

1. A R Bunn and F P Lees, *Chemical Engineering Research and Design*, Vol. 66, September 1988, p. 419
2. P K Andow, *Proceedings of the International Symposium on Preventing Major Chemical Accidents*, American Institute of Chemical Engineers, New York, 1988, p. 1.129
3. T A Kletz, *The Chemical Engineer*, No. 453, October 1988. p. 52

Explosions

Explosions are too big a subject to be discussed adequately here and all I can do is draw attention to a few points that are sometimes overlooked. For more information see *Lees*, Chapter 17.

Most of the materials handled in the oil and chemical industries will not burn or explode by themselves but only when mixed with air (or *oxygen*) in certain proportions. To speak of a petrol *tank* or *tanker* as a bomb is therefore absurd. For an explosion to occur either air has to enter the vessel or the contents have to leak out, vaporize, mix with air and be ignited.

In the open air several tonnes, usually several tens of tonnes, of the explosive material, are usually necessary for an explosion but indoors much less is sufficient. Flammable liquids and gases should therefore be handled in the open air, whenever possible. See *dispersion* and *ventilation*.

During the post-war years the chemical and oil industries expanded rapidly and larger plants containing larger quantities of hazardous materials were built, with the unforeseen result that a number of serious unconfined vapour cloud explosions occurred. *Flixborough* is the best-known example. (See also *change – new problems* and *flashing liquids*.) It was followed by an explosion of papers on the probability of such explosions and methods of estimating their effects but surprisingly little on methods of prevention. See *choice of problems, construction* and *inherently safer designs*.

There are some chemicals (and mixtures) which, like conventional explosives, can decompose violently by themselves, if detonated by heat, sparks or impact (for example, peroxides and ethylene oxide) and they should, of course, be treated with greater respect than the general run of flammable liquids and gases.

An explosion is any violent release of energy and as well as chemical explosions we can have physical ones caused, for example, by the bursting of a vessel (see *BLEVE*) or the rapid vaporization of water (see *foam-over*).

See *defence in depth, dusts, furnaces, static electricity, unexpected hazards* and *windows*.

Explosion venting

The articles on *dispersion* and *ventilation* stress the importance of handling flammable liquids and gases in the open air, so that small leaks are dispersed by natural ventilation and cannot form an explosive mixture of vapour and air. If these materials must be handled indoors, for example, because the climate is very cold, then the damage caused by an explosion can be limited by explosion venting, that is, by making the walls of light plastic panels, loosely fixed, so that they will blow off as soon as the pressure starts to rise and prevent it rising to its full value. Explosion venting will minimize damage but will not protect any people who are present so it is very much a second best compared with good ventilation[1,2,3]. See *construction*.

Explosion venting is widely used to prevent or minimize damage to equipment such as driers in which flammable dusts are handled. The vent panels should have a clear path to the open air, free from bends or restrictions, and other equipment should not be placed in the path of the escaping flames.

1. *Guide to Dust Explosion Prevention and Protection*, Part 1, 1984, by C Schofield, Part 2, 1987, by C Schofield and J A Abbott, Part 3, 1988 by G A Lunn, Institution of Chemical Engineers, Rugby
2. G A Lunn *Venting Gas and Dust Explosions – A Review*, Institution of Chemical Engineers, Rugby, 1984
3. I Swift, *Journal of Loss Prevention in the Processs Industries*, Vol. 2, No. 1, January 1989, p. 5

Extrapolation

Extrapolation can be useful, but we should be aware of its limitations. In Figure 16 the horizontal axis is a measure of the dose or action and the vertical axis is a measure of the response or effect.

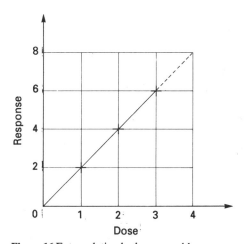

Figure 16 Extrapolation looks reasonable . . .

When the dose is 1, the response is 2.
When the dose is 2, the response is 4.
When the dose is 3, the response is 6.

Can we therefore say that when the dose is 4, the reponse will be 8? Many people would say that we can. To estimate the effects of low concentrations of radiation or toxic chemicals, we measure the effects of high concentrations and then extrapolate.

However, in Figure 17 I have added meanings to the figures. The horizontal axis gives the number of engines that have failed on a four-engine aircraft and the vertical axis gives the delay in arrival at the destination.

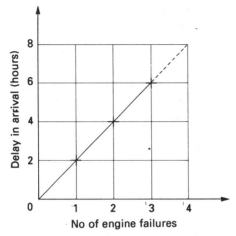

Figure 17 Until we know the meaning of the figures

If one engine fails the plane will be 2 hours late.
If two engines fail the plane will be 4 hours late.
If three engines fail the plane will be 6 hours late.
But if all four engines fail the plane will not be 8 hours late!

S J Gould has shown that over the years Hershey chocolate bars have gradually got smaller. By extrapolating the figures he has calculated the date at which the bars will have zero weight[1].

Before you extrapolate, remember that it can lead to a wrong conclusion.

See *(limitations of) experience* and *radioactivity*.

1. S J Gould, *Hen's Teeth and Horses's Toes*, Penguin Books, 1984, p. 313

Extravagance

'Accidents would fall if companies spent more money on safety'. This is often said but is not always true. Sometimes expensive safety equipment is installed but is not used correctly. The companies concerned have 'more brass than wit', to use a Lancashire phrase.

For example, a dirty water *tank* was fitted with an extravagant collection of equipment to prevent it overflowing: a high level indicator, a high level *alarm* and, in case the alarm was ignored or out of order, an independent high level trip which closed a valve in the inlet line to the tank. Having spent so much money to prevent the tank being overfilled the owners decided that it was not necessary to fit an overflow on the tank, or perhaps they did not realise that the vent, through which air escapes when the tank is filling, was not big enough for the liquid rate.

Despite all this expensive safety equipment the tank was overfilled and the roof lifted. It was then discovered that the level indicator, the high level alarm and the high level trip had been installed incorrectly by the *construction* team four years earlier. This had not been detected by the *inspections* or by the *tests* which were carried out, or should have been carried out, after construction. In addition, although the alarm and trip were scheduled for testing at intervals, these intervals were too long and the tests were not thorough. (See *protective equipment*).

The owners' money would have been more wisely spent if they had left off the expensive trip, installed an overflow and checked the indicator and alarm thoroughly after construction and then at regular intervals, say monthly.

In contrast, many years ago I worked in a factory where they had more wit than brass. A water tank was located on the roof of the foreman's office; the roof leaked. If the tank overflowed the foreman got wet. In those days foremen had more authority than today and the operator did not let the tank overflow very often. Not high technology, but more effective than the equipment on the first tank.

So before you grumble about shortage of money for safety equipment, make sure that you are using correctly the equipment you have got and spending wisely the money you have got.

In addition to the ways I have described, money is often wasted because we carry out safety studies too late in design when all we can do is to add on equipment to control the hazards. If we carried out safety studies earlier we might be able to avoid the hazards by a change in design. See *early involvement in design* and *inherently safer designs*.

See also *emergency equipment* and *penny-pinching*.

Eyebolts

See *threads*.

Eye protection

Eye injuries make up a large proportion of minor accidents. During the construction of a new plant 850 people attended the medical centre for treatment, 423 for dust in their eye. In a chemical works, all the accidents that occurred in a three-week period were investigated more thoroughly than usual. There were 32 injuries and five *dangerous occurrences*; nine of the injuries were dirt or chemicals in the eye. See *LFA*, Chapter 15.

Most eye injuries are trivial but an occasional one is serious. A few of the eye injuries investigated during the three-week study occurred because the injured man did not wear goggles when he should have done but the majority occurred when the risk of injury was not foreseen (and could not reasonably have been foreseen) and the injured man was not expected to wear goggles. If we wish to reduce the toll of eye injuries, the only practicable way of doing so is to insist that everyone working in a factory, laboratory or workshop wears goggles at all times, plain glass lenses being fitted if sight-correcting lenses are not required.

Many companies, while agreeing with this in principle, have shied away from the effort necessary in persuading the workforce to agree and in enforcing the rule. If it cannot be enforced everywhere the wearing of goggles should be enforced in laboratories and workshops. Safety glass goggles are desirable in workshops but safety glass is heavier than normal glass and is not essential elsewhere (except for concave lenses which are thin in the middle).

See *protective clothing*.

Factories Acts, Factory Inspectors

In the UK the first factory act, the Health and Morals of Apprentices Act, was passed in 1802. It restricted the hours of children in textile mills to 12 a day and forbade night work. It also laid down that they should have two suits of clothes per year, proper beds (with no more than two children per bed, in separate quarters for males and females) and some elementary schooling and religious instruction. In addition, factories were to be periodically limewashed and all infectious diseases were to be reported. However, no provision was made for enforcing the act and it was not until 1833 that the first four factory inspectors were appointed. (See *evangelicalism* and *fencing*.)

Today the European Economic Community is busy passing laws on health, safety and the environment but is making no provision for their enforcement.

Later Factories Acts extended the range of industries covered and extended the protection to all workers. Successive acts became more and more complicated. In 1968 the Factory Inspectorate put forward proposals for detailed changes to the 1961 Act. So many comments were received that a committee was appointed, chaired by Lord Robens, to consider whether or not detailed regulation was the best approach. They proposed instead a new method of working which was brought into effect in 1974 by the *Health and Safety at Work Act* (HSW Act).

From the beginning the Factory Inspectors have tried to work by persuation and have used *prosecution* in only a few cases; they still consider education more important than enforcement.

What should we do if a Factory Inspector asks us to do something which we think is unnecessary or even unsafe?:

1. Make sure that his advice is in fact wrong. Factory Inspectors are not fools and most of their advice is very sensible.
2. If he is quoting an *absolute requirement* we must do what the law says whether or not we think it will make the plant safer and whether or not the cost is disproportionate to the risk, but very few regulations affect us in this way. The whole HSW Act and much of the Factories Acts are qualified by the phrase 'so far as is *reasonably practicable*'.
3. If you still think the Factory Inspector is wrong, say so, courteously, and why you think so. Quote any relevant *codes of practice* or experience. Explain that you support his objective, a safer plant, but that you think you can achieve it in another way.
4. If the Factory Inspector wants to insist he can issue an Improvement or Prohibition Notice and then you have to obey (unless you appeal to an Industrial Tribunal) but he is unlikely to do so when there is a difference of opinion on technical matters. Improvement and Prohibition Notices are intended to stop people dragging their heels on necessary changes, not as a means of settling technical arguments.

I cannot recall a case in the UK where a Factory Inspector made a company do something unsafe. At the worst they have made companies take actions which the company thought unnecessary. Abroad this is not always the case. In one country the authorities made a company install a blast wall round a sphere containing LPG to prevent it being damaged by missiles from an *explosion* elsewhere, a most unlikely event. The blast walls prevented the dispersion of leaks by the wind and made an explosion more likely. In other cases the authorities have insisted on conventional bunds around LPG tanks thus making a *BLEVE* more likely. See *iatrogenesis, Seveso* and *ventilation*.

For more information see *Her Majesty's Inspectors of Factories 1833–1983*[1].

1. *Her Majesty's Inspectors of Factories 1833–1983 – Essays to Commemorate 150 Years of Health and Safety Inspection*, HMSO, London, 1983

Fail-safe

See *redundancy*.

Failures

We learn from our failures, not our successes. To quote Robert Stephenson, son of George Stephenson, writing in 1856[1]:

> ... nothing was so instructive to the younger Members of the Profession, as records of accidents in large works, and of the means employed in repairing the damage. A faithful account of those accidents, and of the means by which the consequences were met, was really more valuable than a description of the most successful works. The older Engineers derived their most useful store of experience from the observations of those casualties which had ocurred to their own and other works, and it was most important that they should be faithfully recorded in the archives of the Institution.

See *publication*.

A single failure is often dismissed as an anomaly. Only when a second or third has occurred are we willing to admit that there might be something fundamentally wrong.

In May 1953 a de Havilland Comet, the first commercial jet aircraft, disintegrated in the air over India. The official Indian report said that the explosion was due to a thunderstorm of exceptional severity or to the pilot's over-control in reaction to it. In January 1954 a second Comet exploded over the Mediterranean. It was difficult to recover much of the wreckage and therefore there was no evidence to suggest that the design was faulty. In April 1954 a third Comet exploded, also over the Mediterranean, but over deeper water. Attempts were then made to recover the wreckage from the second Comet; these were successful and investigation showed that all three failures were due to fatigue[2].

In the chemical industry in the 1960s many companies experienced serious fires and explosions. The causes of each were investigated and action taken but it was several years before people realized that there had been a step change. A new generation of plants, larger than those that had been built before, operating at higher temperatures and pressures and containing larger inventories, had introduced new hazards and required a new approach; *loss prevention* rather than the traditional safety approach. See *change – new problems* and *ICI*.

Finally, a quotation from La Rochefoucauld: 'Success has many fathers; failure is an orphan'.

1. Quoted by H Petroski, *To Engineer is Human*, St Martin's Press, New York, 1986, p. 223
2. Quoted by H Petroski, *To Engineer is Human*, St Martin's Press, New York, 1986, Chapter 14.

Failure rates

See *reliability*.

False alarms

False alarms can be general ones about the risks from an industry or specific ones about the risks from individual substances or types of equipment. The former have been described by Robert Murray, former medical adviser to the UK Trades Union Congress[1]:

> One of the essential items of equipment of a false prophet is a small scientific fact with which to bait his seductive hook. Another is a chair in some remote speciality. Yet another is an absence of any conscience and finally he has to have a sublime ignorance of the nature of industry. Thus armed he swings into action and the media, ever hungry for sensation, welcome him with open arms. An incident, a small fact, an orgy of unbridled speculation and a frisson of fear and alarm runs around whipping up evermore speculation and strident calls for urgent action on the part of whatever scapegoat is handiest. Avonmouth, Hebden Bridge, *Flixborough, Seveso* – the names are now household words. I believe that the situation is serious and that even the scientists are becoming the victims of emotion – not to mention the politicians, industry and the unfortunate public who forget that their life expectancy has been markedly improved when they are confronted with some hitherto hidden and preferably sinister risk.

Avonmouth was the scene of a large oil fire, Hebden Bridge the site of an *asbestos* factory. (See *culture*.)

An example of a specific false alarm: a colleague of mine was concerned that the flash point of a product was lower than the figure usually quoted but he had confused the two sorts of flash point, closed-cup and open-cup.

More seriously, in 1970, under pressure from the US Government, several companies stopped making NTA (the sodium salt of nitriloacetic acid). A research worker had found that NTA increased the incidence of birth defects in laboratory animals treated with mercury and cadmium, both of which were themselves known to cause birth defects. The research worker had had trouble getting NTA into solution and had used an additive. Later it was found that it was the additive, not NTA, that enhanced the action of mercury and cadmium. By this time the damage was done, NTA had a bad name and its production was never restarted[2].

See *aversion to risk, cause and effect, perception of risk, perspective, 'they'* and *water*..

1. R Murray, *Health and Safety in the Oil Industry*, Heyden, 1977, p. 27
2. R Stevenson, *Chemistry in Britain*, Vol. 25, No. 1, January 1989, p. 8

Falsification

In *accident investigation* witnesses are sometimes reluctant to tell all they know and information may have to be coaxed out of them. Occasional facts may be distorted but deliberate attempts at wholesale deception are fortunately rare. One of the few cases in which this occurred has been described by Peter Mahon, the judge in charge of the enquiry[1]. An Air New Zealand plane set off on a sight-seeing tour of Antarctica. Unknown to the captain the coordinates of the waypoint to which the plane was flying had been altered in the plane's computer. As a result the plane flew up the wrong valley and crashed into the rock wall at the end, with the loss of all on board. In an attempt to blame the captain the airline put out what the judge described as 'an orchestrated litany of lies'. The book describes in fascinating detail how the judge gradually came to realize that the story he was being told could not be true.

Another case occurred in the US in 1976. An official became suspicious of chemical safety tests submitted by a private testing organization with many well-known firms amongst its clients. Scientific methods and ethics had been abused, it was alleged, to provide manufacturers with positive results, and several employees of the company were prosecuted[2].

At *Bhopal* there seems to have been an attempt to conceal the fact that water was deliberately added to a tank of methyl isocyanate.

See *forecasts, honesty* and *memory*.

1. P Mahon. *Verdict on Erebus*, Collins, Auckland, New Zealand, 1984. (For a shorter account see M Shadbolt, *Readers Digest*, November 1984, p. 164)
2. *Controversial Chemicals*, edited by P Kruus and I M Valeriote, Multiscience Publications, Montreal, Canada, 2nd edition 1984. p. 46

Familiarity

> . . . the horrible thing about all legal officials, even the best, about all judges, detectives, and policemen is not that they are wicked (some of them are good), not that they are stupid (several of them are quite intelligent), it is simply that they have got used to it. Strictly they do not see the prisoner in the dock; all they see is the usual man in the usual place. They do not see the awful court of judgement; they see only their own workshop. – Sir Geoffrey Hoare, circa 1977.

Similarly, accidents do not occur because we are wicked (most of us are good) or because we are stupid (most of us are quite intelligent) but because we have got used to much that we see. We see familiar scenes, not tripping hazards and obstructions. We see our old friend Joe, not a man wearing the wrong protective clothing about to work on a pipeline which may not be properly isolated or identified. We see a place we know well, not leaky equipment in a badly ventilated building.

Many accidents happen because we have got used to the hazards around us.

The destruction of the US space shuttle *Challenger* was due in part to familiarity. A series of successful flights made operations seem almost routine and flights were allowed to continue despite known deficiencies in the design.

One method of jerking people out of their familiarity is to show them colour slides of the hazards they pass every day. They are shocked when they realize what they have failed to see. See *serendipity*.

See *complacency*.

Fatal accident rate

The fatal accident rate (FAR), called the Fatal Accident Frequency Rate (FAFR) until about 1980, was introduced in 1971[1,2] as a convenient measure of the number of fatal accidents that occur in a factory, company, industry or occupation. It is the number of fatalities that occur in a group of a thousand men in a working lifetime, that is, 10^8 hours. While 10^8 hours is just a number, a thousand men for a lifetime gives us a feel for the size of the problem; if we spend our entire career in a factory of this size, it is the number of our fellow workers who will be killed by accidents at work. For all premises covered by the UK *Factories Act*, the FAR is about four but varies from very low figures for light industry to about 60 for *construction* erectors.

Averaged over several decades, the FAR for the chemical industry is about four, made up of about two for chemical risks such as fire and explosion and about the same for ordinary industrial risks, which could occur in any factory. This is the basis for a widely-used *criterion* to help us decide which risks are so big that we should reduce them and which are so small, compared with all the other risks around us, that we should leave them alone, at least for the time being, and spend our resources on bigger ones. (These small risks are often called *acceptable risks* though tolerable risk may be a better name[3].)

The criterion says that if the FAR for the man or men at greatest risk is greater than two the risk should be reduced; if it is below two, the risk can be ignored. Sometimes we are asked to consider a particular risk, such as a fire or explosion, in isolation and do not know the total FAR. We then say that the FAR for this risk should not exceed 0.4. We are thus assuming that there are about five significant risks on a typical plant.

If the man at greatest risk will be killed every time a dangerous incident occurs then there should not be more than 0.4 of these incidents in 10^8 hours or one in 2.5×10^8 hours = 30 000 years. This is the origin of the statement, sometimes seen, that plants should be designed so that explosions do not occur more than once in 30 000 years. However, if there is only a one in ten chance that the man at greatest risk will be killed, the explosion or other dangerous incident can occur once in 3000 years, and so on.

The Health and Safety Executive favours two levels of risk: an upper level (the 'maximum tolerable' risk) which should never be exceeded and a lower level (of 'negligible' risk) which there is no need to go below. In between we should reduce the risk if it is *'reasonably practicable'* to do so[3]. Their suggested maximum risk, for workers in any industry, is FAR 50, which seems rather high.

For *comparison* with the FAR for industrial accidents, if we spend our working life in a factory of 1000 men, about 20 of our fellow workers will be killed by accidents of other sorts, mostly on the roads or at home, and about 370 will die from disease, including about 40 from the effects of *smoking*.

To be meaningful, the FAR usually has to be averaged over several years, even for large companies, as fatal accidents are fortunately rare. However, annual figures for an industry can be meaningful.

For examples of the use of the FAR, see *asbestos* and *ICI*. See also *hazard and risk*.

1. T A Kletz in *Loss Prevention in the Process Industries*, Institution of Chemical Engineers, Rugby, 1971, p. 75
2. T A Kletz, *Hazop and Hazan – Notes on the Identification and Assessment of Hazards*, Institution of Chemical Engineers, Rugby, 2nd edition, 1986
3. *The Tolerability of Risk from Nuclear Power Stations*, HMSO, London, 1988

Fate

In *Disaster*[1] Sheila Tidmarsh points out that politicians and newspapers, when commenting on floods, earthquakes, famines and similar disasters, use phrases such as 'cruel fate', which suggest inevitability and discourage constructive thinking on ways of preventing them or minimizing their consequences.

In the village of Silkstone, near Barnsley, UK there is a memorial to 26 boys and girls, aged 7 to 17, who were drowned when a coalmine was flooded in 1838. It refers to 'an awful visitation by the Almighty. . . On this eventful day the Lord sent forth his thunder, lightning, hail, carrying devastation before them and a by sudden erruption of water into the coal pits of R C Clarke Esquire, 26 human beings whose names are recorded here were suddenly summoned to appear before their Maker'.

Fortunately not everyone was content to blame the Lord. A Royal Commission was appointed and in 1842 Lord Shaftesbury's bill, to prevent children under ten and women working underground, became law.

Describing the designs on the walls of the 16th Century Little Moreton Hall in Cheshire, the guide book says, '. . . the belief by Protestants that individual striving and knowledge, rather than the blind acceptance of Fate, would determine one's destiny, is here plain for all to see. Knowledge of science and the New World were especially admired in the Elizabethan age[2]'.

See *Acts of God* and *evangelicalism*.

1. Sheila Tidmouth, *Disaster*, Penguin Books, 1969.
2. C Rowell, *Little Morton Hall*, National Trust, London, 1986, p. 22

Feedback to design

It is obvious that those who have to start up, operate and maintain new plants should feed back their *experience* and tell the designers how the *design* could be improved; in practice this rarely happens. The start up team are at first too busy; the job is left until the plant has settled down and is then forgotten. Meanwhile the designers have moved on to the next project and are no longer interested. What then can we do to make sure that our expensive experience is not lost? We can:

- Set up a formal system for documenting start-up and operating experience and passing it back to design. When a start-up team is disbanded one or more of them should be retained for a few weeks to write up their experience.
- Include designers in the start team-up.
- Arrange regular transfers of staff between design and operations.
- Include designers in plant *audit* teams.

Fencing

During the nineteenth century unguarded machinery caused many accidents, and laws were passed requiring all moving parts to be fenced (guarded) except prime movers which can be made safe by location. (See *absolute requirements* and *Factories Acts*.) Many manufacturers objected. Charles Dickens, one of the campaigners for tighter laws and more men to enforce them, reported a speech made by a manufacturer in 1855[1]:

> Suppose the mill owners were to go home and set to work to case all their gearing; in many of the mills miles of casing (wooden casing of course) would be required, and the effect would be that, within this casing, a large amount of cotton flake and dust would find its way. This would more or less interfere with the oiling of the machinery, and a spark, communicating to the fibres inside this casing, would inevitably lead to the destruction of the whole mill; the soft fibre would ignite like gunpowder, the fire would pass from shaft to shaft, and it would be found that the moment the fire was put out in one place it would break forth in another and render extinction impossible. The wood casing too, when ignited, would fall in burning fragments and set fire to everything else.

This speaker was followed by a Factory Inspector who said that where casings had been installed, cotton dust did not accumulate in them.

We still hear people speak like this mill owner. Are they just being unreceptive to new ideas, perhaps deliberately, perhaps unconsciously (see *cognitive dissonance*) or are they aware of the risks involved in *modifications* and in *exchanging one problem for another*?

See *poor equipment*.

1. C Dickens, *Household Words*, Vol. XL, 14 April 1855, p. 264. Quoted in *Mill Life in Styal*, Willow Publishing, Altrincham, 1986, p. 33

Fermi estimates

The physicist Enrico Fermi had a reputation for making quick numerical estimates of the answer to a problem or query[1]. For example, how many piano tuners are there in the South Manchester telephone area? The population is about a million, say 250 000 households. If one in five owns a piano which is tuned every five years there will be about 10 000 tunings per year. If each tuner tunes five pianos per day for 250 days per year, i.e., 1250 per year, there will be about eight tuners. But many piano tuners are part-time; they tune other instruments, carry out repairs, sell pianos, so the actual number will be higher, perhaps twelve. This estimate is not accurate, the true figure could easily be five or 25, but it gives us a quick, approximate answer. The Yellow Pages shows that the true figure is 16.

These quick estimates are usually not too far out because errors in our estimates tend to balance out. We are unlikely to over- (or under-) estimate every figure.

When the quantitative methods of *hazard analysis* were first applied in the chemical industry many of the estimates made were of the Fermi type. For example, how many leaks of flammable gas can be tolerated before we have to use Zone 1 electrical equipment (see *electrical area classification*) instead of the cheaper Zone 2?

- Assume a motor or other item of Zone 2 electric equipment is surrounded by flammable gas or vapour for one hour per year.
- Experience shows that it will develops faults which cause sparking or overheating once in a hundred years.
- There are about 10^4 hours per year so a spark will coincide with gas or vapour and there will be a fire or explosion about once in 10^6 years.
- Observation shows that someone is within 3 m of a particular motor for 5% of the time. Assume that anyone within this distance is killed.
- A fatality will therefore occur once in 20×10^6 years. If there are a hundred items of electric equipment on the plant there will a fatality once in 200 000 years, well within the widely-used criterion of once in 30 000 years (*fatal accident rate* 0.4), for any risk considered in isolation[2].
- Zone 2 equipment may therefore be used in areas in which flammable gas or vapour is present for seven (200 000/30 000) hours per year.

In this case the estimates are almost all biased in one direction. It takes time for gas to diffuse into Zone 2 equipment; it is unlikely that everyone within 3 m would be killed and we have assumed that all the risk is concentrated on one man. In many parts of a Zone 2 area leaks are very rare, thus reducing the average risk. We can therefore, as a practical 'rule of thumb', define a Zone 2 area as one in which flammable gas is present for up to ten hours per year[3].

Hazard analysis now often involves complex calculations using a computer model. But there is still a case for Fermi estimates. They may show that the answer is so clearly Yes or No that there is no need to carry out detailed calculations.

See *costs* and *operations research*.

1. H C von Bayer, *The Sciences*, Vol. 28, No. 5, September/October 1988, p. 2
2. T A Kletz, *Hazop and Hazan – Notes on the Identification and Assessment of Hazards*, Institution of Chemical Engineers, Rugby, 2nd edition, 1985, Chapter 3
3. T A Kletz, *Institution of Chemical Engineers Symposium Series No. 34*, 1971, p. 75

Feyzin

See *BLEVE*.

Fin-fans

During a *fire*, fin-fan coolers (air coolers cooled by fans) have been known to draw in flames and hot gases, thus increasing damage, particularly to the fin-fans themselves. It should be possible to isolate the power supplies to fin-fans from a safe distance by duplicating the stop buttons some distance away. (See Figure 18.)

Figure 18 Fin-fans

Another hazard of fin-fans is that even when the power to them has been isolated, they may move under the action of the wind and injure men who are maintaining them. They should be restrained so that they cannot move during maintenance.

Fire

Fire is too big a subject to be discussed adequately here and all I can do is draw attention to a few points. For more information see *Lees*, Chapter 16 and *Loss Prevention Bulletin* No. 082[1].

Liquids and gases can be involved in several different types of fire:

Pool fires

Pool fires occur when a pool of liquid on the ground or on water is ignited. The rate of burning is about 6–13 mm/min and flames are easily displaced by the wind. In both plant and storage areas the ground should be sloped so that pools of flammable liquid do not accumulate under equipment but run off to one side, into drains or a catchment pit. Note that when oil is spilt on water it can spread large distances and catch fire hundreds or even thousands of metres away.

Jet fires

Jet fires occur when a flammable liquid or gas escapes from a puncture or open end. They can be very intense and are comparatively unaffected by the wind. Liquid can 'rain out' and form a pool fire. Vents and relief valve tailpipes which discharge flammable gas should be located so that, if the discharge ignites, it will not impinge on other equipment even though bent 45° by the wind.

Flash fires

Flash fires occur when a cloud of flammable gas is ignited and they may develop into an *explosion*. Flash fires are usually short-lived and may be followed by a pool or jet fire at the source of the release.

Fireballs

Fireballs occur when a quantity of flammable liquid is suddenly released and is immediately ignited, usually following a *BLEVE*. The quantity of material involved is larger than in a flash fire and the fireball may rise in the air.

Sources of *ignition* are often quoted as '*causes*' of fires but it is difficult to eliminate all sources of ignition. To prevent fires we should use *defence in depth*. The first line of defence is to use a non-flammable material instead of a flammable one (*substitution*) or to use so little of the flammable material that it does not matter if it all leaks out and ignites (*intensification*).

The second line of defence is to keep the fuel in the plant and the air out of the plant. The latter is easy as most plants operate under pressure and *leaks* therefore tend to be outwards. *Nitrogen* is widely used to keep the air out of low pressure equipment such as storage *tanks, stacks, centrifuges* and equipment being prepared for *maintenance*. To keep the fuel in the plant we need to *design*, construct, operate and maintain equipment in accordance with good engineering standards. This is easily said but books could be, and have been, written on each of these subjects. (See *construction* and *flammable atmospheres*.)

Other lines of defence are detection of leaks, warnings, isolation, *dispersion*, and elimination of known sources of ignition. If ignition occurs our final lines of defence are fireprotection and fire-fighting.

The main methods of fire protection[2] are *insulation* and water spray.

Insulation has the advantage that it does not have to be commissioned but is immediately available as a barrier to heat input. However, if 10% of the insulation on a vessel is missing the effectiveness of the rest is zero, not 90%.

The first principle of fire fighting was known to Shakespeare: 'A little fire is quickly trodden out, which being suffered, rivers cannot quench.' (Henry VI, Part 3, Act IV, Scene VIII).

1. First Report of Thermal Radiation Working Party of Major Hazards Assessment Panel, *Loss Prevention Bulletin*, No. 082, August 1988, p. 1
2. British Standards Institution, *Code of Practice for Fire Protection in Chemical Plants*, BS 5908

Flame traps

Flame traps or flame arrestors are devices used to prevent the propagation of flames. The commonest type uses an extended metal surface, such as crimped metal, to cool the gas so that combustion is not sustained. Their disadvantage is that the narrow passageways get blocked by dirt. The flame traps should be cleaned regularly but this is often overlooked. Many vessels have been sucked in or over-pressured because the flame traps in their vent lines were blocked with dirt. (See Figure 19 and *will and might*.)

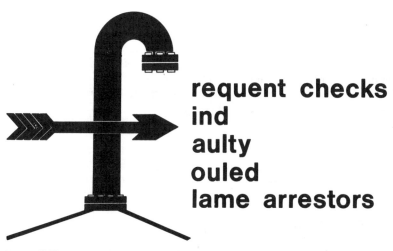

Figure 19 Flame traps

Other flame traps depend on a liquid seal. It is essential to ensure that there is no free gas path through the liquid. A third type of flame trap, uses rapid gas flow through a series of orifices. The velocity of the gas is too high for flash back to occur[1]. Note that the velocity below which flash back will occur is much greater than the 'maximum flame speed' or 'fundamental burning velocity' quoted in many reference books (including *Perry's Handbook*) and depends on the diameter of the orifice[2].

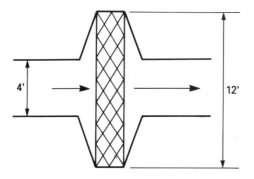

Figure 20 An example of an ineffective flame trap. (Reproduced by permission of the American Institute of Chemical Engineers)

If storage *tanks* containing flammable liquids above their flash points are not inerted, then a flame trap is usually installed in the vent line. It will prevent lightning, or other external source of ignition, igniting the air/vapour mixture in the tank but will not prevent ignition by an internal source such as *static electricity*.

Explosions have occurred because flame traps were not properly designed. For example, the off-gas from an oxidation plant was incinerated to destroy traces of organic compounds. If the plant was upset, the gas could enter the explosive range, so a flame trap was installed in the off-gas line to prevent flash back from the furnace. The line was 1.2 m diameter and the flame trap was 3.65 m diameter. It looked impressive but there were three fatal errors in its design[2]:

- It was made from Pall rings which are known to be ineffective in flame traps.
- The angle of the inlet and exit cones was steep (150°) so flow distribution was not uniform. (See Figure 20.)
- The flame trap was located in the middle of the line, where the flame speed was high, rather than at its end.

Consequently, when the off-gas became flammable, flash back occurred. See *coincidences*.

1. W B Howard, *Loss Prevention Bulletin*, No. 079, February 1988, p. 15
2. W B Howard, *Chemical Engineering Progress*, Vol. 84, No. 9, September 1988, p. 25

Flammable atmospheres

I have often been asked what precautions should be taken when people have to work in flammable atmospheres. The answer is that they should never be asked to do so as, however much care we take to eliminate all known sources of *ignition*, they may still turn up.

On many occasions someone has taken a chance, dashed into a cloud of flammable gas for a moment to isolate a *leak*, and got away with it. However, other people have been badly burnt trying to do so. I will not go so far as to say that no one should ever enter a vapour cloud to isolate a leak. By taking a chance for a moment someone may be able to stop a leak before it spreads and thus prevent a serious fire or explosion. If possible, he should be protected by fire-resistant clothing and water spray. However, we should avoid putting people into situations where they feel a need to take such chances. When experience shows that equipment is liable to leak it should be fitted with remotely-operated *emergency isolation valves*. At the most, someone may be permitted to put his hands, suitably protected, into a cloud of vapour not more that 0.5 m diameter. (See *spark-resistant tools*.)

To prevent *fires* and *explosions* we should never tolerate leaks or rely on the fact that we have tried to eliminate sources of ignition. We should avoid flammable atmospheres as conscientiously as we avoid sources of ignition. Buckets (and other open *containers*) are more dangerous than matches on plants that handle flammable liquids. Matches are dangerous only if struck in a flammable atmosphere and on a well run plant these are rare. If buckets are allowed in a plant they will be used for collecting samples or drainings. A flammable atmosphere will be present over the liquid and sooner or later it will ignite.

In some industries it is difficult to avoid open vessels. If so, good ventilation is necessary to keep the concentration of vapour below the flammable limit (and also below any toxic limit; see *COSHH Regulations*).

No one should be asked to enter a vessel if the concentration of flammable vapour inside is above 20% of the lower flammable limit as the concentration in some parts of the vessel may be higher than at the points where measurements have been taken. If a flammable liquid has to be taken into a vessel for carrying out tests, do not take in more than the amount which, if spilt, will bring the concentration of vapour above 20% of the lower flammable limit. See *entry*.

For electrical equipment suitable for use in flammable atmospheres see *electrical area classification*.

See *LFA*, Chapter 4 and *Health and Safety at Work*[1].

1. *Health and Safety at Work*, Vol. 7, No. 10, October 1985, p. 18

Flare stacks

See *stacks*.

Flashing liquids

Most vapour cloud *explosions* and *BLEVE*s and most large toxic incidents are due to the release of flashing liquids, that is, liquids under pressure above their atmospheric temperature boiling points.

If there is a *leak* of liquid below its boiling point little vapour is formed and it will not spread very far. A *fire* is possible but an explosion is unlikely, especially if the leak is in the open air. If there is a leak of vapour, the leak rate through a hole of a given size will be small and the vapour may disperse by jet mixing. Again, the vapour will not spread very far and an explosion in the open air is unlikely. However, if there is a leak of liquid under pressure, above its boiling point, the liquid will leak at a high liquid rate and then turn to vapour and spray. The fraction that will turn to vapour can be calculated by heat balance but the vapour may carry much, perhaps all, of the rest of the liquid with it as spray, which is just as explosive and just as toxic as the vapour.

Typical flashing liquids are LPG, *ammonia* and *chlorine*. The explosion at *Flixborough* was due to a leak of flashing liquid (*cyclohexane* at 10 bar and 150 °C) and so was the toxic release at *Bhopal* where the liquid (methyl isocyanate) was not normally flashing but addition of water to the storage tank caused a runaway reaction which raised the temperature and pressure.

Davenport[1] has listed 71 vapour cloud explosions. Of these explosions 50 (71%) were due to leaks of flashing liquids, 13 (18%) to leaks of vapour and eight (11%) to leaks of liquids below their boiling points, though some were close to them. Two were the result of rail tankers bursting, spraying the liquid out with great force, in one case the liquid was discharged by a *foam-over* and in another case a storage tank was overfilled, forming a vapour cloud 450–600 m long and 60–90 m wide[2].

1. J A Davenport, *4th International Symposium on Loss Prevention and Safety Promotion in the Process Industries*, Vol. 1, Institution of Chemical Engineers, Rugby, 1983, p. C1
2. T A Kletz, *The Chemical Engineer*, No. 426, June 1986, p. 63

Flexes

See *hoses*.

Flixborough

The *explosion* at Flixborough on 1 June 1974 was a milestone in the history of the UK chemical industry[1]. The destruction of the plant in one almighty explosion, the death of 28 men on site and extensive damage and injuries (though no deaths) in the surrounding villages showed that the hazards of the chemical industry were greater than had been generally believed by the public at large. In response to public concern the Government set up an enquiry into the immediate causes and also an 'Advisory Committee on Major Hazards' to consider wider questions. Their three reports led to far-reaching changes in the procedures for the control of major industrial hazards. (See *CIMAH Regulations*.) The long-term effects of the explosion thus extended far beyond the factory fence.

The plant on which the explosion occurred oxidized *cyclohexane* (a hydrocarbon similar to petrol in its physical properties) with air to a mixture of cyclohexanone and cyclohexanol, usually known as KA (ketone/alcohol) mixture and used in the manufacture of nylon. The reaction was slow and the conversion had to be kept low to avoid the production of unwanted by-products. Therefore, the inventory in the plant was large, many hundreds of tonnes. The reaction took place in the liquid phase in six reaction vessels, each holding about 20 tonnes. Unconverted raw material was recovered in a distillation section and recycled. Similar processes were operated, with variations in detail, in many plants throughout the world. (See *oxidation*.)

One of the *reactors* developed a crack and was removed for repair. In order to maintain production a temporary pipe was installed in its place. Because the reactors were mounted on a sort of staircase, so that liquid would overflow from one to another, this temporary pipe was not straight but contained two bends. Bellows, 28 inches in diameter, were installed between each reactor and these were left at each end of the temporary pipe.

The men who were charged with the task of making and installing the temporary pipe were men of great practical experience and drive; they constructed and installed it and had the plant back on line in a few days. However, they were not professionally qualified and did not realize that the pipe and its supports should have been designed by an expert in piping design. Their only drawing was a full-size sketch in chalk on the workshop floor; the only support was a scaffolding structure on which the pipe rested.

The temporary pipe performed satisfactorily for two months until a slight rise in pressure occurred. The pressure was still well below the relief valve set point but nevertheless it caused the temporary pipe to twist. The bending moment was strong enough to tear the bellows and two 28 inch holes appeared in the plant. The cyclohexane was at a gauge pressure of about 10 bar (150 p.s.i.) and a temperature of about 150 °C. It was thus under pressure above its normal boiling point (81 °C) (see *flashing liquids*) and a massive leak occurred. About 30–50 tonnes escaped in the 50 seconds that elapsed before ignition occurred. The source of ignition was probably a *furnace* some distance away.

The resulting vapour cloud explosion, one of the worst that has ever occurred, destroyed the oxidation unit and caused extensive damage to the rest of the site. In particular the company office block, about 100 m away, was destroyed and had the explosion occurred during office hours, and not at 5 pm on a Saturday, the death toll might have been 128 instead of 28.

The main lessons that can be learned from the disaster are:

- Avoid, if possible, large inventories of hazardous materials. See *inherently safer design*.
- Have a system for the control of plant *modifications*. They should be made to the same standard as the original plant. See *corrosion*.
- Employ professionally qualified people. See *Aberfan* and *knowledge of what we don't know*.
- *Layout* and locate plants and strengthen control buildings so as to minimize the effects of explosions.

See *alertness, operations research* and *Tay Bridge*.
For further details see *LFA*, Chapter 8 and *Lees*, Appendix A.1.

1. *The Flixborough Disaster*, HMSO, London, 1975

Foam-over, froth-over

Foam-overs occur when oil, above a water layer, is heated above 100 °C. The heat gradually travels through the oil to the water layer. When the water boils, the steam lifts up the oil, reducing the pressure and causing the water to boil with increased vigour. The mixture of steam and oil may blow

Figure 21 Foam-over

the roof off the storage tank. In one incident a structure 25 m tall was covered with oil. Witnesses said that the tank had burst and it was an explosion, but a physical one rather than a chemical one.

Foam-overs also occur if oil, above 100 °C, is added to a tank containing a water layer (see Figure 21).

The term *boil-over* is used if the tank is on fire and hot residues from the burning layer travel down to the water layer. The term slop-over is often used if water from fire hoses vaporizes as it enters a burning tank[1].

To prevent foam-overs, which are very common in the oil industry, keep incoming oil below 100 °C and fit a high temperature alarm on the oil inlet line. Alternatively, keep the tank contents well *above* 100 °C so that any small quantities of water that enter are vaporized and a water layer never builds up. The tank should be drained regularly and circulated before any fresh oil is added. Addition of fresh oil should start at a low rate and not more than a tenth of the tank capacity should be added at a time.

The temperature of a heavy oil tank should never be allowed to fluctuate above and below 100 °C.

1. *Loss Prevention Bulletin*, No. 057, June 1984, p. 26, No. 059, October 1984, p. 35, No. 061, February 1985, p. 36

Follow-up

Many people fail to follow up their actions to see if they had the desired effect. Magistrates, for example, seem unable to find out whether or not

"Before you tackle today's problems, check yesterday's problems and see if they're any worse."

Figure 22 Follow-up

their 'clients' responded to treatment or continued in a life of crime. *Safety professionals* should follow up the advice they give, first, to see if it has been taken and, if so, to see if it worked and if there were any unforeseen snags. Often we are so busy dealing with current problems that there is no time to follow up yesterday's. (See Figure 22.) It is impracticable to follow up every word of advice that we give, but we should try to follow up all major problems and a sample of others.

Accidents reports should say when the recommendations will be complete and should be brought forward at that time.

Design engineers should follow up their designs to see how they work and what the operators think of them. See *feedback to design*.

Folly

We cannot install enough safety devices to correct for improper operation or maintenance of a facility. Safety devices are intended primarily as a protection against unusual conditions which operating personnel cannot cope with rapidly enough through normal procedure[1].

The law (in the UK) does not support this view. A person 'is not, of course, bound to anticipate folly in all its forms but he cannot put aside the teachings of experience as to the form those follies commonly take[2]'. (See *employers' responsibility*.)

Folly is not common in industry. Most errors are the result of a slip or a moment's forgetfulness (see *human failing*) but the same principle applies: We have to take into account the teachings of experience as to the form that errors commonly take.

When we have made an error ourselves, especially an error of judgement we are inclined to blame *fate* rather than ourselves.

'To the gods I owe this woeful fate', laments Priam (of Troy), forgetting that he could have removed the cause by sending Helen home at any time or by yielding her when Menelaus and Odysseus came to demand her delivery[3]'.

1. K FitzPatrick, *Safety in High Pressure Polyethylene Plants*, American Institute of Chemical Engineers, 1973, p. 39
2. A House of Lords judgement quoted in *The Guardian*, 7 February 1966
3. B Tuchman, *The March of Folly*, Knoff, New York, 1984, p. 46

Food

The food industry has a surprisingly high accident rate but most of the accidents are due to machinery or handling (human or mechanical). However, there have been many process accidents such as *dust* explosions, particularly in flour mills and silos, and solvent *fires* and *explosions*. As in the rest of the process industry, *tanks* have overflowed and heaters and refrigeration plants have failed because *protective equipment* was absent or neglected.

My impression, from reading published *accident reports*, is that the industry (or at least, parts of it), though sophisticated in other respects, lags behind the rest of the process industry so far as technical safety is concerned. (See *amateurism*.) For example, three explosions occurred within a day in an extractor in which soya beans were treated with hexane[1]. The extractor had been emptied for *entry* and *maintenance*, but was not purged free of hexane and the hexane inlet valve was left open. An unprotected electric tool ignited the first explosion and started a smouldering fire in the deposits that had built up over the years. This fire ignited the second and third explosions which occurred when the hexane concentration built up to to the lower flammable limit (LFL).

The recommendation made in the published report was for the installation of steam lines so that any future smouldering fires could be extinguished. There was no mention of the precautions normally taken in the chemical industry before entry or hot work is permitted, namely:

- Isolate the vessel by slip-plates or disconnection, not by valves which may leak or be left open.
- Purge the vessel of flammable vapour and test to make sure that the concentration is less than 20% of the lower flammable limit. Particular care is necessary if the vessel contains deposits which may give off vapour when disturbed.

A review of dust explosions in the food industry[2] shows a similar lack of awareness.

1. C L Kingsbaker, *JAOCS*, Vol. 60, No. 2, Febuary 1983, p. 197A
2. *Science*, Vol. 222, 4 November 1983, p. 485

Force

Force is often confused with *pressure*, probably because we measure it in similar units. We often measure pressure in pounds per square inch (p.s.i.)

but that is a bit of a mouthful, so instead of saying that a pressure is 10 pounds per square inch we say we say it is 10 pounds. Since 10 pounds is not very much we then assume that 10 pounds per square inch is not very much, forgetting that a force of 10 pounds is exerted on every square inch.

A driver opened the manhole on a road *tanker* while there was *air* inside at a gauge pressure of 10 pounds per square inch. Either he forgot to vent the pressure first or did not realize the importance of doing so. He was blown off the top of the tanker and killed. Many people were surprised that a 'puff of air' at a pressure of 'only 10 pounds' was sufficient and wondered if a chemical *explosion* had occurred. If the manhole was 18 inches diameter then the force acting on it was over a ton. No wonder it was flung open with great violence.

The use of SI units, in which force and pressure are measured in different units, may help to avoid confusion between them.

Forecasts

When we try to forecast the effects of new risks we often overestimate them. Thus, Beckerman[1] writes:

> In retrospect the hysteria which accompanied the early days of the anti-growth movement probably did some harm. For example, associated with it were some excessive conservationist reactions which had most unfortunate results. In Ceylon, for example, where malaria had been almost eradicated, the banning of DDT led to a rapid rise in the malaria death rate. In Sweden, DDT had to be quickly reintroduced when insects started to destroy forests.
>
> One typical unpleasant effect was the movement, in Britain, to encourage school children to monitor pollution levels in their local rivers and lakes, for although there have been no cases of fatal illness from water pollution in Britain for more than 50 years, every year more than 100 children are drowned playing in or near water.

There is no doubt that DDT was overused but nevertheless it has not caused a single human death.

Imagine that railways, escalators, bicycles, pressure cookers, gas fires, glass, razor blades, barbed wire and cigarette lighters have just been invented. Think of the outcry they would arouse and the forecasts of death and destruction that would be made. Would we allow trains to drive at over 100 m.p.h. past people standing on a platform, separated from the train by only a yellow line?[2]

On the other hand, the effects of some new hazards have been underestimated. The motor car, for example, has caused more deaths than its critics ever feared. Lawless[3] gives other examples. He describes 46 technological changes, from thalidomide to plastic turf, that had unforeseen and undesirable side-effects and concludes that in about 40% of the cases the side-effects might reasonably have been foreseen. In 25% of them early warning signs might have been noticed and acted on.

Crystal gazing is not an exact science. We should be humble about our forecasts and willing to admit that they may be widely out. See *extrapolation, false alarms* and *foresight*.

1. W Beckerman, *The Times Higher Education Supplement*, No. 370, 23 November 1979, p. 14
2. H S Eisner, *Science and Public Policy*, June 1979, p. 146
3. E W Lawless, *Technology and Social Shock*, Rutgers University Press, New Brunswick, New Jersey, 1977

Foresight

Some accidents occur because people fail to take precautions against clearly foreseeable events. For example:

- Before carrying out a pressure test we should assume that the equipment might fail and therefore take suitable precautions, keeping people out of the way. The chance of failure is low but if we were certain that the equipment would not fail we would not need to test it.
- Relief valves are designed to lift and so when they do they should not create a hazard by discharging material over people or plant (unless we can show that the chance of lifting is so small that the risk can be accepted). If we were sure a relief valve would never lift we would not need to install it. (See *Seveso*.) Similarly, vent pipes are designed to vent. (See *small companies*.)
- After changing a *chlorine cylinder* two men opened the valves to make sure there were no leaks on the connecting pipework. They did not expect to find any so they did not wear breathing apparatus. There were some small leaks and they were affected by the chlorine.

  If they were sure there were no leaks they did not need to test. If there was a need to test then leaks were possible.
- Before carrying out experiments we should list possible outcomes and decide what actions to take. See *Chernobyl*.
- Three railway research workers were carrying out some measurements on the track using a heavy piece of equipment. A lookout man was present to warn them when a train approached. They had not planned how to they would remove the equipment in the event of an approaching train and failed to do so in time[1].

See *modifications*.

1. *Railway Safety 1986*, HMSO, London, 1987, p. 13

Forgetfulness

See *human failing* and *time-span of forgetfulness*.

Forgotten knowledge

See *lost knowledge*.

Fractional dead time

See *reliability* and *tests*.

Friendly plants

On the last day of a safety conference a young delegate asked a simple question. She said, 'For three days we have been told how to stop accidents. If we know how to stop them, why do they still occur?'.

We know what we ought to do to prevent accidents but we do do not always do it. Being human we do not always perform as well as we can or should, at home or at work, in the car or in bed. We may be able to keep up a tip-top performance for an hour or so while playing a game or a piece of music, but it is hard to keep it up all day, everyday.

Designers have a second chance. They can go over their designs again looking for errors and there are techniques such as *hazard and operability studies* to help them find them. *Construction* teams can do the same, though they do not always do so. But people who operate and maintain plants usually have to get it right first time, and it is difficult to do so for every hour of every day, without fail. When a plant came back on line after a shutdown there was a leak of flammable liquid from a joint. Afterwards the manager said, 'We unbolted and remade 2000 joints during the shutdown and got one wrong, but that is the only one anyone has heard about'.

So what can we do? Whenever possible we should design our plants so that they are 'user-friendly', to borrow a computer term, so that they can tolerate some departure from ideal operation or maintenance or equipment failure without breaking down or injuring people. On the plant where the joint leaked they replaced the gaskets in half the joints – all those exposed to liquid – with spiral-wound ones. These gaskets have the great advantage that if the joint is not made correctly, or becomes slack for some reason, the leak rate is very much smaller than from a joint fitted with a plain gasket. (See *(are things as) black as they seem?)*

Spiral-wound gaskets are 'friendly' equipment. Bellows are an example of 'unfriendly' equipment. They have to be installed with great care as they cannot withstand much distortion. If hazardous materials are being handled it is good practice to avoid the use of bellows and build expansion loops into the pipework instead. (See *Flixborough.*) Hoses are another example of unfriendly equipment.

Friendly plants contain low inventories of hazardous materials or safer materials instead; they are simple, with few opportunities for error; they are less dependent on added-on safety systems which may fail or be neglected, and they respond more slowly and less steeply to changes. Therefore, those errors that are made or those failures that do occur do not produce extensive knock-on effects. Bolted joints are friendlier than quick-release couplings, as they are more tolerant of maintenance errors or ill-treatment. Rising spindle *valves* are friendlier than valves without rising spindles, and spectacle plates are friendlier than slip-plates, as their positions are less likely to be misread.

The *Chernobyl* nuclear reactor was very unfriendly as, at low rates, if it got too hot the rate of heat production increased and it got even hotter, faster. In all other commercial designs, if the reactor gets too hot, the rate of heat production falls and the operators have more time to correct faults. In fast reactors heat production stops completely. (See *nuclear power*.)

Instead of designing plants, identifying hazards and adding on equipment to control the hazards, or expecting operators to control them, we should make more effort to choose basic designs, and design details, that are user-friendly.

When designers are comparing different designs they should not only look at their cost, performance, ease of maintenance and so on but also at their friendliness. It may be worth paying a bit more for a design that can tolerate some misuse. In practice, friendly designs are often cheaper.

See *inherently safer design, human failing, simplicity* and reference 1.

1. T A Kletz, *Chemical Engineering Progress*, Vol. 85, No. 7, July 1989 p. 18

Froth-over

See *foam-over*.

Fugitive emissions

These are the small, continuous or frequent, emissions from plant equipment which produce low concentrations of chemicals in the atmosphere of the workplace and which can result in long-term toxic effects. (See *cancer* and *under-reporting*.) Many loss prevention engineers concentrate on the prevention of large releases and leave the fugitive emissions to their toxicological colleagues. But while the toxicologist can measure the amount of impurity present and can tell us what concentration is tolerable (see *COSHH Regulations*), stopping the *leak* is a problem for engineers.

Most fugitive emissions come from *valves*, followed by oil/water separators, *pump* seals, *drains*, compressor seals, flanges and *relief devices*. Emission rates from these various sources have been estimated[1,2].

1. British Occupational Hygiene Society, *Fugitive Emissions of Vapours from Process Equipment*, Science Reviews, 1984
2. Publications available from the US Environmental Protection Agency

Full-scale deflection

Operators often fail to realize that when an instrument is at the top of its scale the true reading may be very much more than the figure indicated.

For example, a low pressure *tank* containing liquefied ethylene was over-pressured (as the relief valve discharged into a vent *stack* which was plugged with ice). The pressure in the tank was normally 0.8 p.s.i.g., the relief valve was set at 1.5 p.s.i.g., the full-scale deflection of the pressure recorder was 2 p.s.i.g. and a split occurred in the tank at a pressure of several p.s.i.g. For 11 hours before the split occurred the operators, on two shifts, wrote down '2 p.s.i.g.' on the record sheet and did not even draw the foreman's attention to the unusual reading[1].

On another occasion an operator thought the level in a tank was steady. It was slowly rising but the differential pressure level indicator reached its full-scale deflection when the level reached the rim of the tank. The level continued to rise into the dome and when it reached 12 inches above the rim, the tank ruptured at the rim[2]. (See *tanks*, Figure 74.)

With digital instruments it may be less apparent than with analogue ones when the full-scale deflection is reached.

Examining some photographs of a fire, an engineer noted that the flames were almost white and used this observation to to make a rough estimate of their temperature. He did not realize that the photographic emulsion had been fully bleached.

1. T A Kletz, *Chemical Engineering Progress*, Vol. 70, No. 4, April 1974, p. 80
2. R E Sanders, American Institute of Chemical Engineers Loss Prevention Symposium, April 1989

Furnaces

The two main hazards of furnaces are explosion of the fuel and bursting of the tubes.

To avoid the former we should follow the following procedure, either manually or automatically. Failure to do so has caused many explosions during the lighting of furnaces, or, more often, during the relighting of hot furnaces. Sometimes the correct procedure has never been followed but more often the operator has taken a short cut:

- When the furnace is not in use the fuel should be isolated positively, that is, by slip-plating, disconnection, water seal or double block and bleed valves. Valves without a bleed, however numerous, are not sufficient.
- If the fuel is a gas or volatile liquid, test the furnace to make sure no fuel is present. If the fuel is a high-boiling liquid, not detectable with a gas detector, and the furnace is hot, sweep it out, by forced or natural draft, for a sufficient period of time to make sure that any fuel present has evaporated.
- Insert a torch or poker or switch on an electric igniter.
- Remove the positive isolation. This should never be done before a torch, poker or operating igniter is in the furnace.
- Open the fuel supply valve and allow the burner to ignite.

If the flame goes out while the furnace is on line a flame failure device should isolate the fuel (unless another burner, using an independent source of fuel, will reignite the first burner when its fuel supply is restored).

Variation in detail is possible, but all furnace operators should understand the reasons for these principles so that they realize the results of not following them.

Tube failures are usually the result of overheating the tubes, often months or years beforehand. Furnace tubes are usually designed for a life of 100 000 hours (11 years). Suppose the design temperature is 500 °C:

- If the tubes are operated at 506 °C they will last 6 years.
- If they are operated at 550 °C they will last 3 months.
- If they are operated at 635 °C they will last 20 hours.

Failure will be by 'creep'; the tube will expand, slowly at first and then more rapidly, and will finally burst. Creep cannot be detected until its later stages but nevertheless once a tube has been overheated some of its creep life has been used up and however gently we treat the tube afterwards the lost life can never be recovered. Furnace tubes are not just lying there enjoying the warmth; they are working out how long they can go before they burst. (See Figure 23.)

If we can get more output from a compressor or distillation column, then we are free to do so. But if we get more output from a furnace by overheating the tubes we pay for it later.

It should be possible to stop the flow to a furnace, by *emergency isolation valves*, if a tube burts.

Figure 23 Furnaces

Gas detection

See *analysis*.

Gas dispersion

See *dispersion*.

Giving up

We dislike giving up. We admire people who press on, despite difficulties.

... when most of us plan a flight from A to B (for example) we programme our subconscious to get to B... All our thoughts and expectations are of a positive nature; we only think about getting there, and work out how to do it. We rarely plan to get half-way there and turn back[1].

The author then describes an accident which occurred to a light aircraft when an inexperienced pilot who wanted to fly over a range of hills pressed on regardless of low cloud.

Computers have a key marked ESCAPE or EXIT. We get used to pressing it so perhaps in the future escaping or exiting from the planned task will not seem as unmanly as in the past.

See *manliness*.

1. *Aviation Safety Digest* (Australia), No. 129, Winter 1986, p. 129

Glass

See *level glasses* and *windows*.

Good practice

Accidents often occur because someone does something which does not break any *code of practice* or set of *instructions* but is not good engineering or operating practice.

There are so many codes, standards, guidance notes and so on available to the design engineer that one might expect them to cover every possible situation, but they do not. Many points are not covered because the are considered too obvious.

For example, a line carrying liquefied gas was protected by a small relief valve which discharged onto the ground. The ground was uneven and after rain the end of the tail pipe was below the surface of a puddle. The puddle froze and the line was over-pressured; it did not burst but an instrument was damaged.

I doubt if any code says that that tail pipes should not discharge into puddles. It is obvious; it is good engineering practice, but the man who installed the tail pipe did not think of it and the *construction* inspector and the operating team did not notice what he had done. The engineering inspector has to be on the look out for errors which no one would dream of forbidding specifically. See *LFA*, Chapter 16.

Another example: The roofs of large fixed roof storage *tanks* (and of floating roof tanks when off float) have to be supported. Hollow tubes are often used. Accidents have occurred because liquid got inside the tubes through corrosion holes and then came out when the tank was emptied for repair. It is good practice to drill holes in the tubes so that liquid can drain out, or to use girders instead of tubes. See *WWW*, §5.5.2.

Similarly, much good operating practice is not written down. Operators are expected to pick it up by a sort of osmosis but some people are not very permeable. Many operators do not realize the power of compressed *air*. (See *pressure*.) Others do not realize that their job is to do something about an unusual reading or observation, not just write it down, even if all they do is tell the foreman. See *alertness, full-scale deflection* and *small companies*.

Grimaldi

John Grimaldi, at one time Director of the Center for Safety at New York University, referred, some years ago, to 'The dearth of intellectual curiosity and scarcity of scholarship which seems to typify the pursuit of safety'. He, at least, is an exception, as the following quotations show:

> The individual, when given a goal which he finds is beyond attainment, simply settles for what is good enough. The accident pattern thus continues . . .
>
> Another problem is the dispersion of effort when people try to put under control *all* possible causes of harmful events. The effect . . . has been compared to collecting feathers in a wind storm.
>
> The inherently high hazard industries, where single events could have severe or catastrophic consequences, are examples of intense safety effort compared to others where the hazards are less likely to be fearsome. The visible presence of pronounced threats to the system are powerful inciters to safety action.
>
> The natural center of safety interest is the *hazard and its control*, rather than the accident which derives from it[1].

See *philosophers' stone* and *production v safety*.

1. J Grimaldi, *Safety and Maintenance*, March 1969, p. 13

Guarding

See *fencing*.

Guidance notes

See *codes of practice* and *good practice*.

'Gut feel'

We are often told that techniques cannot answer all our problems and that sometimes we have to rely on 'gut feel', or managerial judgement to give it a more dignified name. I do not dispute this, but different guts feel differently. If your gut tells you that a risk is large and demands immediate attention and mine says that it is trivial, a dialogue between us is difficult. In contrast, if we can discuss the probability of an accident and the size of the consequences and put numbers to them, however roughly, a dialogue is possible. (See *Hazard Analysis (Hazan)*).

If you feel something in your bones, to use another metaphor, you should try to puzzle out why you feel as you do. Is it past experience of a similar situation, suspicion of technical arguments you cannot fully understand, distrust of an individual's judgement? If you put your feelings into words you are more likely to convince others.

'Gut feel', like female intuition, coincides too often for comfort with what people would like to believe.

Hazard analysis (Hazan)

After we have identified the hazards on a plant, by a *hazard and operability study* or by other means, we have to assess them, that is, decide what to do about them. (See *assessment*.) We may decide to remove or reduce the hazards by a change in design or method of working or we may decide that the hazards are so unlikely or trivial, compared with all the other hazards around us, that we should ignore them, at least for the time being. Sometimes the decision is obvious, sometimes our experience or a *code of practice* tells us what to do, but sometimes we have no experience, there is no code of practice and the arguments for and against living with the hazard seem finely balanced. In such cases we may use the methods of hazard analysis[1], comparing the risk with a target or *criterion* (see *fatal accident rate*) in order to decide on action.

A hazard analysis, therefore, has to answer three questions:

1. How often will the hazard occur?
2. What will be the consequences to employees, the public and to the plant?
3. How do these compare with the target or criterion?

In brief:

- HOW OFTEN?
- HOW BIG?
- SO WHAT?

Whenever possible the answers to 1 and 2 should be based on experience but sometimes there is no experience and we have to estimate the answers. Fault trees are often used to estimate how often the hazard will occur, i.e. its probability.

A hazard analysis may take a few minutes (see *Fermi estimates*) or it may take weeks or even months of detailed calculation.

Other names for hazard analysis are risk analysis, Probabilistic Risk Assessment (PRA) and Quantitative Risk Assessment (QRA). Risk analysis is sometimes used to describe methods of assessing the commercial risks of a project.

A hazard analysis (hazan) is often confused with a hazard and operability study (hazop), but they are quite different:

- Hazop identifies hazards, is qualitative and can be applied to all designs.
- Hazan assesses hazards, is quantitative and should be used only when other methods fail.

If you are asked (or you ask someone) to carry out a hazop (or hazan) make sure that they are using the words in the same sense as you are or the job done may not be the one wanted.

Needless to say, risk quantification must always be applied with good judgement, and it should be viewed as a supplement to, not as a substitute for, engineering design standards, best industry practices, and safe designs based upon long experience. Some companies may not find risk quantification to be suited to their needs, and I do not recommend it to everyone. I will say, however, that our company has found risk quantification to be extremely useful as a means of uncovering risks which are far greater than had been perceived and as a means of identifying the most effective ways to reduce those risks[2].

See *assumptions, asymptote, Canvey Island, confidence limits, operations research, perception of risk, perspective, reliability, sloppy thinking* and *Lees*, Chapter 9.

1. T A Kletz, *Hazop and Hazan – Notes on the Identification and Assessment of Hazards*, Institution of Chemical Engineers, Rugby, 2nd edition, 1986
2. P L Thibaut Brian (a vice-president of Air Products), *Safety Management* (South Africa), Vol. 16, No. 1, January 1989, p. 5

Hazard and operability study (Hazop)

A hazard and operability study (hazop) is the preferred method, in the process industries, of identifying hazards on new or existing plants.

At one time we identified hazards by waiting until an accident had occurred; we then took action to prevent it happening again. This 'every dog is allowed one bite' philosophy was acceptable when plants were small but is no longer viable now that we keep dogs as big as *Flixborough* or *Bhopal*. We need to identify hazards before accidents occur. *Check lists* have the disadvantage that new hazards, not on the list, may be overlooked so we prefer the more open-ended hazop technique. It allows a team of people, familiar with the design, to let their minds go free and think of all the deviations that might occur but it is done in a systematic way in order to reduce the chance of missing something.

Note that hazop identifies operating problems as well as hazards.

Although hazop has been used mainly in the oil and chemical industries it can be applied to many other operations.

The technique is applied to a line diagram, line-by-line. Using the guide word 'NONE' we ask if there could be no flow, or reverse flow, in the first line. If so, we ask if this would be hazardous or would prevent efficient operation. If it would, we ask what change in design or method of operation will prevent no flow, or reverse flow (or protect against the consequences). Using the next guide words, 'MORE OF' and 'LESS OF', we ask if there could be more (or less) flow, pressure or temperature in the line and if this would be hazardous etc. Using the guide words 'PART OF' and 'MORE THAN' we ask about the effects of changes in concentration or the presence of additional substances or phases. The guide word 'OTHER' reminds us to apply our questioning to all states of operation, including start-up, shutdown, catalyst regeneration and so on and we also ask if the equipment can be safely prepared for maintenance. We then study the next line in the same way. All lines should be studied including *service lines* and *drains*.

On a batch plant it is also necessary to question the instructions. If a tonne of A is to be added to a reactor we ask about the effects of not adding A, adding more or less A, adding something else instead of or as well as A, adding part of A (if A is a mixture) and reverse addition of A (that is, *reverse flow* from the reactor to the A container).

156 Hazard and operability study (Hazop)

A hazop is carried out by a team who have been involved in the *design*. It is not a technique for bringing fresh minds to look at a design. It assumes that errors in the design do not arise from lack of knowledge but because the complexity of plant design makes it difficult to apply our knowledge unless we go through the design systematically, line-by-line, deviation-by-deviation.

After we have identified the hazards we have to decide what to do about them. Most of the time the team decides during the meeting, using the members' experience, *codes of practice* and their knowledge of *good practice*. Sometimes the decision is more difficult calling for *hazard analysis* to be used. See *Lees*, Chapter 8 and reference 1.

Audits are widely used for identifying hazards on existing plants. While hazop looks for hazards inherent in the design or method of operation. audits show whether or not the agreed procedures are in fact being followed.

The following example will show the advantages and limitations of hazop: An emulsion of nitroglycerine (NG) and acid was separated in a centrifuge. A *choke* occurred in the NG exit line, due to swelling of the *plastic* pipe. Some NG went down the acid line into the acid tank and settled on top of the acid. Two *explosions* occurred, one in the acid tank and the second in the recycle line out of the tank. (See Figure 24.) The first was probably triggered by vibration and the second was possibly triggered by the sun's heat. Several men were killed.

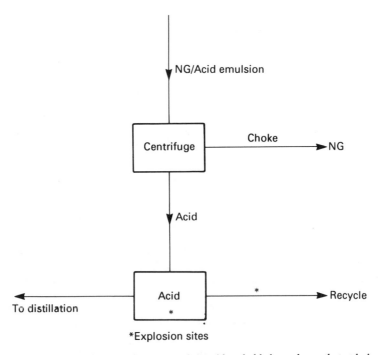

Figure 24 A Hazard and operability study would probably have shown that a choke could cause NG to enter the acid line

If the plant design had been hazoped then consideration of 'no flow' in the NG line or 'more than' in the acid line would probably have disclosed the hazard; chokes are one of the obvious causes of 'no flow' (the others being closed valves, pump failures, line breakage and empty suction vessel). On the other hand, if the hazop team had said that line chokes have never been known to occur and never could occur then the accident might have been dismissed as impossible; this is unlikely. Without a hazop we might never consider the possibility of a choke. Once we think of it, we realize that it might occur.

It is difficult to prevent chokes in the NG line so methods of detecting NG in the acid line were developed.

We believe it is beneficial to involve a large number of our people in our hazard review efforts. While most of them will participate in a formal hazard review only occasionally, this training and experience will, we believe, help them to think logically about process safety in their every day work[2].

The limitation of hazop is that it comes too late for major changes in design. See *friendly plants* and *inherently safer designs*. Similar studies are needed at the conceptual stage of a project (when we are deciding which product to make, by what route, and where to locate the plant) and at the flowsheet stage.

See *expert systems*.

1. T A Kletz, *Hazop and Hazan – Notes on the Identification and Assessment of Hazards*, Institution of Chemical Engineers, Rugby, 2nd edition, 1986
2. P L Thibaut Brian (a vice-president of Air Products), *Safety Management* (South Africa), Vol. 16, No. 1, January 1989, p. 5

Hazard and risk

A hazard is a physical situation with a potential for human injury, damage to property, damage to the environment or some combination of these.

The term major hazard is often used to describe a hazard with the potential to kill a number of people while a major hazard site is one where large quantities of hazardous chemicals are processed or stored. These terms came into widespread use after *Flixborough*. See *CIMAH Regulations*.

A risk is the probability that the injury or damage will occur. It may be expressed as a frequency (the number of times it occurs, on average, in a year or other period of time) or as the probability that the injury or damage will occur during a period of time or will follow a previous event.

Risks are often expressed as:

- The number of deaths or injuries per year in a group of employees (see *fatal accident rate*) or in the surrounding population.
- The probability that someone – any person in a defined group such as the employees on a plant or the people living nearby – will be killed or injured in a year.
- The probability that a particular person (perhaps the employee or member of the public at greatest risk) will be killed or injured in a year.

158 Hazardous substances

- The probability that a certain number of people will be killed or injured. This is called the societal risk to distinguish it from an individual risk.

Frequencies and probabilities are sometimes confused. A frequency is a rate (so many times per hour or per year) while a probability is a dimensionless number between 0 and 1. Nonsense answers have been produced in *hazard analysis* because the analyst was unsure of the difference[1].

See *criteria, perspective, underestimated hazards, unexpected hazard, unlikely hazards* and reference 2.

1. T A Kletz, *Hazop and Hazan – Notes on the Identification and Assessment of Hazards*, Institution of Chemical Engineers, Rugby, 2nd edition, 1986, Chapter 3
2. *Nomenclature for Hazard and Risk Assessment in the Process Industries*, Institution of Chemical Engineers, Rugby, 1985

Hazardous substances

Some substances, those that are flammable, explosive, toxic or corrosive, are obviously hazardous. Here I want to draw attention to some which are less obviously hazardous but are involved in more than their fair share of accidents, i.e. *air, heavy oils, nitrogen, steam* and *water*. Why is this? Partly because people are unaware of the damage and injury these subtances can cause but also because looking after the services is seen as less interesting and rewarding than looking after the manufacturing operations. Nobody wants to do it and the job gets starved of resources or given to the less able people. Similarly, designers would rather design the heart of the plant than the off-plots and other peripherals. When carrying out *audits* the services (see *service lines*) should always be looked at.

See *WWW*, Chapter 12.

For obviously hazardous substances see *ammonia, arsenic, asbestos, benzene, BLEVE, chlorine, 'Controversial Chemicals', corrosive chemicals, cyanides, cyclohexane, dioxin, explosions, fire, flammable atmospheres, flashing liquids, lead, oxygen, solvents* and *toxicity*.

Health and Safety at Work Act

The item on *Factories Acts* should be read first. While the Factories Act laid down an increasing number of detailed requirements, the UK Health and Safety at Work Act, which came into force in 1974, laid down instead a general requirement on all employers to provide a safe plant and system of work and adequate (see *adequacy*) *instruction, training* and supervision, so far as is *'reasonably practicable'*. In the first place it is the responsibility of the employer to decide what is a safe plant etc. (those who create a hazard know more about it than the *Factory Inspector*) (see *Summerland*) but if the Factory Inspector does not agree he will say so and, if necessary, issue an Improvement or Prohibition Notice.

Regulations made under the Act are 'inductive', that is, they say what has to be done but do not say how, though guidance and a *code of practice* may be available. If there is such a code (or another generally accepted one) then an employer is expected to follow it unless he can show that it is inapplicable or he is doing something as safe or safer.

The Health and Safety at Work Act is enforced by the Health and Safety Executive, a body which includes the Factory Inspectorate, the Mines and Quarries Inspectorate, the Nuclear Installations Inspectorate and the Agricultural Inspectorate, while policy is decided by the Health and Safety Commission, which consists of representatives of the employers' organizations, trade unions and local authorities.

The arguments in favour of the Health and Safety at Work Act, compared with the old Factories Acts, are set out in the report which led to the Act[1].

In brief, they are:

1. Codes can be changed more easily than regulations when new problems arise or new solutions are found to old problems. It takes time to change regulations.
2. Employers are not compelled to follow rules which are inapplicable or out of date. They do not have to follow the letter of the law if another course of action is as safe or safer.
3. Employers (and inspectors) cannot say, as I have heard them say in other countries, 'The plant must be safe because it complies with the regulations.' Instead they must continually be on the lookout for new hazards and new ways of controlling familiar hazards.
4. It is not practicable to write detailed *rules and regulations* for complex and changing technologies especially when there may be only one or two plants in the country using these technologies. On the other hand, regulations may be appropriate when a straightforward activity, such as *tanker* labelling, is carried out by many people.
5. The Health and Safety at Work Act is a more powerful weapon in the hands of the inspector than the Factories Act. He does not have to look for a regulation that has been broken. It is sufficient to show that the plant or method of working is not safe, so far as is reasonably practicable, or does not comply with a generally accepted code.

Much of the opposition to the Health and Safety at Work Act came from people who thought it would be a soft option for the employer, compared with the Factories Act. As I have tried to show, this is not the case.

The UK system is not common. Most countries still try to control accidents at work by detailed regulations. Comparing Sweden, which has a similar system to the UK, with the US, Kelman writes[2]:

> American inspections are designed more as formal searches for violations of regulations; Swedish inspections are designed as informal, personal missions to give advice and information, establish friendship ties between inspector and inspected, and promote local labour-management cooperation.

In contrast, Burger writes[3]:

> German safety programmes seem to be heavily influenced by scores of statutory regulations with which industry has to comply. . . When a programme has to conform rigidly to regulations by the authorities

160 Heavy oils

and/or trade unions there is little motivation to structure it on safety policies or objectives. There is a much stronger urge to satisfy the requirements . . .

See *employers' responsibilities, law, (legal) responsibilities, management* and *United Kingdom*.

1. *Safety and Health at Work: Report of the Committee 1970-1972* (Chairman: Lord Robens), HMSO, London, 1972.
2. S Kelman, *Regulating America, Regulating Sweden*, MIT Press, Cambridge, Massachusetts, 1981, p. 203
3. S Burger, *Safety Management* (South Africa), February 1988, p. 4

Heavy oils

Heavy oils, those with a flash point above ambient temperature, are more hazardous than many people realize and are therefore treated with less respect than obviously *hazardous substances* such as petrol. Heavy oils, widely used as fuel oils, solvents, lubricants and heat transfer fluids, will not burn or explode when cold but will do so once they are heated above their flash points.

Many fires and explosions have occurred when welding has been carried out on *tanks* which have contained heavy oils (or materials which polymerize to form heavy oils). It is almost impossible to remove all traces by cleaning and the welding vaporizes the oil and then ignites it. (See Figure 25.) If welding has to be carried out on such tanks they should be

Figure 25 Heavy oils

inerted with *nitrogen* or with fire fighting foam made with nitrogen (but not foam made with air). The volume to be inerted can be reduced by filling the tank with water up to the level at which welding has to be carried out.

Spillages of heavy oils, at temperatures below their flash points, can be ignited easily if a piece of material such as a rag, which forms a wick, falls into them. Heavy oil mists, like all flammable mists, can be ignited below the flash point of the bulk liquid. See *WWW*, §12.4 and, for another hazard of heavy oils, see *foam-over*.

See *unexpected hazards*.

Higee

See *distillation* and *innovation*.

Hired equipment

When equipment is hired it is important to be clear who will be responsible for its testing and maintenance. The hirer of a *nitrogen* vaporizer thought that the owner would test the instrumentation and the owner thought that the hirer would do so. Result: nobody tested it.

A company hired a van for use by *contractors*. The catch on the rear door was faulty. After a man had fallen out of the back of the van and been injured it was discovered that each of the three parties involved, the company, the van owner and the contractor, thought that one of the others was responsible for maintenance.

Before hiring equipment we should make sure that it conforms to our usual safety standards or, if it does not, that we are aware of the fact and have nevertheless taken a considered decision to hire it. See *packaged deals*.

History of safety

People have been concerned since the earliest times that others should not be hurt. For example, The *Bible* states:

> When you build a new house, be sure to put a railing round the edge of the roof. Then you will not be responsible if someone falls off and is killed. – Deuteronomy 22: 8

In the East roofs were, and still are, often used as extra living space. The sixteenth century *Code of Jewish Law*[1], a summary of rules going back to the second century, states that the railing should be 10 hand-breadths high. This is 3 ft 4 in (100 cm), if one hand-breath is taken as 4 inches (10 cm). According to the Construction Regulations, made under the UK *Factories Act*, every workplace in which someone is liable to fall more than 6 ft (1.8 m) must be securely fenced with a fence between 3 ft and 3 ft 9 in (90–115 cm) high, so this law has not changed very much.

Even earlier than the Bible the Code of Laws of the Babylonian King Hammurabi, written about 1700 BC, said:

If a builder builds a house for a man and does not make its construction firm and the house which he has built collapses and causes the death of the owner of the house that builder shall be put to death.

If it causes the death of a slave of the owner of the house, he shall give to the owner of the house a slave of equal value.

If a builder builds a house for a man and does not make its construction meet the requirements and a wall falls in, that builder shall strengthen the wall at his own expense.

The industrial revolution led to legislation on *boilers*, the UK *Factories Act*, and similar legislation in other countries (though in some countries not until much later). (See *evangelicalism*.)

The growth of safety as a subject worthy of systematic study in industry is more recent. In 1922 the Chief Labour Officer of *Brunner Mond* visited the *United States*. He reported that Solvay established a fire service at their Syracuse works in 1890 and that in about 1909 they realized the importance of studying accidents that had occurred with a view to preventing their recurrence. They soon realized that it would be valuable to study all plant conditions in detail and correct those found to be dangerous without waiting for an accident to occur. Starting the following year, outside experts were called in to carry out what we would now call *audits* of the fire hazards and plant conditions. The first full time safety officer was appointed in 1916.

See *Du Pont, loss prevention, retribution* and *'safety first'*.

1. S Gantzfried, *Code of Jewish Law*, Hebrew Publishing Company, New York, 1927

Honesty

While a welder was burning a hole in a pipe, 6 inches diameter, with walls ½ inch thick, in a workshop, a sudden noise made him jerk, his gun touched the pool of molten metal and a splash of metal hit him on the forehead. Fortunately, his injuries were not serious. If the pipe wall had been more than ½ inch thick a hole would have been drilled in the pipe first and the foreman said, on the accident report, that this technique should be used in future for all pipes. However, this would be troublesome: a *crane* would be needed to move the pipe to the drilling bay and back and this could cause delay. In practice, nothing was done.

The welder had cut thousands of holes before without incident, the chance of injury was small and it may have been reasonable to do nothing and accept the slight risk of another incident. In practice this is what the foreman, and the engineer in charge, decided to do. However, they were not willing to say so and instead they recommended action that they had no intention of enforcing.

After an accident many people feel that they have to recommend something even though they know it will be difficult and they have no intention of carrying out the recommendation with any thoroughness. It would be better to be honest and say that the chance of another incident is so low that the risk should be accepted. The law does not expect us to do everything possible to prevent accidents, only that which is *'reasonably practicable'*.

In another case a man went up a ladder to fix an additional support to a pipeline and, while doing so, burnt his hand, not seriously, on a hot pipe which was not insulated. The foreman realized that in all probability no one would ever go near the pipe again, but he felt he had to recommend something and he recommended that the hot pipe be insulated. In this case the recommendation was carried out.

See *falsification*.

Hoses

Flexible hoses are widely used and have been involved in many accidents. When used correctly, they are safe but they cannot tolerate misuse to the same extent as a steel pipe (or an articulated arm) and so they should be used only when essential, for example, for filling and emptying *tankers* and for making temporary connections. Whenever possible we should use *friendly plants* and equipment which can tolerate failure or human error. If a temporary connection looks like becoming permanent it should be replaced by a permanent pipe. Articulated arms should be used instead of hoses for filling *chlorine* and *ammonia* tankers.

Common reasons for hose failure are:

1. Using the wrong sort of hose. Does everyone know the sorts that are suitable for each duty? Do you have enough of them? Must you have so many different sorts in use? It may be cheaper in the long run to standardize on fewer types even though they are more expensive. (See *action replays*.)
2. Using a damaged hose. All hoses should be tested and inspected regularly and marked to show that they have been passed as fit for use. The colour of the test mark should be changed after each round of testing. Do not leave hoses lying on the ground so that they get run over; provide racks for them and make sure the hose is well-supported. If it is supported at only one or two points it may be damaged[1]. (See *oxygen*.)
3. Making connections incorrectly, in particular, securing screwed joints by only one or two *threads*, combining different threads, leaving out gaskets or securing hoses by Jubilee clips (as used in motor car cooling systems). These clips are unsuitable for industrial use; bolted clamps should be used instead.
4. Disconnecting the hose before the pressure has been blown off. All hoses or hose connections should be fitted with small vent valves for blowing off the pressure. (See Figure 79, p. 326)
5. Allowing a process material to get back into a hose containing a service such as *steam, nitrogen* or compressed *air*. The valve at the process end of the hose should be closed first and, in addition, the hose should be chosen so that it can withstand the process material. (See Figure 26.)

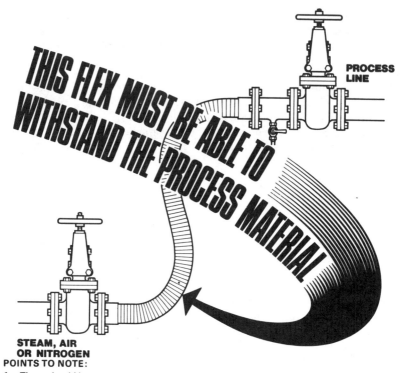

POINTS TO NOTE:
1 There should be a non-return valve in the service line.
2 The flex should be able to withstand the process pressure as well as the process material.
3 If the service valve is really located below the process valve, then an additional drain point should be fitted next to the service valve.

Figure 26 Hoses

6. Using a hose at a higher pressure than necessary. For example, when emptying a tanker, using a hose to provide suction to a fixed pump is better than connecting the hose to the delivery of the tanker's pump.
7. Not wearing suitable *protective clothing*.

A road tanker delivered concentrated sulphuric acid to a chemical works. The tanker carried five hoses, including only one suitable for sulphuric acid, but it was fitted with the wrong couplings, so the driver used another hose, under pressure. It burst, showering him with acid. He was not wearing protective clothing and died in hospital three months later[2].

See *EVHE*, §10.4.1–2.

1. *Loss Prevention Bulletin*, No. 083, October 1988, p. 1
2. *Hazardous Cargo Bulletin*, March 1986, p. 41

Hubris

The Supplement to the *Shorter Oxford Dictionary* defines hubris as: 'Presumption, originally towards the gods; pride, excessive self-confidence.'

A company safety newsletter described an accident that had occurred elsewhere. A pipeline had been welded to two supports which were bolted to a concrete foundation. When the pipe expanded and vibrated, it could not move and a piece was torn out of it. There was a serious fire[1].

A colleague telephoned the editor to say that such an incident could not occur in a well-run company. No one would design or construct pipework that way. The editor should, he suggested, describe accidents that might occur in his organization and that had a message for his colleagues, not accidents which could not conceivably occur.

Soon afterwards a similar accident did occur. A reflux line was clamped firmly to brackets which were welded onto the side of a distillation column. At start-up the difference in expansion between the hot column and the cold reflux line was sufficient to tear one of the brackets from the column, causing a leak of flammable vapour.

1. V G Geisler, *Loss Prevention*, Vol. 12, 1979, p. 9

Human failing

In some companies 80–90% of the accidents that occur are said to be due to human failing, that is, a failure by the injured person or fellow-workers rather than a failure by the manager, supervisor or designer. (It seems that managers and designers do not fail, or are not human!) There is apparently nothing that managers or supervisors can do except tell the people concerned to be more careful. In fact, as I shall try to show, most accidents can be prevented by better management or better design. I do not say that they are caused by bad management; if I did people would become defensive, but most of us are ready to admit that we could do something better.

To say that accidents are due to human failing or human error is not so much untrue as unhelpful. It does not lead to effective action, only to advice to take more care. Instead of asking what is the *cause* of an accident we should ask what we can do to prevent it happening again. We may then think of ways of improving the training, supervision, design and so on.

So-called human failings can be classified as follows:

Failings that could be prevented by better training or instructions

Someone does not know what to do or, worse still, thinks he knows but does not. Sometimes he lacks elementary knowledge, sometimes sophisticated knowledge, such as methods of diagnosing faults. Sometimes there are no instructions; sometimes they are not clear. The action required is obvious (see *discretion*).

Failings due to a lack of motivation

Someone knows what to do but decides not to do it. An operator may decide not to wear the correct protective clothing; a manager may decide to keep the plant running to complete an urgent order, despite a serious leak. These failings can be prevented by better management and supervision, in particular, by making sure that the correct procedure is not unduly difficult or inconvenient to follow, by explaining the need for the rules, by checking up from time to time to see that they are followed and, in particular, by not turning a *blind-eye*.

According to Rook[1], it is 'always worthwhile to expend effort on improving training and motivation until most of the workforce is brought up to the general industrial average level. However, it is doubtful whether further improvement can be obtained economically by further training and motivational efforts'.

Swain[1] goes further. 'Motivational appeals have a temporary affect because man adapts. He learns to tune out stimuli which are noise, in the sense of conveying no useful information. Safety campaigns which provide no useful, specific information fall into the noise category; they tend to be tuned out.'

Failings due to a lack of physical or mental ability

Someone is asked to do a job beyond his powers or, more often, beyond anyone's powers. The equipment or method of working should be redesigned.

A momentary slip or aberration

Someone knows what to do, intends to do it, and is able to do it but forgets to do it or does it wrongly. These slips are similar to those of everyday life and are impossible to prevent, though they can be made less probable by reducing *stress* and *distraction* (and *boredom*). Routine tasks are delegated to the lower levels of the brain and are not continuously monitored by the conscious mind. If everything we did required our full attention we would be exhausted soon after we got up; so we put ourselves on autopilot. If anything disturbs the smooth running of the program a slip occurs. We cannot prevent these failings but we can remove opportunities for error by changing the work situation, that is, the design or method of working. Alternatively we can guard against the consequences of errors or provide opportunities for recovery.

To illustrate these four sorts of human failing consider a common situation on process plants: an alarm sounds and an operator has to select the right valve and close it within, say, 10 minutes. If he fails to do so, it may be for one of several reasons.

1. He may not know that he was supposed to close the valve, or which valve.
2. He may decide not to close the valve although he knows that he should.
3. He may be unable to close the valve as it is too stiff or out of reach.
4. He may be busy on other jobs and forget to close the valve or be distracted and close the wrong valve. We cannot prevent these errors by telling him to be more careful. We have either to accept the occasional error or install an automatic system. (Note that this will not remove our dependence on men. We now depend on the men who design, install,

test and maintain the automatic equipment. They may also fail but they probably work under conditions of less stress and have more opportunities to check their work.)

Estimates of human error rates apply only to the last sort of error. We can hardly estimate the probability that someone, or the whole workforce, will be poorly trained, unwilling or incapable.

As another example of the four sorts of error, consider spelling mistakes:

1. If I write 'recieve' or 'seperate', it is probably due to ignorance. If I write 'wieght' it may be the result of following a rule that is wrong, i.e. 'i before e, except after c'. Better training may prevent the errors but a spelling-checker program for my word processor would be cheaper and more effective.
2. If I write 'thru' or 'gray', it is probably due to a deliberate decision to use the American spelling. To prevent the errors someone will have to persuade me to use the English spelling.
3. If I write 'Llanfairpwllgwyngyllgogerychwyrndrobwlllantysiliogogogoch' (a village in Wales), it is probably because it is beyond my mental capacity to spell it correctly from memory. (Can you spot the error?)
4. If I write 'opne' or 'thsi' it is probably a slip, overlooked when I checked my typing. Telling me to be more careful will not help. A spelling-checker program will.

(1) is an error in knowledge-based (or rule-based) behaviour, (2) could be described as a deliberate decision rather than an error and (4) is an error in skill-based behaviour.

To sum up, do not try to change people. Accept them as we find them and try to change situations.

For many examples of accidents which at first sight were due to human failing but could have been prevented by changing the work situation, that is the design or method of working, see *EVHE*. See also *automation, carelessness, (personal) responsibility* and *(visit accident) sites*.

1. Quoted by A D Swain in *Selected Readings in Safety*, edited by J T Widner, Academy Press, Macon, Georgia, 1973, p. 371

Humour

Some people feel that humour has no place in safety. It is too serious a subject and we should not laugh about actions or omissions that might lead to death or injury. But though we should be serious about safety that does not mean we have to be solemn. If a little humour helps to get our message across and helps people to remember it, then it serves a good cause. W S Gilbert wrote, in *The Yeomen of the Guard*:

> He who'd make his fellow creatures wise,
> Should always gild the philosophic pill.

Figure 27

Figure 28

Figure 29

Figure 30

Furthermore, there is many a true word said in jest. Humour can tell us something about our underlying motives and meanings. There is a story about a mother who gave her son a red shirt and and a blue shirt. The next day he wore the red shirt. She said, 'So you don't like the blue one!' This story strikes a chord: families work by making each other feel guilty. We may never have said so in so many words, but we recognize the truth in the story[1].

The illustrations, in Figures 27–34 show some cartoons that have appealed to me. I leave the reader to decide their message, if any.

1. E de Bono, *Word Power*, Penguin Books, 1979, p. 118

Figure 31

Figure 32 Like this doctor, safety professionals are often called in too late. See *friendly plants* and *inherently safer design*

Figure 33

Figure 34

Hydrogen

Hydrogen differs from other flammable gases in several respects. Because it is very light it disperses readily and large gas clouds are unlikely to form in the open air. On the other hand only very small quantities are necessary for an *explosion*. Several have been reported in which only 100–500 kg were involved (compared with several tonnes or more needed for other gases). The efficiency (the fraction of the heat of combustion appearing as a shock wave) was 1–2%[1].

Indoors, hydrogen will disperse quickly if there are openings in the roof but if there are no openings it can accumulate below the roof and an explosion can occur[2].

If hydrogen is flared then as soon as the flow stops the hydrogen will be displaced by air. A very large purge gas rate is necesssary to prevent air entering the *stack*. Whenever possible, therefore, hydrogen should not be flared with other gases but sent to an independent vent stack.

1. J R Hawksley, *Loss Prevention Bulletin*, No. 068, April 1986, p. 1
2. J A MacDiarmid and G J T North, *Plant/Operations Progress*, Vol. 8, No. 2, April 1989, p. 96

Hydrogenation

See *wolves in sheep's clothing*.

Iatrogenesis

Medical scientists use the word 'iatrogenetic' to refer to disabilities that result from medical treatment. Do we need a word to describe design faults that are the result of following the *safety professional*'s advice?

Modifications to plants and processes often have unforeseen side-effects, and safety professionals sometimes get their priorities wrong and spend too much time and money dealing with minor hazards but I cannot think of many cases when they have actually advised someone to take action that was clearly wrong. However, in a few cases those most powerful of safety professionals, *Factory Inspectors*, have made companies make what seemed to me to be wrong decisions. Here are two examples, neither from the UK:

- A company was made to construct a blast wall round a liquefied flammable gas storage sphere to protect it from missiles from an *explosion* on a neighbouring plant, an unlikely event. The blast wall prevented natural *ventilation* of the tank's surroundings, small leaks could not be dispersed and an explosive atmosphere could be formed inside the walls. They created a bigger hazard than the one they were designed to prevent.
- On several occasions companies have been made to install compressors inside buildings, to reduce the noise level outside, preventing dispersal of leaks by natural ventilation and increasing the chance of an explosion. There are other ways of reducing the noise from compressors.

For other examples see *caution* and *exchanging one problem for another*.

ICI

ICI (Imperial Chemical Industries p.l.c.) was formed in 1926 by the amalgamation of *Brunner Mond* and three other chemical companies, and inherited many of Brunner Mond's traditions and practices, including a high commitment to safety. The company's safety record, as measured by the *fatal accident rate*, continued to improve until the 1960s, when it deteriorated. (See Figure 35.) A new generation of plants had been built, larger than those built before, containing higher inventories of flammable liquids and gases under more extreme conditions, but methods of design and operation had not changed. The result was a series of *fires* and *explosions*. Senior managers realized, though it took them a few years to do so, that there had been a step change, and that a new approach was needed. Technical people with operating experience were transferred to safety (I was one of the first) and *loss prevention* replaced the traditional safety methods. Figure 35 shows the results achieved.

See *airlines* and *change – new problems*.

Ideas

See *innovation*.

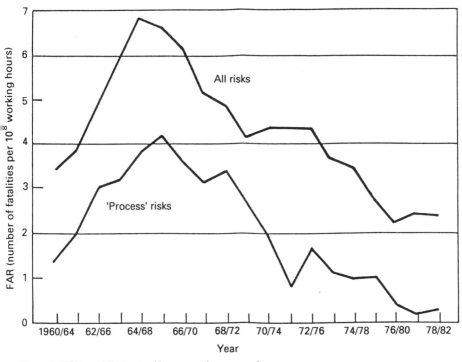

Figure 35 ICI's accident rates (5-year moving average)

Identification of equipment

Many accidents have occurred because a maintenance worker opened up the wrong pipeline, pump, tank etc. Pointing out the equipment will not prevent these accidents. The repairman goes for his tools or for a cup of tea, comes back and breaks into the wrong pipeline or pump. Chalk marks are no better; they are not weatherproof and old marks sometimes remain leading the repairman to the wrong chalk mark.

Describing the equipment is even worse: 'The pump you repaired last week has failed again'; 'The one that usually gives us trouble is at it again'; phrases like these lead to accidents. To identify equipment correctly we should mark it with a number or name and put this number or name on a permit-to-work which is given to the repairman. If there is no permanent number or name then a numbered label or tag should be fixed to the equipment and the number put on the permit. If a pipeline is to be broken the tag should be fixed at the point at which it is to be broken. (See Figure 36.)

Numbering systems should be simple and logical. It is asking for trouble to number a row of pumps as in Figure 37. The man who was asked to repair pump No. 7 assumed, reasonably enough, that it would be the 7th one along and in fact started to dismantle No. 6.

Tag along with us

They make the job exact and neat... Please return when job complete.

Figure 36 Identification of equipment

There were seven pumps in a row

A fitter was given a permit to do a job on No. 7. He assumed No. 7 was the end one and dismantled it. Hot oil came out.

The pumps were actually numbered:

Equipment which is given to maintenance must be labelled. If there is no permanent label then a numbered tag must be tied on.

Figure 37 Identification of equipment

On one plant the pump numbers were painted on the coupling guards. Before long the guards on two adjacent pumps were removed at the same time and put back wrongly. Pump numbers should be painted on pump bodies or, better still, on the plinths.

Labels seem to have a high vapour pressure as they vanish after a while. They are a sort of protective equipment and, as with all protective equipment, we should inspect them at regular intervals to make sure thay are still in working order.

Seeing that equipment is correctly labelled and checking from time to time that the labels are still there seems a dull job, giving us little opportunity to stretch our intellectual muscles or utilize our technical skills and knowledge. Nevertheless, it is as important as more demanding tasks. One of the signs of a good manager is that he sees to the tedious jobs as well as those that are fun. If you want to judge a manager or a management team do not just look at the technical problems they have solved. Look at their labels.

Colour coding of pipelines can often avoid confusion but much effort is needed to make sure the colours are correct. It is better to have no colour coding than have lines coloured incorrectly.

For other examples of accidents caused by poor labelling see *access, King's Cross* and *WWW*, §1.2 and Chapter 4.

Identification of hazards

Many accidents occur because we fail to foresee a hazard, though afterwards we say, 'we ought to have thought of that'. In *hazard analysis* the biggest source of error is failing to see some of the hazards. We quantify with great accuracy those that we have foreseen but overlook bigger ones. Hazard identification is thus of the greatest importance. *Audits* can identify hazards on existing plants, *hazard and operability studies* (the preferred method) and *check lists* can identify them on both new and existing plants[1].

Other methods of hazard identification, including tests for runaway chemical reactions, are detailed in *Lees*, Chapter 8.

When we have identified the hazards our problems are not over. If many accidents occur because we have not foreseen the hazard, many more occur because we fail to take the precautions we know we ought to take or follow the procedures we know we ought to follow.

1. T A Kletz, *Hazop and Hazan – Notes on the Identification and Assessment of Hazards*, Institution of Chemical Engineers, 2nd edition, 1986

Ignition

When a *leak* of flammable gas or liquid catches fire the report on the incident often gives the source of *ignition* as the *cause*. For example, an operator had to drain water from a tank of aviation fuel. He opened two drain valves and left the job; oil came out and caught *fire*. The report said, 'The probable direct cause of the fire was sparking when water reached the poorly protected cable joints'[1].

Cable joints should, of course, be properly protected but it is more important to prevent the spillage; there are possible sources of ignition everywhere. To prevent similar fires, operators should be trained not to leave the job while tanks are being drained, one of the drain valves should be spring-loaded so that it has to be held open, the drain valves should be small and, if the oil is very volatile, a remotely-operated *emergency isolation valve* should be fitted in the drain line.

Obviously, we should do what we can to remove sources of ignition from areas in which flammable gases or liquids are handled (in the UK this is required by the Highly Flammable Liquids and Liquefied Petroleum Gases Regulations 1972) but it is difficult, if not impossible, to remove all sources of ignition. We should therefore try to prevent the formation of flammable mixtures of gas or vapour and air, except in clearly defined circumstances where the risk is accepted.

It is difficult to remove all sources of ignition because the amount of energy needed to ignite a mixture of flammable gas or vapour and air is so small, 0.2 mJ . This is the amount of energy released when a 1 penny or 1 cent coin falls 5 mm. Of course, dropping a 1 penny or 1 cent coin 5 mm will not ignite vapour as the energy is too spread out, but the same amount of energy concentrated into a spark or a speck of hot metal, will do so.

Which is the more dangerous to bring into a plant handling highly flammable liquids: a bucket or a box of matches?

The bucket is the more dangerous as it will probably be used for collecting drainings or samples; a flammable mixture will be present above the liquid surface and sooner or later a source of ignition will be present. In contrast, matches are dangerous only when they are struck in the presence of a leak, and on a well-run plant leaks are few and far between.

I am not suggesting the indiscriminate use of matches on plants handling flammable liquids but I do suggest that we keep out buckets as conscientiously as we keep out matches.

What are these sources of ignition that are so hard to eliminate? D J Lewis lists 116[2]. Here are a few selected at random:

- Hot bearings.
- Hot and/or burning particles from rotating metal items fouling casings.
- Electric sparks when live cables work loose at connections due to vibration.
- Atmospheric corona discharges (St Elmo's fire).
- Spontaneous combustion of oil-soaked rags, oil-soaked lagging on hot pipes, used polishing cloths etc. (See *auto-ignition temperature*.)
- Water ingress into electric cables.
- Chewing or gnawing of cables by rats, mice or other animals.
- Sparking during the connection of batteries to electric equipment.
- Pyrophoric materials inside vessels.

In the UK, the view that we should expect sources of ignition to arise is supported by the law. A contractor's labourer was carrying a drum of lacquer across a shipyard when he tripped over some electric cables and dropped the drum. The drum broke and cut the cables and a short circuit ignited the laquer; the ensuing damage was serious. The judge ruled that the contractor was responsible. Trailing cables are a normal hazard in a

shipyard and the labourer should have taken more care. 'The labourer, if he had applied his mind to the matter, could reasonably have foreseen that dropping the drum would create a fire hazard, even if he could not reasonably have foreseen the particular source of ignition which arose'[3].
See *containers, diesel engines* and *fire*.

1. *Petroleum Review*, April 1982, p. 34
2. Course on Loss Prevention in the Process Industries, Center for Professional Advancement, The Hague, March 1987
3. *Industrial Safety*, May 1974, p. 261

Ignorance

See *knowledge of what we don't know*.

Impatience

See *patience*.

Improvement notice

See *prosecution*.

Incentive schemes

Under incentive schemes people are given money or gifts if they achieve a safety target, such as working for so many hours without a lost-time accident. The money is often given to charity.

The best incentive is not to get hurt, so why give gifts as well?

The theory is that people pay more attention to the comments of their mates than to those of their bosses. If their mates say, 'Don't do that; we'll lose the award', a man (or woman) will take more notice than if the boss says, 'Don't do that; you'll get hurt.' Incentives are therefore given to groups, not individuals.

There is no doubt that incentive schemes can reduce the *lost-time accident rate*, as pressure from fellow workers encourages people with slight injuries to continue at work. Whether the schemes actually reduce accidents is more doubtful.

An incentive scheme may be a useful way of arousing interest in safety where it is lacking. It should be considered part of a safety programme and not an end in itself. After a while schemes lose their impact and should be dropped. This may be difficult unless it is made clear at the outset that the scheme has a limited life.

If an incentive scheme is to produce results the rewards must be worthwhile. People have proudly displayed the ballpoint pens – specially engraved – which have commemorated a safety achievement. But the pens were not the incentive for a good safety record; they were appreciated as a symbol that management recognized the achievement. If the safety record is poor, an offer of pens will not improve it.

Inconsistency

See *old plants and modern standards* and *variety*.

India

See *Bhopal*.

Indices of woe

If we reduce the inventory of liquefied petroleum gas (LPG) in a plant from 100 tonnes to 10 tonnes it is obvious that we have a safer plant and we do not need a special technique to tell us so. But suppose we replace 100 tonnes of LPG with 10 tonnes of chlorine; will we have a safer plant? Several ways of answering this question have been suggested.

Marshall's mortality index

This is the average number of people killed by the explosion of a tonne of LPG or the release of a tonne of chlorine or ammonia[1]. Marshall shows that the historical record is:

- Chlorine   0.30
- Ammonia   0.02
- Liquefied flammable gases   0.60
- Unstable substances   1.50

From these figures we could deduce that 10 tonnes of chlorine is 20 times safer than 100 tonnes of LPG.

However, it is not quite as simple as this. The probability of a leak may not be the same on the two plants. One may contain more leakage sources, such as pumps or drain points, or leaks may be more likely because operating conditions are more extreme. On some plants leaks of LPG are more likely to ignite. Consequently, more sophisticated indices have been devised including the Dow Index[2] and its development the Mond Index[3].

Dow and Mond indices

Numbers are assigned to the properties and inventories of the various materials in the plant and to the operating conditions and these numbers are then combined to give a unified index of woe (a term suggested by A V Cohen) – a single figure which measures the hazard rating of the whole plant or section of plant. We can then use this number to compare different designs or to help us decide how much protective equipment to install. The numbers have no physical meaning but they do give a rough indication of relative hazard just as the star rating of a hotel gives us a rough indication of its standard. If a hotel has five stars we know that it is suitable for a business trip, but not for the family holiday.

Risk to local inhabitants

A third method is to calculate this risk. The calculations are complex as we need to estimate:

- The frequency of equipment failure
- The amount that leaks out
- The fraction that forms vapour and spray
- How far it will spread, on average, in each direction
- If it is toxic, the direct effect on people; if it is flammable, the effect on people of the heat radiation if it ignites and of the over-pressure if it explodes.

Many computer packages are now available to carry out the calculations for us. Their absolute accuracy may not be high but their relative accuracy is greater, and alternative designs can be compared. Their advantage, compared with the Dow and Mond Indices, is that the results have a physical meaning.

Indices measuring financial risk

These have been developed by several insurance companies. The best known are Instantaneous Fractional Annual Loss (IFAL)[4] and the IC Insurance Index (see *Lees*, §5.6.). IFAL is a measure of the average rate of loss and is therefore the financial equivalent of the indices described above. The IC Insurance Index is more like the Dow and Mond indices as it is based on arbitrary factors which measure the quality of the hardware, the software and the fire-fighting facilities.

1. V C Marshall in *Hazardous Materials Spills Handbook*, edited by G F Bennett *et al.*, McGraw-Hill, 1982, p. 5
2. *Dow's Fire and Explosion Index Hazard Classification Guide*, 6th edition, American Institute of Chemical Engineers, 1987
3. B J Tyler, *Plant/Operations Progress*, Vol. 4, No. 3, July 1985, p. 172
4. H B Whitehouse, *The Assessment and Control of Major Hazards*, Institution of Chemical Engineers Symposium Series No. 93, 1985, p. 309

Industrial hygiene

This is the study of the conditions that cause occupational disease and the actions needed to prevent it. See *asbestos, cancer, COSHH Regulations, fugitive emissions* and *under-reporting*.

Inert gas

See *nitrogen*.

Influences

See *decisions*.

Information

See *books on process safety* and *publication*.

Inherently safer design

To make a plant safer we usually add on *protective equipment* to control the hazards. In contrast, in an inherently safer design we try to avoid the hazards, for example, by using so little hazardous material that it does not matter if it all leaks out (*intensification*); using a safer material instead (*substitution*); using the hazardous material in a less hazardous form (*attenuation*). Inherently safer designs are often cheaper than conventional ones, as there is less need for added-on protective equipment and, if intensification is possible, the smaller inventory requires smaller equipment. Large inventories are often due to low conversion so that large quantities of raw material get a 'free ride' and have to be recovered and recycled. If we can increase conversion then energy usage will be less.

Flixborough showed the desirability of making plants inherently safer but progress has been slow, for reasons that are logistic rather than technical. Safety studies take place late in plant design. If we are going to make fundamental changes they will have to take place early in design. But when we recognize the need for fundamental redesign early in the life of a project, there is rarely time to carry out the necessary development, as the plant is wanted soon. We can break out of this impass only by thinking about the plant after next. While we are designing the next plant we are conscious of many changes we would like to make but do not have time to develop. We should start work on them at once, ready for the plant after next[1,2].

See *actions, friendly plants, hazard and operability studies, innovation*

1. T A Kletz, *Cheaper, Safer Plants – Notes on Inherently Safer and Simpler Plants*, Institution of Chemical Engineers, Rugby, 2nd edition, 1985
2. R Malpas in *Research and Innovation in the 1990's*, edited by B Atkinson, Institution of Chemical Engineers, Rugby, 1986, p. 28

Innovation

Loss prevention engineers, when they first appeared on the scene, were more innovative than traditional *safety* officers but the difference is less marked now that loss prevention has matured. The problems that beset the innovator in loss prevention are much the same as those of innovators elsewhere and are illustrated by the story of ICI's Higee *distillation* equipment. An ICI engineer, Colin Ramshaw, realized that in the Sherwood equation, which describes the behaviour of a packed bed, the expression g (the acceleration due to gravity) is not a constant to keep the units right but a variable. If it can be increased, by rotating the packed bed, then the gas and liquid flows can be increased or the packing size can be reduced and its area increased. *Distillation*, Figure 8, shows the resulting equipment. Vapour is fed to the outside of the packed bed – a disc with a hole in the middle – and travels inwards. Liquid is fed to the inside and travels outwards. The radius of the packed bed corresponds to the height of a normal column while the height corresponds to the diameter of a normal column. Two units are required, one for the stripping section and one for the fractionation section. The machine rotates at about the speed of a centrifuge so there is nothing innovative in the mechanical engineering[1,2].

Higees have to be attached to a conventional condenser and reboiler so the next stage was to intensify these operations[3].

Tests have shown that Higee comes up to expectation and that the inventory can be reduced by a factor of up to a thousand (an example of *intensification*) with a corresponding increase in safety, if the contents are hazardous. The cost is half that of a normal column, less if expensive grades of steel have to be used. Nevertheless very few Higees are in use, even within ICI. Individual project managers are reluctant to install them, in case there are any unforeseen snags which might delay the entire project. Understandably, they prefer familiar equipment. Nevertheless, it is surprising that each of the larger companies have not installed one or two, in order to gain experience, the cost being underwritten by the whole company rather than an individual project or business area.

See *cognitive dissonance*.

1. C Ramshaw, *The Chemical Engineer*, No. 389, February 1983, p. 13 and No. 388, January 1983, p. 7
2. R Fowler, *The Chemical Engineer*, No. 456, January 1989, p. 35
3. T Johnson, *The Chemical Engineer*, No. 431, December 1986, p. 36

Inhibitors

See *(accidental) purification*.

Inspection

All equipment which is liable to deteriorate in service should be inspected, and/or tested, at regular intervals to make sure that it is still fit for use. Most companies inspect pressure *vessels* and test *relief devices* at regular intervals. Many now test trips and *alarms*. Equipment which should be inspected and/or tested regularly, but often is not, includes pressure pipework, open vents and *non-return* (check) *valves*. (See *protective equipment* and *tests*.) The inspection and test schedule should state what is to be inspected or tested and how often and what should be looked for. The manager or engineer responsible should be reminded when a test or inspection is due and senior managers should be informed if it has not been carried out by the due date.

The following incidents show how poor inspection may fail to detect faults:

1. *Corrosion* of a *distillation* column was suspected. However, ultrasonic testing showed that the thickness was still adequate. Some months later,

when the column was out of use, the insulation was removed and part of the column was found to be so thin that it could be flexed by hand.

The thin spot was immediately opposite the vapour return line but the thickness meaurements had been made on the other side of the column where access was easier.

A baffle is sometimes installed near the vapour return line and corrosion is then most likely near the edges of the baffle. (See Figure 38.)

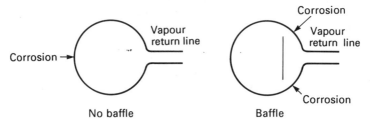

Figure 38 Corrosion was not detected as tests were not carried out at the places where it was most likely to occur

2. A company had a thorough system of vessel inspection. If several vessels were on similar duty one vessel in the group was inspected every two years and if no corrosion was found the other vessels were not inspected until their turn came round. The maximum period between inspections was twelve years.

   However, what is a 'similar duty'? After an absorption tower on a nitric acid plant had leaked it was realized that it operated at 100–125°C while the other towers in the inspection group operated at 90°C. The higher temperature increased the rate of corrosion. Similarly, a change was made to the transmission of a two-rotor helicopter. The manufacturer decided to test the new design on the aft transmission as it had had slightly more problems in the past. The new design passed the tests but failed in service on the forward transmission. The helicopter crashed, killing 45 people[1].
3. If tests are being carried out on a vessel they are often made on the points of a grid. Lines of weakness, such as welds may then be missed. The grid should be tilted so that the test points are not all above or below each other. (See Figure 39).

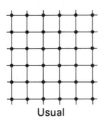 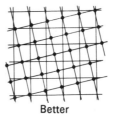

Figure 39 If test points are on a grid it should be tilted so that lines of weakness are not missed

Finally, an example to show how ingenuity can make inspection easier. The insides of spheres are difficult to inspect. If the sphere can be filled with water it can be inspected from a boat as the water level is gradually lowered. The inspector should, of course, wear a lifejacket.
See *'normal accidents'* and *Tay Bridge*.

1. Air Accidents Investigation Branch, *Report on the Accident to Boeing 234LR, G-BWFC 2.5 miles east of Sumburgh, Shetland Isles on 6 November 1986*, HMSO, London, 1989, p. 36

Instructions

Accidents have occurred because people who were constructing or installing equipment have made minor changes to their instructions, for what they thought were good reasons. Here are some examples:

1. Instead of drilling a hole in one of the rails carrying a *crane*, someone burnt a hole with an oxyacetylene cutter. The heat made the area around the hole brittle, the rail broke and the crane collapsed[1]. We do not know why a hole was burnt and not drilled; the man who burnt it may have thought he was saving time – or perhaps it was just easier to do it that way.
2. Compressors handling flammable gases should be installed in the open so that leaks can disperse easily by natural *ventilation*. However, in cold climates they may have to be placed inside a building. If there is a leak an explosion can occur and so the buildings are often designed with light walls which will blow off as soon as the pressure starts to rise, and prevent it rising further. (See *explosion venting*.) A *construction manager* looked at the drawings for one such building. He saw that the walls were to be made from plastic sheets secured by very weak clips. He did not know why such weak clips had been chosen. He decided that they were too weak to hold the sheets securely and he used stronger fasteners instead. When an explosion occurred the walls did not blow off until the pressure was much higher than the designers intended and damage was much greater than it should have been.
3. Storage tanks for flammable liquids are usually made with a weak weld between the walls and the roof, but a normal, full-strength weld between the wall and the floor. If an explosion occurs in the tank, the roof weld will fail rather than the floor weld and the liquid will be retained in the tank. (See *domino effects*.) On one occasion a welder decided that the standard of welding requested for the roof was poor, decided to make a proper job of it and installed a full-strength weld.
4. A joint in a helicopter blade was skimmed with a metal tool instead of a plastic one. The metal tool scratched the blade and started a fatigue crack. The helicopter crashed[2].

What can we do to prevent incidents like these?

- If we construct, install or repair equipment, we should always follow instructions precisely. Suggest changes, by all means, but do not make them until the consequences of the change have been thought through and the design organization has agreed them. (See *modifications*.)

- If we design equipment, we should explain to those who will have to construct, install, use and repair it why we have made it the way we have, especially if it has any unusual features, and we should check to make sure that the unusual features have been built and installed correctly. (See *team working*.)

See *breaking the rules*, *contradictory instructions* and *rules and regulations*.

1. B Argent, *Institute of University Safety Officers Symposium Series No. 5*, 1985, p. 21
2. F Jones, *Air Crash*, Comet Books, London, 1986, p. 142

Instruments

Some years ago the low oil pressure warning light in my car lit up. I pulled into a garage where the attendant said, 'It will be a fault in the switch. Why don't you carry on?' An hour later the AA confirmed that he was correct.

Unfortunately the garage attendant's attitude is common in industry. Because *alarms* are sometimes false or unusual instrument readings are often wrong, we assume that they are always false or wrong and instead of looking for a fault on the plant we send for the instrument mechanic or just ignore the warning. For example:

- A nightwatchman in a warehouse saw the sprinklers starting to operate. 'The stock will be ruined', he thought, and turned off the water supply at the main. The warehouse was destroyed by fire[1].
- An alarm sounded to warn the operators that a tank was nearly full. 'Impossible', they said and checked the stock sheet to make sure. Twenty minutes later the tank overflowed. Someone had moved a batch of liquid into the tank without telling them and the stock was higher than they thought. (See Figure 40.)
- A plant was shut down during the night and allowed to cool. When the manager came in the next morning the temperature in the plant was reading 930°C though it should have been less than 300°C. 'The temperature indicator is out of order', he was told. The manager went round the plant and found that the valve on a compressed *air* line, which should have been disconnected, was open. Air was entering the plant and causing the high temperature.

For another example see *symptoms*.

Never ignore unusual readings or alarms. They might be correct. When instruments tell us what we would like to believe, then we are more likely to believe them.

Some instruments cannot do what we want them to do[2].

All instruments, especially trips and alarms, should be tested regularly or they may not work when required. See *tests*.

Instruments should measure directly what we need to know, not some other property from which the information can be inferred. See *Three Mile Island*.

We are dependent on instruments, especially trips and alarms for plant safety and more and more of them are installed as the years go by; we could not manage without them. Nevertheless, whenever possible, we

Figure 40 Instruments

should try to avoid hazards by a change in design rather than control them by adding on more and more instruments. It is better to reduce the inventory of hazardous material in a plant (*intensification*), so that it does not matter if it all leaks out, or use a safer material instead (*substitution*), than to add on intruments for detecting leaks, isolating the leaking equipment and so on.

Control engineers do not object to this point of view and are not worried that *inherently safer design* will put them out of a job. They know the limitations of their trips and alarms. Process managers have greater faith in them than control engineers.

See *analysis, centrifuges* and *protective equipment.*

1. J A Fletcher and H M Douglas, *Total Loss Control*, Associated Business Programmes, London, 1971
2. T A Kletz, *Chemical Engineering Progress*, Vol. 76, No. 7, July 1980, p. 68

Insularity

It is commonplace to say that responsibility for safety starts at the top, but do those at the top really believe it?

186   Insulation

Like all good managers, they accept responsibility for the actions of their staff. They urge their staff to do better and give them money, support, manpower, encouragement, all that they ask for (well, almost), but it rarely occurs to them that their own actions or inactions may directly cause or prevent an accident. This can be illustrated by an example.

Some years ago a serious *explosion* occurred in a factory handling high pressure ethylene. A badly-made joint had leaked and the ethylene was ignited by an unknown source. After the explosion many recommendations were made, and carried out, to improve the standard of joint-making: better training, tools and inspection.

Poor joint-making had been tolerated for a long time before the explosion as all sources of *ignition* had been eliminated and thus *leaks* could not ignite, or so it was believed. The factory was part of a large group but the individual parts of it were technically independent. The other factories in the group had never believed that leaks of flammable gas will not ignite. They knew from their own experience that sources of ignition are liable to arise, even though we do everything we can to remove known sources, and therefore strenuous efforts must be made to prevent leaks. Unfortunately, the managers of the ethylene factory had hardly any technical contact with the other factories; handling flammable gases at high pressure was, they believed, a specialized technology and little could be learnt from those who handled them at low pressure. The factory was a monastery, a group of people isolating themselves from the outside world; the explosion blew down the monastery walls.

If the management of the factory where the explosion occurred had been less insular and more willing to compare experiences with other factories in the group, or if the managers of the group had allowed the component parts less autonomy, the explosion might never have occurred. It is doubtful if the senior managers of the factory or the group ever realized or accepted this or discussed the need for a change in policy. The leak was due to a badly-made joint and so joints must be made correctly in future. No expense was spared to achieve this aim but the underlying weaknesses in the management system went largely unrecognized.

For more details of this incident see *LFA*, Chapter 4. See also *amateurism*.

Insulation

Equipment may be insulated to keep it hot (or cold) or to prevent damage by *fire*. Different materials are used for fire protection but most thermal insulants provide some fire protection, if they are securely attached.

Insulation has the great advantage compared with water cooling that it does not have to be commissioned and is immediately available as a barrier to heat input. It should therefore be used to protect pressure *vessels* against *BLEVEs*.

There are many materials available for fire protection, including intumescent paints which swell when exposed to fire to produce an insulating char and materials which absorb heat by sublimation. However,

Insulation 187

Figure 41 Insulation

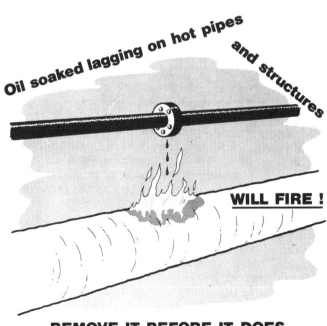

Figure 42 Insulation

many fire protection engineers still recommend the more traditional materials such as lightweight concrete. I have known vermiculite concrete with a nominal 2 hour fire rating to withstand 8 hours fire exposure without damage.

Some *reactors* have to be insulated on the inside to prevent the metal getting too hot.

Note that passive *protective equipment* such as insulation requires *inspection* just as much as active equipment. If 10% of the fire insulation is missing the rest is useless. See Figure 41.

Oil soaked insulation should be removed before it ignites spontaneously. See Figure 42 and *auto-ignition temperature*.

Insurance

The principles and purpose of insurance were well described in the preamble to a 1601 Act of Parliament[1] which is believed to have been written by Francis Bacon. (The spelling has been modernized.)

> It has been time out of mind an usage amongst merchants, both of this realm and of foreign nations, when they make any great adventures, specially into foreign parts, to give some consideration of money to other persons (which commonly are in no small number) to have from them assurance made of their goods, merchandise, ships and things adventured, or some part thereof, at such rates and in such sort as the parties assurers and the parties assured can agree, which course of dealing is commonly called a policy of assurance, by means of which policies of assurance it comes to pass that upon the loss or perishing of any ship there follows not the undoing of any man, but the loss lights rather easily upon many than heavily upon few. . .

In large organizations we cannot say, 'Don't worry. The insurers will pay'; in the long run the insurers will get back in premiums all that they pay in claims plus enough to cover their profit and expenses. All that the insurance does is spread the loss between one year and another and between one part of the organization and another.

Before the late 1960s the UK insurance companies took little interest in the technical problems of the oil and chemical industries. Premium income covered outgoings and kept them happy. The losses of the late 1960s (in one year ICI got back from the insurance companies all the premiums it had paid since the end of the War) made them adopt a more sophisticated approach. They engaged technical staff better able to distinguish between good risks and bad and greatly increased the premiums paid by the companies they considered bad risks. They were surprised to find that in one case the consequential losses from a fire were 30 times the material damage and in 1974 *Flixborough* showed them that the 'exposure' (the maximum they have to be prepared to pay) was higher than they had realized.

See *United States*.

1. Quoted in *Engineering Risk*, Institution of Professional Engineers of New Zealand, 1983, p. 42

Intangibles

See *selling safety*.

Intensification

Intensification is the preferred method of achieving *inherently safer design*[1]. The best way of avoiding a leak of hazardous material is to use so little that it does not matter if it all leaks out. For example, many batch reactors contain large quantities of hazardous chemicals as the reactions are said to be slow. But often the chemistry is not slow. Once the molecules come together they react quickly. The slow reaction is due to poor mixing and a small well-mixed continuous reactor may give the same output. Another example: many plants which used to carry large stocks of hazardous intermediates now find that they can manage without them. Instead of 100 tonnes in a tank there may be only a few kilograms in a pipeline.

If intensification is not possible then the alternatives are *substitution* (using a safer material) and *attenuation* (using the hazardous material in a safer form). Intensification is the first choice as it produces a bigger reduction in capital cost. If less material is present then vessels, pipes, structures, foundations etc. are smaller and cheaper. The incentive to develop intensified reactors, distillation columns, heat exchangers and so on has come mainly from a desire to reduce costs.

Events at *Flixborough* and *Bhopal* encouraged the use of intensification. At Flixborough the inventory was large as the reaction was slow, with low conversion, so that most of the raw material had to be recycled. It got a free ride through the plant. At Bhopal the material that leaked – methyl isocyanate (MIC) – was an intermediate, not a product or raw material, and while it was convenient to store it, it was not essential to do so. Since Bhopal many companies have reduced their stocks of MIC and other intermediates.

See *innovation, production or safety?* and *LFA*, Chapters 8 and 10

1. T A Kletz, *Cheaper, Safer Plants – Notes on Inherently Safer and Simpler Plants*, Institution of Chemical Engineers, Rugby, 2nd edition, 1985

Interference

See *alertness*.

Isle of Man

See *Summerland*.

Ionizing radiations

See *radioactivity*.

Isolation

Many people have been killed or injured because they opened up pipes or vessels which contained flammable or toxic materials, often under pressure; sometimes valves had not been closed; sometimes they had leaked.

Before maintenance work is carried out on any equipment it should be freed from hazardous materials and also isolated from them, by disconnection or slip-plating (blinding), unless the job to be done is so quick that disconnection would take as long and be as hazardous as the main job. Valves leak and should not be relied on for isolation except for quick jobs.

When using valves for isolation, including the closure of valves so that slip-plates can be inserted, they should be locked shut with a padlock and chain or equally effective device. A notice hung on the valve is not sufficient. (See Figure 43.)

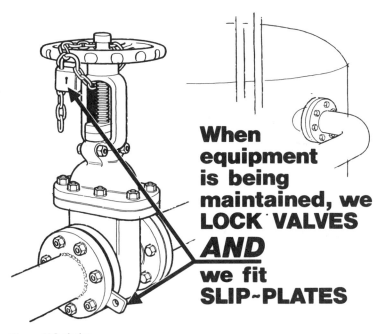

Figure 43 Isolation

These rules should apply to all materials except cold *water* at low pressure. People have been injured by hot water, *steam, nitrogen* and compressed *air*.

Electricity should be isolated by locking off or removal of fuses before anyone is allowed to work on or near electrically-driven machinery. In 1981 a youth lost both feet because someone switched on a stirrer while he was cleaning the inside of a mincing machine.

If fuses are withdrawn, how do you know they are the correct ones? Are they clearly and correctly labelled? As a check always try to start the machine after removing the fuses. Do not leave the fuses near the fuse box for anyone to replace. If anyone is going inside a vessel containing moving parts it is good practice to disconnect the cables.

Another form of hazard is potential energy. Before working on a machine, put it in a position of minimum potential energy. Springs should be released from compression or extension. The forks on a forklift truck should be lowered. On some machines the position of minimum energy is not so obvious. Do not work underneath heavy suspended objects – or above objects which may move upwards. In an aircraft factory a man was working above a fighter plane which was nearly complete. The ejector seat went off and he was killed.

See *amateurism, (personal) responsibility, WWW,* §1.1 and *LFA*, Chapter 1.

Investigation

See *accident investigation*.

Joints

See *(are things as) black as they seem?*

Joint ventures

See *Bhopal.*

Judgement

See *discretion.*

King's Cross

A *fire* at King's Cross underground station, London on 18 November 1987 killed 31 people and injured many more. The immediate cause was a lighted match, dropped by a passenger on an escalator, which set fire to an accumulation of grease and dust on the running track. A metal cleat which should have prevented matches falling through the space between the treads and the skirting board was missing and the running tracks were not cleaned regularly.

No water was applied to the fire. It spread to the wooden treads, skirting boards and balustrades and after 20 minutes a sudden erruption of flame ocurred into the ticket hall above the escalator. A water spray system was provided under the escalator but it was not actuated automatically and the acting inspector on duty did not know the location of the water valves, which were unlabelled; he walked right past them. To quote from the official report[1], '. . . his lack of training and unfamiliarity with water fog equipment meant that his preocccupation with the fire and smoke led him to forget about the system or the merits of its use' (p. 62). In general, the London Underground staff, who were promoted largely on the basis of seniority, had little or no training in emergency procedures, and their reactions were haphazard and uncoordinated. They were 'woefully inequipped to meet the emergency that arose' (p. 67).

Although the combination of a match and dust was the immediate cause of the fire the underlying cause was the view, accepted by all concerned, including the highest levels of management, that occasional fires on escalators and other equipment were inevitable and could be extinguished before they caused serious damage or injury. From 1958 to 1987 there were an average of 20 fires per year, which were called 'smoulderings' to make them seem less serious (p. 45). Some had caused considerable damage and passsengers had suffered from smoke inhalation but no one had been killed (p. 93). The view thus grew that no fire could be become serious and fires were treated almost casually. See *(limitations of) experience*. The recommendations made after previous fires were not followed up. Yet escalator fires could have been prevented, or reduced in number and size, by replacing wooden escalators with metal ones, by regular cleaning, by using non-flammable grease, by replacing missing cleats, by installing smoke detectors which automatically switched on the water spray, by better training in fire-fighting and by calling the Fire Brigade whenever a fire was detected, not just when it seemed to be getting out of control; there was no *defence in depth*.

The attitude of London Underground to escalator fires was similar to that of some chemical companies towards sources of *ignition*: if we eliminate all known sources of ignition, *leaks* cannot ignite and we need not worry if the plant leaks. (See *insularity*.) Similarly, London Underground believed that escalator fires could be exinguished easily and did not worry about them; both groups were wrong.

The *management* of safety in London Underground was criticized in the official report[1]. There was no clear definition of responsibility, no auditing, no interest at senior levels. One is left with the impression that London Underground ran trains very competently and professionally but was less interested in peripheral matters such as stations. In the same way, many process plants give *service lines* less than their fair share of attention and they are involved in a disproportionate number of incidents; rot starts at the edges.

1. D Fennell, *Investigation into the King's Cross Underground Fire*, HMSO, London, 1988.

Knock-on effects

See *domino effects*.

Knowledge

See *lost knowledge*.

Knowledge of what we don't know

The temporary pipe at *Flixborough* which failed, releasing many tonnes of hot hydrocarbon, was designed and built by men who had great practical experience but little theoretical knowledge. ('Designed' is hardly the word as the only drawing was a full sized sketch in chalk on the workshop floor.) They did not know how to design large pipes suitable for use at high temperatures and pressures – few engineers do; piping design is a specialized branch of mechanical engineering – but in addition they did not even know that an expert should have been consulted. They did not know what they did not know. See *(legal) responsibilty*.

As manpower is reduced and individual engineers are asked to do more, could another Flixborough arise? For example, in many plants there is no longer an electrical engineer; the control engineer is also responsible for electrical work. There is an electrical engineer available for consultation somewhere in the company, but will the control engineer know when the electrical engineer should be consulted?

If individual maintenance engineers are going to be asked to carry out a wider range of work, should some engineers be trained as generalist engineers instead of everyone becoming a mechanical, chemical, civil, electrical or some other sort of engineer? Is there a need for 'general practitioner' engineers as well as specialists? Like general practitioner doctors they would know enough to recognize when a specialist ought to be called in. A few universities try to produce such engineers but they are looked down on by chartered engineers and cannot join the engineering institutions except as technician members.

'It is becoming easy to take on design work outside the engineer's area of experience simply because a software package is available[1].'

Another example: most managers say that there is no need to employ human factors specialists but after an accident a lot of errors in job design, which no one had noticed, come to light and everyone says we ought to have spotted them. See *human failing*. For example, after an explosion in the *reactor* of a *batch process* it was found that:

1. The operators did not know the exact function of some controls.
2. A charge meter was unreliable and often by-passed, but there was no way of emptying the reactor if too much raw material was charged.
3. The raw material was supposed to be charged at 20–25°C but it was stored at 10°C and there was no way of heating it.
4. The water supplies to two condensers were in series but each condenser had its own valve.

Would a human factors specialist have spotted these errors, or do we just need a competent technologist with time to look at the process in detail?

Never hesitate to speak up if you think a colleague or an expert has overlooked something. See *alertness* and *welding*.

See *lost knowledge*.

1. H Petroski, *To Engineer is Human*, St Martin's Press, New York, 1982, p. 203.

Labels

See *identification of equipment*.

Language

Many years ago F R Farmer said, 'The language of safety is often second class, evasive and fails to convince or hold the interest of senior management[1]'.

At the time this was true. Most articles and talks on safety consisted of familiar exhortations by people who lacked imagination. There has been an enormous improvement in the quality of *safety professionals* in recent years and a consequent improvement in their output. Management get what they pay for. But if there has been an improvement in the quality of the ideas expressed, the language used is often cumbersome. When writing accident reports, safety bulletins and safety instructions we often use expressions that we would never use in everyday speech. At home we would never ask someone to establish a level in a cup of tea but we talk of establishing a level in a tank. Here are a few more such phrases:

- Degradation of visual information.
- Unfavourable mortality experience.
- Video and audio indications showed no abnormality.
- Several bags of sand/salt should be retained to alleviate local ground ice hazard.
- . . . provided no other activity which could adversely affect the permitted task has been initiated.
- On completion of the permit workscope (meaning, when the job is finished).
- Enhanced facilities (meaning, better service).

The writers of instructions do not fill in a permit before letting someone start, carry out and finish a job. Instead they complete the necessary paperwork prior to authorizing someone to commence (or initiate), implement and complete an activity.

Perhaps the writers think this language sounds more dignified than ordinary speech. Whatever the reason the effect is to make the reader switch off. We should always try to write as simply and clearly as possible, especially in safety. If our the message does not get through, someone may get hurt. See *communication*.

In writing about an unfavourable mortality experience the writer was using a euphemism, writing to soften the impact of something unpleasant by renaming it. We do not like to say that someone was killed or is dead so we talk about a fatality or mortality; it makes death seem a little less deadly. There is less excuse for those who make fires seem less serious by calling them smoulderings. See *King's Cross*.

1. Quoted by C Sinclair, *Innovation and Human Risk*, Centre for the Study of Industrial Innovation, London, 1972

Large plants

During the 1960s and 1970s there were large increases in plant size. Companies built one large single-stream plant instead of two or more smaller ones. People asked if two small plants would be safer than one large one? The answer is no, because two small plants will have twice as many leaks as one large one and the leaks will not be much smaller. If the large plant uses 4 inch pipes the half-size plants will use 3 inch ones and a leak from a joint on a 4 inch pipe will not be much larger than a leak from a joint on a 3 inch pipe.

Suppose the large plant costs £10M. Then the half-size plants will cost about £6.5M each or £13M for two. By building one large plant we save about £3M. Some of this saving can be spent on extra safety precautions. We get a safer plant and still save money.

To quote from J A Lofthouse, a former ICI director, writing about 1970:

There are great dangers in putting all the eggs in one basket but in a competitive world, if a number of people take that risk and get away with it, the remainder will be forced to follow this lead. At least it is a wise precaution to ensure that it is a good, stout, well-made basket and not a flimsy paper bag and that a reliable and trustworthy human being is employed to carry it.

Of course, if we can build a small plant with the same output as a large one it will be safer and many companies are now trying to do this. See *intensification*.

(better) Late than never

This is not always a sound philosophy. The following has happened on several occasions: one liquid was being added, slowly, to another with which it reacted, when the operator realized, after a while, that the stirrer or circulation pump was not running. He switched it on, and a runaway *reaction* occurred. The two liquids had formed separate layers which reacted violently when they were mixed[1]. (See also *WWW*, §3.2.8.)

While a *maintenance* worker was changing the suction valve on a pump he realized that the pump was still running, so he asked an operator to switch it off. Some liquid which was held up in the pump ran backwards out of the suction valve and sprayed the two men. The liquid was corrosive and they were both injured, one seriously[2].

Corrosion was found on the outside of a pipeline, so it was wrapped. This made the corrosion worse and the pipeline failed catastrophically. If it had been left alone it would probably have leaked before failure.

1. *Loss Prevention Bulletin*, No. 029, October 1981, p. 124 and No. 078, December 1987, p. 26
2. *Loss Prevention Bulletin*, No. 085, February 1989, p. 3

Law

Today no one disputes the need for laws to protect employees and the public from the hazards of work but this was not always the case. (See *boilers* and *evangelicalism*.) There are, however, differences of opinion on the form the laws should take. At one extreme, in many countries, including the *United States*, the Government writes a book of rules that looks like a telephone directory, though it is rather less interesting to read. It has to be followed to the letter whether or not it is inapplicable or out-of-date, and whether or not there is a better way. In contrast, in the UK, since the *Health and Safety at Work Act* came into force in 1974, instead of detailed rules there is a general obligation on employers to provide a safe plant and system of work and *adequate* instruction, training and supervision, so far as is *'reasonably practicable'*. If a *Factory Inspector* does not agree that the plant is safe etc. he will say so and, if necessary, issue an Improvement or Prohibition Notice, but in the first place it is up to the employer to decide on the action needed to achieve a safe plant etc. Factory Inspectors are willing to give advice, and regard this as more important than *prosecution*, but they cannot be expected to know the problems of an industry better than those who work there. Those regulations that are made in the UK are mainly 'inductive', that is, they define the objective, but do not say how it has to be achieved. This is usually covered by a *code of practice*; if the code is not being followed this is *prima facie* evidence that the plant is not safe. However, the employer can argue that the code is not applicable to his problems or that he is doing something else that is as safe or safer.

Whatever system of legislation is adopted, complying with the law should be a minor part of a company's safety activities. In my 14 years as a safety adviser with ICI I never bothered much about the law. I tried to do whatever was necessary to protect employees, members of the public and the company's investment and as a result we did far more than was required by the law. There are now more laws, even in the UK, than there used to be and as a result there is a danger that people will say, 'Just do what the law requires.' As a result of the mass of legislation in the US, some companies list compliance with the law as the first subject to be covered in a safety *audit* and include the *safety professionals* in a Regulatory Affairs Department, headed by a lawyer. Obeying the law can become more important than having a safe plant. (See *morality*.)

The law follows good practice; it cannot lead because until a company has shown that something can be done, the law can hardly ask companies to do it. Until someone had invented a windscreen wiper for cars and it had been shown to work, the law could hardly make them obligatory. Today the law brings the stragglers into line more quickly and energetically than it used to and this should be welcomed by those companies that are already acting voluntarily and thus imposing on themselves a tax that some competitors do not have to pay.

For an account of the development of UK law and its enforcement see *Her Majesty's Inspectors of Factories 1833–1983*[1].

This item has been concerned with the criminal law. If it is broken a company or its employees can be prosecuted. In addition, an employee or member of the public who is injured as the result of an accident can sue the employer, under the civil law, for damages. (See *Abbeystead* and *protective clothing*.)

See *(learning from others') experience, modifications after an accident* and *'reasonable care'*.

1. *Her Majestys Inspectors of Factories 1933–1983 – Essays to Commemorate 150 Years of Health and Safety Inspection*, HMSO, London, 1983

Layered accident investigations

See *accident investigation* and *chains*.

Layout and location

The importance of laying out plants in a safe way was recognized hundreds of years ago. The wood of the yew tree was needed for bows; the leaves are poisonous to sheep; yew trees were therefore grown in churchyards where the sheep could not get at them.

Even earlier, the Mishna, the Jewish Code of Law written about 200 AD, stated: '. . . tanneries may not remain within a space of 50 cubits (about 20 m) from the town. A tannery may be set up only on the east side of the town.' In Israel the prevailing wind is from the west.

In 1917 an explosion in a munitions plant in a densely-populated part of London killed 69 people and injured over 400. (See *Brunner Mond*.) In modern times events at *Flixborough* and *Bhopal* greatly increased awareness of the importance of plant layout and location. It is now agreed that plants handling hazardous chemicals should not be built close to concentrations of people and that people should not be allowed to settle close to existing plants. Sophisticated computer programs are available for calculating the probability of a release of toxic or flammable materials and the consequent risk to life at various distances from the plant. As a rough rule of thumb plants handling large quantities (say, more than 50 tonnes) of *flashing* flammable *liquids*, such as liquefied petroleum gas, should be located at least 700 m from housing.

Within the plant, the erection of *fire* breaks, at least 15 m wide, as in a forest, can prevent the spread of fire. Protection against *explosion* damage can be by distance – locating equipment or buildings some distance away – or by building blast resistant buildings and structures. See *domino effects, Lees*, Chapter 10 and references 1 and 2.

1. *Process Plant Layout*, edited by J C Mecklenburgh, Godwin, London, 1985
2. *Loss Prevention*, Vol. 13, 1980, p. 147

Lead

The average concentration of lead in the blood of the UK population is about a quarter of that at which umistakeable signs of lead poisoning appear. No other toxin is universally present at such high levels and it is therefore difficult to disagree with the view of the Royal Commission on

Environmental Pollution[1] that no opportunity of reducing the concentration of lead should be missed. Their recommendation that lead should be removed from petrol, at a cost of £70M per year (1982 prices), was accepted by the Government although petrol accounts for only 20% of body lead and little or no action has been taken to reduce other intakes. (See *cause and effect*.)

Adults get most of their lead from food but children get most of theirs, it is believed, from eating dust. Some of the lead in dust comes from petrol but existing dust will be around for a long time. Can we prevent children eating so much dust? It is easy to say 'impossible' but if we offered £1M (about a hundredth of the present annual cost of removing lead from petrol) in prizes to those who can devise ways, some ideas might be produced.

The lead story shows what happens when we tackle the problems that are easy to tackle rather than those that most need tackling and when we look at problems qualitatively rather than quantitatively.

1. *Lead in the Environment, Royal Commission on Environmental Pollution: Ninth Report*, HMSO, London, 1983

Leaks

Most of the materials handled in the oil and chemical industries will not burn or explode by themselves but only when mixed with *air* (or *oxygen*) in certain proportions. To prevent *fires* and *explosions* we need to keep the air out of the plant and the fuel in the plant.

The former is easy as most equipment operates under pressure. *Nitrogen* is widely used for blanketing low pressure equipment, such as *centrifuges, tanks* and *stacks* and for sweeping out equipment which is to be, or has been, under *maintenance*.

The main problem, therefore, is keeping the fuel inside the plant, that is, preventing leaks.

The explosion at *Flixborough* in 1974 was followed by an plethora of papers on leaks but, surprisingly, most of these considered, theoretically or experimentally, the behaviour of the leaking material, the probability of a leak and the protection of people and plants from its effects. Very few have considered the reasons why leaks occur. Yet if we knew why leaks occur, and could take action to make them less likely, we could worry less about the behaviour of the leaking material.

Most large leaks are the result of pipe failures which in turn are the result of *construction* teams failing to follow the design in detail or not following *good* engineering *practice* when details were left to their discretion. The most effective action we can take to prevent large leaks is to specify designs in detail and then inspect after construction, much more thoroughly than in the past, to see that the design has been followed and that good engineering practice has been followed when details were not specified. This initial inspection of pipework is much more important than ongoing inspection throughout the life of the plant.

Figures 44–54 summarize the actions to take when a leak occurs. They all apply to leaks of flammable liquids or gases but only some of them apply

Don't wait for a leak to occur...

imagine there is one, and discuss what you would do.

Figure 44

If equipment is liable to leak...

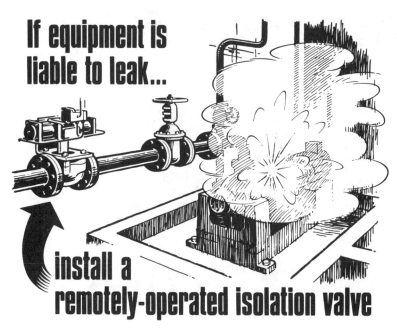

install a remotely-operated isolation valve

Figure 45

Figure 46

Figure 47

Figure 48

Figure 49

204 Leaks

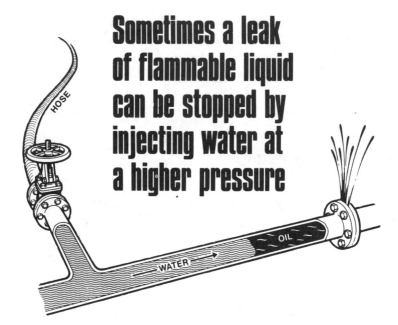

Figure 50

Figure 51

Leaks 205

If the only valve that will stop a leak of flammable gas is near the leak...

Do not rush into the leak — the gas or vapour may ignite

Figure 52

If the only valve that will stop a leak of flammable gas is near the leak...

force back the vapour with water spray

Figure 53

206 Leak tests

Figure 54

to leaks of toxic gases or liquids. If there is a leak of toxic gas or vapour and breathing apparatus is not immediately available it is usually better to seek shelter in a nearby unventilated building, than to try to leave the area.

Structural failures, as distinct from pipe failures, seem to be caused mainly by overloading[1].

See *choice of problems, defence in depth, LFA*, Chapter 16 and *WWW*, Chapter 7. For small leaks see *fugitive emissions*.

1. *Journal of Occupational Accidents*, September 1981, p. 1630

Leak tests

See *pressure tests*.

Learning

See *(learning from) experience*.

Level glasses

Failures of level glasses, sight glasses and glass rotameters have caused many incidents[1,2]. The explosion of a tonne of ethylene, released from a broken level glass completely destroyed a plant[1]. Men have been blinded by exploding sight glasses which were installed under strain so that they were broken by a light blow. One company suggests that sight glasses should be 'avoided like the plague'. Level glasses, sight glasses and glass rotameters should be avoided whenever possible and never used with *flashing* flammable or toxic *liquids*, that is, liquids under pressure above their normal boiling points, such as liquefied petroleum gas, *chlorine* or *ammonia*.

Level glasses, if used, should be fitted with ball check cocks which prevent a massive leak if the glass breaks. Unfortunately, the balls have sometimes been removed by people who did not understand their purpose, or made inoperative because the cocks were not fully open. See *WWW*, §7.1.4.

Level glasses containing hot or corrosive liquids should be provided with a protective screen to prevent injury should the glass break.

There is, of course, no objection to the used of magnetic level glasses, frost plugs and other glass-free types. (A frost plug is used with very cold liquids. It is an insulated metal tube with protrusions along its length which stick through the insulation. Frost forms on those protrusions that are below the liquid level.)

In the UK the law requires boiler steam drums to be fitted with level glasses.

1. C T Adcock and J D Weldon, *Chemical Engineering Progress*, Vol. 63, No. 8, August 1967, p. 54
2. *Loss Prevention Bulletin*, No. 080, April 1988, p. 19 and No. 083, August 1988, p. 25

Life

See *(loss of) expectation of life*.

Limitations

See *(limitations of) experience*.

Liquefied petroleum gass

See *aerosol cans, BLEVE, drums, entry, iatrogenesis, indices of woe, layout and location, level glasses, sloppy thinking, smoke screens, timebombs* and *time span of forgetfulness*.

Location

See *layout and location*.

Loss

If *loss prevention* is our aim we need to know how losses occur. A report from one of the major US *insurance* companies[1] shows that the money they paid out for 1985–87 was split as follows:

Risk covered	Percentage of annual outlay
Fire	42%
Explosion	24%
Liquid damage, flood, collapse, theft, transit	18%
Windstorm	4%
Sprinkler leakage	2%
Lightning	1%
Others	9%

Athough explosions represented only 6% of the claims, they made up 24% of the dollar loss. Half the explosion damage was due to explosions inside equipment, 10% to runaway *reactions*, 7% to explosions outside equipment and 2% to molten materials.

Altogether, 42% of the losses, but only 13% of the incidents, occurred in storage areas.

In individual companies and factories the split may be very different. Do you know how your company's insured losses (and your uninsured losses and consequential losses) are made up? I suggest that you circulate the figures every quarter so that everyone knows where the profits are going down the drain.

1. *The Sentinel* (published by Industrial Risk Insurers, Hartford, Connecticut), Third Quarter 1988, p 13.

Loss control

See *damage control*.

Loss prevention

Loss prevention is not the same as safety. It is not just a new, better-sounding name, intended to give safety a new image, like calling dustmen refuse collectors. Loss prevention differs from the traditional safety approach in the following ways:

- There is more concern with those accidents that arise out of the technology.
- There is more emphasis on foreseeing hazards and taking action before accidents occur.
- There is more emphasis on a systematic rather than a trial and error approach, particularly on systematic methods of identifying hazards and of estimating the probability that they will occur and their consequences. See *hazard and operability studies* and *hazard analysis*.

- There is concern with accidents that cause damage to plant and loss of profit but do not injure anyone, as well as those that do cause injury.
- Traditional practices and standards are looked at more critically. See *myths*.

See *Lees*, Introduction and *safety professionals*.

There are some forms of loss which are not usually considered part of loss prevention, for example, operating below maximum efficiency or taking too long over a shutdown. This may not seem very logical but is accepted practice. In the US, loss prevention is sometimes used to describe the prevention of losses from theft, vandalism and fire.

Loss prevention can be applied in any industry, especially highly technical ones, but the term, and the approach, have been particularly widely used in the oil and chemical industries which are exposed to the major hazards of *fires, explosions* and toxic releases. (See *toxicity*.) Loss prevention in these industries is the same as process safety.

There is, of course, a big overlap between loss prevention and safety but investigating the reason why a relief valve lifted, for example, is not usually considered one of the responsibilities of a traditional safety offficer.

I do not wish to imply that loss prevention is more important or worthwhile than the traditional safety approach. Both are needed and they complement each other. Far more people are killed and injured, even in the high technology industries, by simple mechanical accidents than by the technology. The only people killed by the *nuclear power* industry in the UK are those who have fallen off the structures during construction.

See *books on process safety and loss prevention* and *operations research*.

Lost knowledge

The Tasmanian aboriginies, before they became extinct, are said to have lost the skill of making fire. In industry accidents often occur because, with changes in staff, the correct method of plant operation has been forgotten[1,2]. Here are some examples:

1. A *hose* should be vented before disconnection. After someone, disconnecting a hose, has been sprayed with the contents or injured by a sudden movement of the hose it is found that many of the hose connections have no vent valves. There is a campaign to fit them. After a few years the accident is forgotten. Hoses are used again without vent valves and there is another accident[1].
2. Sample bottles should never be carried in the hand, only in carriers, closed ones if the chemicals are corrosive or very toxic. After someone has been injured, by broken glass or chemicals, proper carriers are issued. What has happened to them two years later?
3. More serious accidents are remembered for longer but, after about ten years, most of the staff have changed and the accident is forgotten and can recur. A man was killed because a *vent* was choked. It cleared suddenly; the back pressure caused the vent pipe to move and it hit the operator on the head. The material being vented was solid at ambient

temperature and the vent pipe should have been heated – and properly secured.

Fifteen years later another vent 100 m away, although under the control of a different department, also choked and this time the vessel burst and killed two men[2].

4. In December 1969 explosions occurred in the cargo tanks of three oil tankers while they were being washed out with high pressure water jets. The source of *ignition* was a discharge of *static electricity* from the cloud of water droplets.

   In the 1840s the discharge of static electricity from water droplets was well known and was the basis of the Armstrong machine for producing high electrostatic voltages. Steam jets were allowed to impinge on an isolated metal comb. 'This illustrates how easy it is for a body of fundamental knowledge, once well disseminated, to become forgotten with the passage of time . . . we should be keen students of the past if we want to avoid costly mistakes in the future[3]'.

5. When new plant is installed the operating instructions usually cover all the circumstances that the start-up team can think of – and the instructions are often added to during commissioning. Once the plant settles down some problems may not recur for many years. When they do no one can remember how they were dealt with in the past and the old instructions cannot be found. Hard-won experience has been lost.

For other examples see *repeated accidents* and *time span of forgetfulness*.

How can we prevent the loss of knowledge that has occurred in cases like these?

1. By putting old accident reports in a *black book* which is compulsory reading for newcomers and which others dip into from time to time.
2. By using our safety bulletins to remind us of past accidents.
3. By discussing from time to time those accidents of the past from which most can be learnt.
4. By indexing old reports so that they can easily be found[4]. (I received a letter some years ago from a major international company. The writer said, '. . . I looked for records of accidents in our company for your response and was quite surprised to find that they have been piled up in the stacks for years without any arrangement. No one in the safety section has been aware of the value of them. We have to depend on the memories of individual persons to find out what happened'.)
5. By making sure that instructions are always up-to-date and in good condition and that the reasons for them are explained in an Appendix.

See *novelty* and *old days*.

1. T A Kletz, 'Accidents that will occur during the coming year', *Loss Prevention*, Vol. 10, 1976, p. 151
2. T A Kletz, 'Organisations have no memory', *Loss Prevention*, Vol. 13, 1980, p. 1
3. A F Anderson, *Electronics and Power*, January 1978 (quoted in *Loss Prevention Bulletin*, No. 021, June 1978)
4. T A Kletz and R W Fawcett, 'One organisation's memory', *Plant/Operations Progress*, Vol. 1, No. 1, Jan 1982, p. 7

Lost-time accident rate

A lost-time accident (LTA) is one which results in absence from work beyond the day or shift on which it occurred. (In some companies and countries several days absence is needed before the accident is counted.) The definition is widely used to distinguish between trivial and 'serious' accidents though most LTAs are not really serious, as the term is popularly understood; the injured person is back at work, fully recovered, after a few days or weeks.

The LTA rate is the number of LTAs in 100 000 working hours. (Some companies use 200 000 or a million hours instead.) A 100 000 hours is a working life-time so an LTA rate of 2 means that on average each employee can expect to have two LTAs in his or her working life.

In the process industries most companies seem satisfied if they can keep their LTA rate below 1 but *Du Pont* achieve well below 0.1.

The LTA rate is by far the most widely used statistic for comparing industries, firms, factories and individual units with each other and with their own past performance. Every manager and safety officer tries to get his figures as low as possible. This can be done by preventing accidents (hard) or by preventing lost time (easier). Many organizations therefore try to persuade injured men to continue at work. Easy jobs are found for them and if necessary cars are sent out to bring them into work. If this proves impossible the next step is to show that the accident should not 'count'. Perhaps the injured man hurt his back at home, or never hurt it at all, or his injuries are so slight that he could perfectly well come into work. The LTA rate often measures the willingness of injured men to continue at work rather than the actual accident rate and is an index of morale rather than safety. As such it still has some value.

For some alternative figures that might be reported as a measure of safety see *accident statistics*. Another method, suggested by the UK Health and Safety Executive is to divide LTAs into three groups, by severity. Group 1 includes those causing serious injuries such as fractures, dislocations and amputations, Group 2 other injuries causing hospitalization or more than 28 days absence and Group 3 the rest; only a small proportion of LTAs come into Group 1.

LPG

See *liquefied petroleum gas*.

Lying

See *falsification*.

Machinery

See *absolute requirements* and *mechanical accidents*.

Magic charm to prevent accidents

Used correctly, the magic charm shown in Figure 55 will reduce the chance of an accident on your plant.

Figure 55 Used correctly this magic charm will reduce the number of accidents on your plant

You must carry the charm every day to every part of the plant under your control – into every building, up every structure, past every pump, past every instrument on the panel, not forgetting the 'holes and corners' – behind the buildings, along the pipebridge walkways and so on.

For the best results, carry the charm round at a different time each day, and occasionally at night and during the weekend.

If you follow these instructions then by this time next year you will be having fewer accidents – and better output and efficiency.

Photocopies will work just as well if you do not want to cut up your copy of this book.

Maintenance

Maintenance is a major source of accidents, not so much the maintenance in itself but the preparation of equipment for maintenance. Sometimes the procedures are inadequate; sometimes the procedures are satisfactory but are not followed. In particular:

- The equipment to be repaired may not be isolated from sources of danger such as pressure or hazardous chemicals. See *isolation*.
- The equipment may not be identified so that the wrong equipment is opened up. See *identification of equipment*.
- The equipment may not have been freed from dangerous materials and tested to make sure that it is free.
- The maintenance workers may not have been warned of the residual hazards or advised what precautions to take; or they may not have taken precautions, even after advice. See *trapped pressure*.

See also *entry*, *WWW*, Chapter 1 and *LFA*, Chapters 1 and 5.

To prevent similar accidents we should:

- Draw up a good procedure for the preparation of equipment for maintenance[1]. See *Lees*, Chapter 21.
- Persuade those concerned that the procedure is necessary, for example, by discussing some of the accidents that have occurred because procedures were poor or were not followed. Many foremen and lead operators find permit-to-work procedures tiresome. Their job, they feel, is to run the plant, not fill in forms. We have to convince them that accidents occur when proper procedures are not followed.
- Check from time to time to make sure that the procedure is being followed. If managers take no interest foremen and operators assume that it is not important. If managers take an interest and point out any shortcomings, then foremen and operators will see that everything is correct before the manager comes round. A friendly word before an accident is far more effective than disciplining people after it has occurred. I suggest that first line managers should look at the permit-to-work book everyday and should compare a permit with the actual job several times per week. See *audits*.

Since maintenance causes so many accidents, a plant which needs less maintenance is a safer plant. The nuclear industry builds very reliable plants that need little or no maintenance, as it has to be carried out by remote control and is very expensive (or impossible). Other industries cannot afford nuclear standards of *reliability*, but some movement towards nuclear standards might be justified.
See *amalgamation*.

1. T A Kletz in *Safety and Accident Prevention in Chemical Operations*, edited by H H Fawcett and W S Wood, Wiley, New York, 1982, Chapter 36.

Major hazard

See *hazard and risk*.

Management

The safety record of a company depends on the quality of the management rather than the inherent hazards of the materials and equipment handled. *Construction* sites and mines have, on the whole, a poor safety record but when chemical companies have their own construction sections they have a record typical of the chemical industry, not the construction industry. When *ICI* operated an anhydrite mine it was one of the safest units in the company.

Safety achievement is, in fact, far more dependent upon management commitment and attention than upon the type of industrial activity, the geographic location, or all the other factors so often invoked to attempt to justify mediocre performance[1].

The importance of management is now recognized by the Health and Safety Executive:

> ... in contrast to earlier legislation the *Health and Safety at Work Act* of 1974 creates a different dimension for the enforcing body ... when a physical defect or a shortcoming is seen, the inspector will not only recognize the omission or defect but will go further and determine not only why the organization has allowed such a development, but also what is the weakness that has failed to monitor the situation. The onus is now on top management to create and monitor a system which effectively controls and regulates the whole of the working environment[2].

Although senior managers, in general, recognize that safety is ultimately their responsibility they tend to discharge that responsibility by exhorting their staff to do better. They rarely identify the major problems and the actions required and ask for regular reports on progress, as they do when output, efficiency or product quality are causing problems. See *amateurism, insularity* and *platitudes*.

Safety, at all levels, cannot be managed from an office, only by walking round the plant. Unfortunately, many managers use the level of paperwork as a reason, or excuse, for staying in the warmth and comfort of their offices. While there is room for a good deal of difference in management style I suggest that any first level manager (supervisor in most US companies) who finds he is spending less than 3 hours per day out on the plant should ask himself if he is spending enough time there.

What should we look for while walking round? Anything that looks unusual (what does not look right is usually not right); anything that has changed since the last visit and a few things picked at random. Pick a repair job and check that the permit-to-work has been made out correctly; try a shower to see if it works; look in an eyewash bottle cabinet; ask why an *alarm* light is red and what is being done about it; look where others do not, behind and underneath equipment. Everyone will soon know that you have been round and next time they will try to get it right before your visit.

Walking round the the site will improve output, efficiency and product quality as well as safety. (See *magic charm*.) A quotation from the historian, Barbara Tuchman, writing about American independence, shows what happens when those in charge never visit the site[3]:

> The Grenville, Rockingham, Chatham-Grafton and North ministries went through a full decade of mounting conflict with the colonies without any of them sending a representative, much less a minister, across the Atlantic to make acquaintance, to discuss, to find out what was spoiling, even endangering, the relationship and how it might be better managed.

See *human failing, philosopher's stone, risk management, King's Cross, (visit accident) sites* and *Zeebrugge*.

1. P L T Brian, *Safety Management (South Africa)*, Vol. 16, No 1, January 1989, p. 5
2. *Health and Safety – Manufacturing and Service Industries, 1976*, HMSO, London, 1978, p. 11
3. B Tuchman, *The March of Folly*, Knopf, New York, 1984, p. 229

Manliness

Safety seems to be opposed to many of the 'masculine' aspects of human nature such as heroism, bravery and taking-a-chance, and more in line with 'feminine' characteristics such as care and compassion. Thus operators who wear all the protective clothing they are supposed to wear are seen, by managers as well as other workers, as good, responsible and often elderly men, conscientious plodders who are rather out of touch, rather fastidious and compulsive, slow but reliable men who can safely be left alone[1].

Can we overcome this attitude? There are some suggestions in the item on *safety*. We can also show how failure to control so-called 'manliness' has caused accidents. At the third attempt a pilot succeeded in landing a plane at a fog-shrouded airport. A second plane, of the same type, approached and was told that the first was already on the ground. The pilot made two abortive approaches. The knowledge that the first pilot had succeeded challenged the second pilot and he made a third attempt, killing 41 people[2].

Perhaps we can also show that safety is not opposed to the 'masculine' virtues of wisdom and logicality. However, like President Johnson at the time of the Vietnam war, some men prefer strength to wisdom[3]:

In the nervous tension of his sudden accession, Johnson felt he had to be 'strong', to show himself in command, especially to overshadow the aura of the Kennedys, both the dead and the living. He did not feel a comparable impulse to be wise.

See *giving up*.

1. *Personnel Management*, February 1976, p. 25
2. M Allnut in *Pilot Error*, edited by R and L Hurst, Aronson, New York, 2nd edition, 1982, p. 15
3. B Tuchman, *The March of Folly*, Knopf, New York, 1984, p. 311

Materials of construction

In the chemical industry many incidents have occurred because the wrong materials of construction were used. The most spectacular incident occurred when the exit pipe from a high pressure converter on an ammonia plant was made from mild steel instead of 1¼% Cr, ½% Mo. Hydrogen attack occurred and a hole appeared at a bend. The hydrogen leaked out and the reaction force pushed the converter over.

Steel washers inside one of Britain's nuclear power stations have corroded because they were made of mild steel instead of the stainless steel specified.

As a result of these and other incidents many companies now insist that if use of the wrong grade of steel can affect the integrity of the plant, all steel arriving on site must be checked for composition before it is used. The checks can be carried out very easily with a spectrographic analyser such as a Metascop. It is important that all steel is checked: nuts, bolts, washers, flanges and welding rods as well as pipes. Complete components such as valves should be checked as well as raw materials.

Many incidents have demonstrated the need for this thorough checking. For example, vendors often send without warning what they regard as superior material. If a vendor is asked to supply 20 carbon steel flanges and he has only 19 in stock, he is quite likely to add a twentieth in 2¼% Cr. If challenged, he may be indignant because he supplied 'superior' (that is, more expensive) material at the original price. He does not realize that a more expensive material may not be suitable for some applications or, if suitable, may require a different welding technique[1].

Even if the grade of steel is stamped on a piece of equipment it is still necessary to check it. An alloy steel heat exchanger was found to be fitted with two large carbon steel flanges. The flanges were stamped as alloy.

Even if the composition of the steel is correct, it may have received the wrong heat treatment. Pipe hangers have failed because they were too hard. Pumps have leaked because they were assembled using bolts of the wrong hardness.

Stainless steel is non-magnetic so magnets are often used to check that a steel article is made of stainless steel. However, a magnet will not distinguish between different grades of stainless steel. During the night a valve had to be changed on a plant which handles acids. The fitter could not find a suitable valve in the workshop but on looking round he found one on another unit. He tested it with a magnet and finding it to be non-magnetic he assumed that it would be suitable and installed it. The valve was actually made of Hastelloy. Four days later it had corroded and there was a spillage of acids.

See *plastics, smoke screens* and *WWW*, Chapter 16.

1. G C Vincent and C W Gent, *Ammonia Plant Safety*, Vol. 20, 1978, p. 22.

Measurement of safety

See *accident statistics* and *criteria*.

Meccano or dolls?

As children most engineers played with meccano rather than dolls. We were interested in the way things worked, otherwise we would not have become engineers. We are more at ease finding technical solutions to problems than dealing with the human aspects. But plants consist of people as well as equipment and many of our problems have human as well as technical *causes*. When engineers are asked to list the reasons why a car will not start, they list all the technical reasons (tank empty, petrol pump failed, flat battery, dirty plugs and so on) but only a few list the human reasons (such as too much or too little choke, wrong ignition key, automatic car not in park or neutral, due in each case either to a slip or to poor training).

In *accident investigation* , we usually identify the technical problems and devise solutions but often brush aside the human problems with phrases such as 'operator told to take more care' or 'operator requires extra training'. I have seen this phrase used when the accident was not due to lack of training at all but to a moment's forgetfulness; extra training would not prevent the accident happening again. We do not install *emergency equipment* unless we expect it to work when needed but we tell people to take more care without asking ourselves if they are likely to heed our advice or if more care is the answer. See *human failing*.

Mechanical accidents

Most of the accidents that occur in the oil and chemical industries are mechanical ones that could occur in any industry. For example:

- An operator left a drum of hot water at the foot of a cat ladder. He came down the ladder into the drum and scalded his leg.
- A scaffolding pole fell 14 m and punctured a cable 0.8 m below the ground.
- A lorry driver, while sheeting his load, threw a rope over the top. A loop on the end fell over the head of a passing cyclist and pulled him off his bicycle.
- Welding rods left inside scaffolding poles have slid out and injured people when the scaffolding was dismantled.
- A large slip-plate fell while being lifted as the weld between the lug and the body was below standard.
- Ropes have broken while lifting heavy loads as no one knew the weight of the load.
- Tools have broken at the point where identification marks have been welded on. Indentification marks should be engraved, not welded.
- The top casing of a turbine was located on the bottom casing by two ⅜ inch (10 mm) dowels. There were blind holes in the bottom casing and open holes in the top casing. (See Figure 56.) When the top casing was

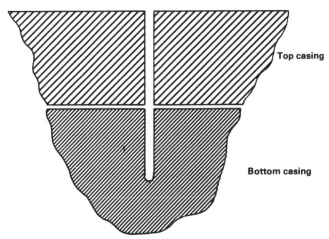

Figure 56 Mechanical accidents can occur in the process industries. When the top casing was removed a dowel stuck to it. When the top casing was replaced and bolted down the dowel was ejected

removed, one of the dowels stuck to it. When the casing was replaced the dowel did not slide right into the lower hole. When the cover was tightened the dowel was ejected and hit the fitter above the eye.

Before putting a peg into a hole, make sure the hole is clear, and deep enough. It would have been better to screw the dowel into the lower casing, or machine a groove in the dowel.

See *leaks, (visit accident) sites* and *traps*.

Memory

The men on the job are a major source of information on accidents. Occasionally one or more of them may lie, to avoid embarrassment or to protect colleagues. (See *falsification*.) A far more important reason for distortion is the poorness of the human memory, especially under stress. Allnut writes[1]:

> Closely controlled laboratory experiments confirm the everyday experience that memories are distorted in the direction of simplicity and coherence. If a story is repeated many times from one witness to another, it is progressively simplified until only the bare bones remain. Similarly, the human desire for order out of chaos means that *non-sequiturs* in a story are gradually eliminated until a clear and logical account of events emerges. However, this final account usually bears very little resemblance to the original message.
>
> Distortion can be minimised, although not eliminated, when statements are taken from the parties concerned as soon as possible after the accident[1].

Shehadeh refers to '. . . the psychological mechanism that makes people believe the concise, documented account as opposed to the confused, incoherent verbal one[2]'.

1. M Allnut in *Pilot Error*, edited by R and L Hurst, Aronson, New York, 2nd edition, 1982, p. 19.
2. R Shehadeh, *The Third Way*, Quartet Books, London, 1982, p. 69.

Methods of working

Safety by design should always be our aim but very often there is no way of making a plant safe by design, or the only way is too expensive, and we have to rely on safe methods of working – software rather than hardware. Even when we can make a plant safe by design it is not always possible to achieve an *inherently safer design* and we have to add on *protective equipment* such as trips, *alarms*, fire *insulation* and *relief devices*. All this safety equipment should be inspected and/or tested regularly so we are still dependent on methods of working. (See *inspection* and *tests*.)

Unfortunately, procedures are subject to a form of corrosion more rapid than that which affects the steelwork and can vanish without trace in a few weeks or months once managers loose interest. A continuing *management* effort, sometimes called 'grey hairs', is needed to make sure that procedures are followed. This management effort should be three-pronged. We should:

1. Design procedures that, as far as possible, are simple and easy to follow. If the safe way of working is difficult people will not follow it.
2. Persuade people to follow the procedures. We do not live in a society where people will follow the rules just because they have been told to do so. They must be convinced that the rules are necessary and reasonable.
3. Check up from time to time to make sure that the procedures are being followed. See *blind-eyes* and *management*.

For examples see *maintenance* and *modifications*.

Methyl isocyanate

See *Bhopal*.

Mexico city

See *BLEVES*.

Minimum standards

Some companies call their safety standards 'minimum standards'. Individual factories and design engineers are allowed to do more, but may not do less.

I do not like this arrangement. The resources we can spend on safety are large but not unlimited; only so much time, money and effort is available. If we spend too much on one plant or problem, there is less left to spend on others. We should try to identify the biggest risks and deal with them first.

Safety professionals spend most of their time persuading people to take action to make the plant safer. If they find that someone is going too far they should be prepared to say so and not rejoice that more money is being spent on safety. They should say, 'There is no need to do that; the risk is slight. It would be better to spend the time and money on a bigger risk.' They should be prepared to say this after an accident when *over reaction* is common.

See *'reasonably practicable'*.

Misquotations

See *vessels*.

Modern standards

See *old plants and modern standards*.

Modifications

Changes to plants and processes have caused many accidents. We have a problem. We think of a solution and are then so pleased we have solved the problem that we do not see the unwanted side-effects. Here are some examples:

- The vacuum in an experimental unit was too hard – 20 mm Hg instead of the 80 mm required – so a compressed air supply was connected to the air bleed. (See Figure 57.) The gauge pressure of the compressed air was 8 bar (120 p.s.i.). No one asked if the equipment could withstand this pressure or what would happen if the vacuum pump stopped.

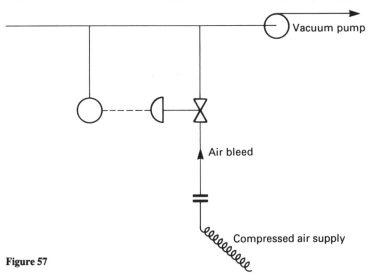

Figure 57

- In carbon dioxide absorption plants balanced control valves are usually used to let down the potassium carbonate solution from high to low pressure. One plant used a motorized ball valve instead. When the jet from this type of valve impinges on a surface it produces a ring-shaped corrosion groove. A disc of metal was blown out of a bend downstream of the valve.

  Sensing the loss of pressure, the controller opened the ball valve fully, discharging hot potassium carbonate solution out of the hole. Unfortunately the pipe was opposite the control room window. The window was broken and all the operators were killed. It was shift change time and more operators than usual were present[1].

- A reactor was cooled by circulating brine through the jacket. The brine system was shut down for repair so town water was connected to the jacket. The gauge pressure of the town water (9 bar) was greater than the design pressure of the jacket's inner wall, which gave way[2].

  The works modification approval form, which had been completed by the manager and engineer, asked 20 questions, one of which was, 'Does the proposal introduce or alter any potential cause of over/underpressurising the system or part of it?' They had answered 'No'.

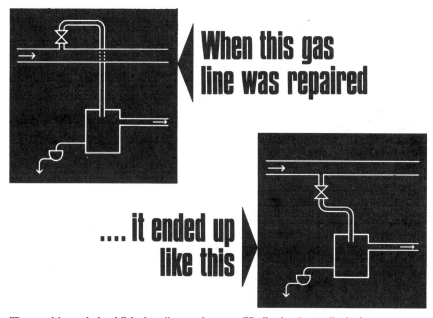

Figure 58 Modifications

- When a plant is repaired, there is a temptation to improve it. The man who reassembled the fuel gas line shown in Figure 58 did not know why the branch came off the top of the line and thought it would be simpler to take it from the bottom. Liquid was carried over into a furnace and extinguished the burners. Maintenance workers should be encouraged to suggest modifications but unless they have been authorized they should put the plant back exactly as it was. See *(personal) responsibility*.
- The modifications described so far have been plant modifications. Here is a process modification. Instead of letting a reactor age for 30 minutes and then cooling it, the usual procedure, it was allowed to cool by heat loss to the surroundings; it exploded. The cover of the reactor rose 50 m into the air and landed 75 m away[3].
- The duck pond at a company guesthouse was full of weed. The company water chemist was asked for advice. He added a herbicide which was also a detergent; it wetted the ducks' feathers and they sank.

Sometimes a modification is followed by a chain of further modifications, as we try to put right the unforeseen results of each step[4]. Often it is the simple hazards that are overlooked rather than the esoteric or unlikely ones.

For other incidents see *anaesthetics, Flixborough, exchanging one problem for another* and *(managerial) responsibility*. See also *Lees*, Chapter 21, *WWW*, Chapter 2, *LFA*, Chapters 1, 7, 8 and 14 and *Chemical Engineering Progress*[5,6].

Figure 59 Modifications

To prevent similar incidents:
1. Do not allow any modification, temporary or permanent, however trivial, to plant or process, to take place until it has been approved by a professionally qualified person, say, the first level of professional management. (See Figure 59.)
2. Provide the approver with a guide sheet or check list to help him examine the proposed change. However, this system will be useless if the form is completed in a perfunctory manner. (See the third incident above.) For large modifications a *hazard and operability study* is recommended. A form for use with minor modifications is described in *Chemical Engineering Progress*[5] and in *Lees*, Chapter 21.
3. Check, when the modification is complete, that it has been carried out correctly and that it 'looks right'.
4. Carry out a training programme to convince those concerned that a modification control procedure is necessary and check from time to time to make sure that it is being followed.

Effective control of modifications requires a cultural change as well as a formal procedure. For many years the hero in industry was the go-getter who got things done quickly and brushed difficulties and reservations aside. Now the hero is, or should be, the man who says, 'I know it's urgent but we can still spend an hour going over the proposal'.

1. M Schofield, *The Chemical Engineer*, No. 446, March 1988, p. 39
2. *Chemical Safety Summary*, Vol. 56, No. 221, 1985, p. 6
3. L Silver, *Loss Prevention*, Vol. 1, 1967, p. 61
4. T A Kletz, *Plant/Operations Progress*, Vol. 5, No. 3, July 1986, p. 136
5. T A Kletz, *Chemical Engineering Progress*, Vol. 72, No. 11, November 1976, p. 48
6. R E Sanders, *Chemical Engineering Progress*, Vol. 79, No. 2, February 1983, p. 73

Modifications after an accident

If we modify the plant after an accident, is this evidence that the original design was unsafe? According to M Dewis, the short answer is 'No'.

English law, . . . although not entirely clear would appear reluctant to stigmatise machinery as defective or unreasonably dangerous merely on the ground that after an injury has occurred to an operative, remedial measures were taken of an accident-prevention nature. In Hart v Lancashire and Yorkshire Rail Co. the court distilled the essence of the matter as follows: '. . . it is not because the defendants have become wiser and done something subsequently to the accident, that their doing so is to be evidence of any antecedent negligence on their part in that respect'. Here the rail company had altered their way of changing points after an accident had occurred. It was held that this was not admissible to establish that the accident had been caused by negligence[1].

1. M Dewis, *Occupational and Health*, Vol. 8, No. 6, June 1978, p. 6

Money

See *costs* and *time and money*.

Morality

In an article on television's responsibility[1], Sir Robin Day wrote:

Many years ago, a wise jurist – Lord Moulton – said that there were three great domains of human action. The first is where our actions are limited or forbidden by law. Then there is the domain of free personal choice. But between those two lies the domain of 'obedience to the unenforceable', where people do right although there is no one to make them do right but themselves. And the extent to which there is obedience to the unenforceable is the measure of true civilisation.

Obedience to the unenforceable means self-imposed restraint. It means a sense of right and wrong. It means not what we can do but what we ought to do.

Lord Moulton's views apply to safety as well as television. Some things we are required to do (or not do) by *law*. There is an area where it does not matter what we do; then there is a very large area where we install equipment and procedures, to prevent accidents, though no law requires us to do so. Unfortunately, as the first domain increases, i.e. as the law requires us to do more and more, there is a tendency for people to think that that is all they have to do.

1. *Daily Telegraph*, 7 August 1980

Motivation

In a paper entitled *We ain't farmin' as well as we know how*[1] W B Howard tells the story of a young man with a degree in agriculture who tried to tell an old farmer how to improve his farming. The old man listened for a while and then said, 'Son, we ain't farmin' now as well as we know how'.

Howard then described some accidents which occured because people took a deliberate decision not to perform as well as they could, and should, have done. For example, a vessel was left full of reaction product for many hours, the pressure rose and the vessel exploded. It was then found that:

- An untrained operator, who did not realize the importance of the presure rise, had been left in charge of the plant.
- No tests had been carried out to to see what happened when the material was left standing.

The explosion was not due to equipment failure or to human error as normally understood, that is, forgetfulness. (See *human failing*.) It was due to conscious decisions to postpone testing and to put an untrained operator in charge.

After an explosion in a pipeline it was found that four trips and *alarms*, each of which could have prevented the accident had all failed. No one knew they had failed because a decision had been made not to test them regularly.

An example from the UK: after a disgruntled employee had blown the whistle, a Factory Inspector visited a plant where liquefied flammable gases were stored in six tanks each fitted with a high level trip and alarm. He found that five of the alarms were out of order and no one knew how long they had been in that condition; he also found that the trips could not be tested, so no one knew whether they were in working order or not. (See *talebearing*.)

The organization concerned employed a large safety staff in its head office. Many of them spent their time working out equipment reliability. They would have been better employed down on the site persuading the managers to test their trips and alarms and persuading designers to install trips that could be tested.

In *Normal Accidents* C Perrow writes, '. . . humans do not exist to give their all to organizations run by someone else... At some point the cost of extracting obedience exceeds the benefits of organized activity'[2].

Even when people do give their all, opinions will differ on what is best for the organization.

Note that in the incidents described people did not consider safety unimportant, but they did not realize the need for the actions they failed to carry out. They did not realize the importance of trip testing or operator training and the effects they can have on accident rates. Accidents, by and large, do not occur because we put a low value on human life, but because we lack competence or do not pursue an objective with sufficient energy, drive and initiative. Our safety record is poorer than it need be for the same reasons that our output, profit and human relations are poorer than they need be. See *blame, King's Cross, management* and *'Normal Accidents'*.

1. W B Howard, *Plant/Operations Progress*, Vol. 3, No. 3, July 1984, p. 147
2. C Perrow, *Normal Accidents*, Basic Books, New York, 1987, p. 339

Myths

During my years in the chemical industry, as manager and safety adviser, I came to realize that many chemical engineers accept unquestionably many statements of doubtful accuracy. I have set some of them down in a book entitled *Myths of the Chemical Industry or 44 Things a Chemical Engineer ought NOT to Know.* (See Introduction to this book.) I said why they were wrong and I described some of the resulting accidents and wrong decisions. Some examples of these widely accepted but incorrect beliefs are:

- A relief valve, properly designed and maintained, will prevent a vessel bursting.
- Accidents are due to *human failing* so we should eliminate the human element when we can.
- A *pressure* of 10 p.s.i. is small and will not cause injury.
- Major disasters in the chemical industry are becoming more frequent.
- We can rely on the advice of the accountant where money is concerned.

These myths are not completely untrue – there is a measure of truth in them – but neither are they completely true. They are deeply ingrained; when our reasons for believing them are shown to be invalid we look for other reasons. See *cognitive dissonnance.*

Near-miss

See *dangerous occurrence*.

'Need to know'

Some organizations operate on a need-to-know basis. Staff are given the information they need to do their job, but no more. They are protected from a deluge of time-wasting, unnecessary paper, but the system has snags. New ideas come from a creative leap that connects disjointed ideas, from the ability to construct fruitful analogies between fields[1,2]. If our knowledge is restricted to one field of technology or company organization we will have few new ideas.

For example, at *Abbeystead* if one of those concerned with the design of the pumping station had known, however vaguely, that methane might be present in ground water, and had been given an opportunity to bring forward this knowledge, the accident might not have occurred. (*Hazard and operability studies* can stimulate people to bring out knowledge stored away in the depths of their memories.) Engineers need to build up a ragbag of bits and pieces of knowledge that may one day be useful. They will never do this if they are restricted (or restrict themselves) to what they need to know.

We don't know what we need to know until we know what there is to know.

Creative people often 'belong to the wrong trade union'. If a civil engineer, say, has a new idea in, say, control engineering, the control engineers may be reluctant to accept it.

1. A Koestler, *Act of Creation*.
2. S J Gould, *The Panda's Thumb*, Penguin Books, 1983, p. 57

Never

See *(better) late than never*.

New hazards

See *underestimated hazards*.

New processes

Bryan Harvey, Chief Inspector of Factories, writing in 1972, said:

> The costs of development and exploitation of a new process or technique must include all of the costs of control, and only when these are taken into account can a proper decision be taken whether a new process is commercially viable. This is not manifestly so at present (although there

are one or two encouraging signs) since the concept involves what is a near revolution in some boardroom thinking[1].

There has been much improvement since these words were written but nevertheless, on a new plant, the cost estimates for added-on safety equipment are still more likely to be overspent than any other items, as nobody sees the problems even until late in design, or even until the plant is running. See *hazard and operability studies* and *inherently safer design*.

1. B Harvey, *Department of Employment Gazette*, August 1972, p. 695

Nitrogen

Nitrogen is widely used in the oil and chemical industries to prevent fires and explosions. Equipment which operates at or near atmospheric pressure, such as *tanks, stacks* and *centrifuges* is blanketed with nitrogen, to prevent air diffusing in; equipment which has contained flammable gas or a flammable liquid above its flash point is swept out with nitrogen before opening up for *maintenance* and is swept out again before flammable gas or liquid is readmitted. Nitrogen has prevented many *fires* and *explosions* and saved many lives.

However, a high price has been paid. Nitrogen has caused many deaths by asphyxiation. In *ICI* it has killed more people than any other substance. (Fires and explosions have killed more but many different substances have been involved.) Other companies' experience is similar; it is our most dangerous gas. (*Water* is probably our most dangerous liquid.) The term *inert gas*, often used to describe nitrogen, is misleading as it leads people to believe that it is harmless; it is not. A high nitrogen, that is, low oxygen, concentration knocks people out like a blow on the head. (See *cylinders*.)

Sometimes people have been overcome by nitrogen because they were unaware of the dangers. It is not necessary to enter a vessel full of nitrogen to be overcome. Putting your head inside for a quick look is sufficient. (See Figure 13, p. 115.) Other people have been affected by nitrogen when they were working near leaking joints. (See Figure 60.)

People have been overcome by nitrogen because it was confused with compressed *air*. Different *hose* fittings should be used for compressed air and nitrogen. On other occasions men have entered a vesssel not knowing it had been connected to a nitrogen supply. Before anyone enters a vessel it should be isolated, tested and a permit issued by a competent person, even when the vessel is still in the hands of the *construction* team and there is apparently no way in which nitrogen or other hazardous materials can enter.

When equipment is blanketed with nitrogen the atmosphere inside it should be tested regularly or continuously to make sure that the blanketing is effective. It is not necessary to remove every trace of oxygen as 10% is needed for an explosion (unless hydrogen is present). High oxygen concentration *alarms* should therefore be set at 5%. See *stacks* and *WWW*, §12.3. Also see *LFA*, Chapter 6 for an account of an explosion that occurred because nitrogen blanketing was considered to be an *optional extra*.

See *tankers*.

Figure 60 Nitrogen

Non-events

If there is a *dangerous occurrence* but no one is injured, many foremen and managers like to treat it as a non-event and pretend it never happened. No report is written and no one is told. Does it matter?

It does matter, because if we draw attention to the incident and circulate a report someone elsewhere may take action to prevent the same accident happening on his plant. If we hush up the incident we may even forget about it ourselves and it may occur again, on the original plant, and this time someone may be injured.

Of course, when someone publicises an accident he does not look very clever and if his manager criticizes him we can hardly *blame* him for hushing up the next incident. Managers should be scientists looking for ways of preventing accidents, not policemen looking for culprits. When possible, *accident reports* that are widely circulated should be anonymous, so that no one knows exactly where they occurred.

Other reasons for reporting all accidents, whether or not someone is injured, are:

- To draw attention to the cost.
- To make us think out clearly what happened and what should be done to prevent it happening again.
- We are more likely to take action if we commit ourselves in writing .

Non-return valves

Non-return (check) valves have a poor reputation amongst engineers who consider them unreliable. However, many non-return valves, once installed, are never looked at throughout the life of the plant. No item of equipment, especially one containing moving parts, can be expected to work for ever without attention. If *reverse flow* through a non-return valve can affect the safety or operability of a plant then the valve should be scheduled for regular inspection, say, once every year or two. If reverse flow does not affect the safety or operability of the plant, do we need the non-return valve?

Non-return valves are designed to protect against gross reverse flow; they will not prevent slight leaks. Sometimes it is necessary to prevent any reverse flow with high reliability. For example, if ethylene oxide is reacted with another substance, reverse flow from the *reactor* into the ethylene oxide storage tank has caused a violent runaway *reaction* in the tank[1]. An elaborate protective system is necessary to prevent such reverse flow[2].

1. R Y Levine, *Loss Prevention*, Vol. 2, 1968, p. 125
2. H G Lawley, *Hydrocarbon Processing*, Vol. 55, No. 4, April 1976, p. 247

Non-sparking tools

See *spark-resistant tools*.

'Normal Accidents'

In a book with this title[1] Charles Perrow argues that accidents in complex systems are so likely that they must be considered normal (as in the expression System Normal, All Fouled Up (SNAFU)). Complex systems, he argues, are *accident-prone*, especially when they are tightly-coupled, that is, have little slack or room for recovery. Errors or neglect in design, *construction*, operation or *maintenance*, component failure or unforeseen interactions are inevitable and will have serious results.

His answer is to scrap those complex systems we can do without, particularly *nuclear power* plants, which are very complex and very tightly-coupled, and try to improve the rest. He does not consider the alternative, the gradual replacement of present designs by *friendlier plants*, which can withstand equipment failure or human error without serious effects on safety (though they are mentioned in passing and called 'forgiving').

Many of the author's descriptions of accidents on petrochemical plants are hard to follow. (He is a sociologist, not an engineer.) The plants seem even more complex than they are and this helps his thesis! The description of *Flixborough* ends with a sneer, 'And, finally, there are the great lessons learnt... It is unlikely that any of these recommendations will find their way into operating practice.' Many of them have. However, these weaknesses should not blind us to the fact that Dr Perrow's diagnosis is sound, though he has prescribed the wrong cure.

A similar view has been put forword by O H Critchley[2] who describes the degradation of complex systems as an increase in entropy and emphasises the value of *inspection*. He writes, 'Disorder and restricted flow of information in the lower tiers of an organization can usually be corrected. When higher *management* is afflicted by unperceived entropy, self-appraisal is difficult'.

See *astonishment*.

1. C Perrow, *Normal Accidents,* Basic Books, New York, 1984
2. O H Critchley, *Radiation Protection in Nuclear Energy,* Vol. 1, International Atomic Energy Agency, Vienna, 1988, p. 91

Nothing

Chemicals cause injury so when equipment is empty it ought to be safe. In fact more accidents seem to occur when *vessels* are empty than when they are full. See *entry* and *WWW*, Chapter 12. Of course, it is not the presence of nothing which causes injury but the traces of hazardous material which have been left behind, and not detected by the tests which should be carried out before anyone is allowed to enter a vessel. Or perhaps the right precautions were not requested, or not taken. On other occasions hazardous material has leaked into the vessel because it was not adequately isolated.

Empty drums have blown up because they were used by welders as a working platform. A trace of flammable liquid was ignited by welding sparks (see Figure 61). On one occasion a brand new drum exploded; the manufacturer had used a flammable solvent to clean the inside and a trace of it remained.

A vacuum can be dangerous if the vessel is not designed to withstand it. *Tanks* and even pressure vessels have collapsed because they were pumped out with the *vent* closed or without a big enough vent or allowed to cool after steaming without a big enough vent.

Another incident occurred when a manhole cover was removed from a vessel for modification. It was replaced by a thin metal sheet, 3 mm (about ⅛ inch) thick, secured by three nuts. To prevent fumes coming out a vacuum was applied to the vessel. The operators intended to pull only a slight vacuum but they pulled a strong one. When two fitters unscrewed the three bolts the metal sheet and the fitters were sucked into the vessel. The men were killed[1].

Many people have been killed or injured because they did not realize the strength of compressed *air*. As this accident shows, vacuum can also be very powerful.

1. J Bond, *Loss Prevention Bulletin,* No. 078, December 1987, p. 15

Novelty

All previous civilizations have valued ancient knowledge. Ours is probably the first to value new knowledge more than old. After I had distributed a five-year-old paper as a course handout one of the students wrote, in his comments on the course, that he expected more up-to-date material.

Novelty 231

A drum was used as a makeshift platform. It exploded when a spark from an oxy-acetylene cutting torch fell through the bung-hole. The operator was seriously injured.

Figure 61 Nothing

There are some branches of loss prevention where progress has been so rapid, during the last few years, that five year old papers really are out of date, for example, gas *dispersion* and the venting of *batch reactors*. But so far as *accident reports* are concerned, many old reports tell much the same story as new ones and can teach us just as many lessons. *WWW*, Chapter 11 (2nd edition only) describes an accident involving *entry* to vessels that occurred in 1910. The same accident was repeated in the same company 45 years later and many similar accidents have occurred in other companies. Other serious accidents have been repeated after ten or more years – time for the staff to change and the accident to be forgotten. See *lost knowledge* and *old days*.

We have paid a high price, in injuries and damage, for the knowledge of past accidents. Value that knowledge and see that it is not forgotten, or the accidents will happen again. Past accidents explain the reasons for many of our operating practices and design features. Once the accidents are forgotten someone changes a procedure or design and another accident occurs.

Never ask a *safety professional* 'What's new?' The answer is very little. Most of the reports he receives are depressingly familiar. However, there are some new problems and some new solutions to old problems. See *change*.

Nuclear power

The most important lesson to be drawn from *Bhopal* is that the material that leaked, methyl isocyanate (MIC), need not have been there at all. it was an intermediate, not a product or raw material, and while it was convenient to store it, it was not essential to do so.

The best way of avoiding a leak of hazardous material is to use a safer material instead (*substitution*), or so little of the hazardous one that it does not matter if it all leaks out (*intensification*). This is far better than trying to prevent or control leaks by adding on protective equipment which might be neglected or might fail. 'What you don't have, can't leak'. See *inherently safer design*.

If we try to apply this philosophy to electricity production, does this mean that we should avoid nuclear energy altogether because an error in design or operation may result in a melt-down and the release of massive quantities of radioactivity to atmosphere? However much protective equipment we add on to a nuclear power station, coincident failure is still possible. More important, as at Bhopal and *Chernobyl*, the operators may not keep all the protective equipment in working order. (See *'Normal Accidents'*.)

However, there are designs of nuclear reactor which are inherently incapable of overheating to anything like the same extent as most of those currently in use. Most commercial reactors, outside the UK, are water-cooled. If the water supply fails, then the reactor will overheat and so emergency water supplies are provided. There is always a possibility that these will fail. At *Three-Mile Island* the operators actually isolated the emergency water supply in error. On gas-cooled reactors, if the forced circulation fails, convection cooling will prevent serious overheating. All existing UK commercial reactors are gas-cooled but a water-cooled reactor is being built at Sizewell.

The UK prototype fast reactor at Dounreay, cooled by liquid sodium, is safer still. If forced circulation is lost, natural circulation will keep the reactor cool. In addition, if the reactor gets too hot, heat production stops.

In the High-Temperature Gas reactor, under development in the US and Germany, a small core is cooled by helium gas. If the helium flow stops, the high surface/volume ratio and the high heat capacity prevent overheating. The reactor is small and a typical power station might contain ten factory-made units. Another inherently safer design is the Swedish Process Inherent Ultimate Safety (PIUS) reactor in which a water-cooled reactor is immersed in boric acid solution. If the water pumps loose pressure, the boric acid solution floods the core, stopping the nuclear reaction (as boron absorbs neutrons) and cooling the core; no make-up water is needed. Development of a small, water-cooled reactor in which convective cooling can prevent overheating – the Safe Integral Reactor (SIR) – has recently been proposed[1].

In the short term the Pressurized Water Reactor (PWR) at Sizewell may well be the right answer for the UK. The CEGB has the resources, the competence and the commitment to see that the added-on safety systems, on which this design is dependent, are properly designed and maintained.

But the advantages of the inherently safer designs are so great that I would guess that we shall be building PWR's for no more than a few decades.

For those countries that lack the resources, *culture* or commitment necessary to maintain complex added-on safety systems, water-cooled reactors are not suitable even today and they should wait until the inherently safer designs become available.

See *publication, radioactivity* and LFA, Chapters 11 and 12.

1. M Hayns, *Atom*, No. 292, June 1989, p. 2

Nuts and bolts

See *threads*.

Occupational disease

See *cancer, COSHH Regulations, fugitive emissions* and *under-reporting*.

Old days

'The past is a foreign country; they do things differently there[1].'

This is true if we go back far enough. In ancient times people took different things for granted; something we should bear in mind when reading old books. But many people think that the past ended the day they were born, or the day they joined the company, and that those who went before them were simple, unsophisticated souls who made mistakes that no one would ever make today. Not so; we make the same mistakes as they did, as every *safety professional* knows. (See *lost knowledge* and *novelty*.) Though we have learnt much we have also forgotten much.

There are no 'sell by' dates on safety messages. We can buy the old ones. But some of the old ways of putting them across might be given a rest. After a while familiar messages become part of the background and are no longer seen or heard.

1. L P Hartley, *The Go-between*, Penguin Books (opening line)

Old equipment

A few years ago a gunsmith was trying to free the corroded trigger of a 200-year-old musket. The gun went off and shot his wife who was in the next room; it had not been used since 1745 and had remained charged since then.

An old gas line which had not been used for 12 years was being brought back into use. A branch had to be welded on to it. Fortunately the line was drilled and the atmosphere checked before welding was allowed to start; the atmosphere inside was still flammable.

These incidents remind us of the need to take care when working on equipment which has been out of use for some time. It may contain toxic or flammable materials or *trapped pressure*; it may contain chemicals which we no longer use, and we may not know how to handle them safely.

When old equipment is taken out of use it should be left clean and gas free. If complete cleaning is impossible, for example, because persistent residues are stuck to the sides, then warning notices should be fixed to the equipment; not a piece of paper but a notice that will last. Leave your knowledge to the next generation, not your problems, but do not assume that your predecessors left everything clean for you.

As well as the chemical hazards of using old equipment there are engineering hazards. Is the grade of steel, the detailed design and the quality of welding suitable for the present day? How much of the creep life or the fatigue life of the equipment has been used up? Expert advice should be sought. Sometimes old equipment needs so much doing to it to bring it up to current standards that it is cheaper to buy new.

See *demolition, penny-pinching, plugs* and next item.

Old plants and modern standards

How far should we go in bringing old plants up to modern standards? Some changes are impossible to make on old plants. We cannot increase the spacing between units or the distance between the plant and the nearest houses. Occasionally it may be necessary to demolish the plant and start again but usually we say that inconsistency is the price of progress; we accept somewhat lower standards of design on old plants and try to compensate by high operating standards.

Sometimes it is as easy, or almost as easy, to introduce a change on an old plant as on a new one, installing gas detectors, for example. We should then make the change as soon as possible, unless the plant is going to be closed.

In between there are many changes which are difficult but not impossible to introduce on a new plant and a compromise may be necessary. For example, after a number of pipe joints fitted with fibre gaskets had leaked, replacement by spiral-wound gaskets was considered. There were thousands of joints and the task appeared daunting. In a compromise, joints exposed to liquid, over half the total, were replaced over a three year period but those exposed to vapour were left as they were.

Some companies have been concerned that improving the standard of new plants is tantamount to admitting that their old plants are unsafe. This need not be so. We should not accept a higher level of risk on our old plants but the methods we use to achieve a safe plant may be different. If we cannot bring the equipment on old plants up to modern standards we can make sure that the *training, instructions, tests, inspections* and *audits* are fully up to standard. Older plants may need our best managers and operators, even though new plants have the glamour and people like to be transferred to them.

See *LFA*, Chapter 17.

Old-timer

The old-time safety officer did not concern himself with the hazards of the technology, which often he did not fully understand. Instead he may have spent most of his time in the office, keeping the *accident statistics* and filling in forms for Government departments. He 'numbered the dead and injured on the industrial battlefield' and shared a teapot with the ambulance driver or first aid man.

Or he may have been a 'sergeant-major' who enforced the rules that others had made, sharing a teapot with the security officer.

Others were 'guardsmen' who saw their job as designing, ordering and fitting protective guards or *protective clothing* and seeing that they were used. They shared a teapot with the workshop foreman.

There were 'lawyers' whose main objective was to see that the letter of the *law* was obeyed. They were very concerned if the statutory *inspection* of a small *air* receiver was a week overdue but did not mind if a *vessel* holding liquefied petroleum gas was never inspected, as no law specifies an inspection frequency. They would have liked to share a teapot with the company lawyer, but he would not reciprocate.

Finally, there was the 'padre' who felt that all problems are human ones and if a man has an accident he must have a personal or career problem. He tried to find out what was wrong and what could be done about it. He shared a teapot with the welfare officer, if there was one, or the personnel officer[1,2]. (See *caring*.)

Each sort of safety officer tended to propagate his own kind. He assumed that only someone with the same background and interests could do the job.

All the tasks I have described are important and necessary. We do need accident statistics, a measure of discipline (see *blind-eyes*), machine guards, protective clothing, knowledge of the law and an understanding of people's problems. But they are not enough. In the high technology industries they are not even the most important tasks awaiting the *safety professional*.

See *loss prevention* and *sleeping beauties*.

1. R Barry, ROSPA National Industrial Society Conference, 1970
2. *Safety*, published by British Steel, November, 1970, p. 9

Operations research

Operations research is 'The technique of the scientific analysis of operations of war'; 'the scientist can encourage numerical thinking on operational matters, and so can help avoid running the war by gusts of emotion'. In a wider sense, it is 'a scientific method for providing executives with a quantitative basis for decisions'. The quotations are from *Studies of War* by P M S Blackett[1]. Part II of the book describes the technique and gives some examples of its use. Thus operations research showed that aircraft would be more effective when used on anti-submarine duties than on bombing Germany and that larger convoys would result in fewer ship losses.

The spirit behind operations research also inspired the early work on *hazard analysis*, see Lees §9.3.4: 'The situation is studied and described by a simple model, the implications of the model are explored, initially using estimated rather than field data, and then if the need for field data is indicated, these are obtained and the decision made.' Hazard analysis is thus the opposite of what I have called *sloppy thinking*; that item includes some examples. Here are two more:

A chemical was carried 200 miles by road for further processing. It was in the form of a harmless aqueous solution but money was being spent to transport water. So someone suggested that an alternative water-free intermediate should be transported instead. The quantity to be transported would be reduced by 80% but unfortunately the alternative material was corrosive and if a tanker was involved in an accident and the contents spilt someone might be injured, even killed. The company at first decided that they should continue to transport the bulkier, harmless material.

However, this decision was changed when I pointed out that ordinary road accidents are far more probable than accidents involving chemicals. Using average figures I showed that transport of the corrosive chemical would save one life every 12 years on average, even though an accident involving a tanker of the corrosive chemical is very slightly more likely to result in a fatality than an accident involving a tanker of the aqueous solution.

After the explosion at *Flixborough*, a BBC reporter described it as 'the price of nylon'. Many people must have wondered if we shoulc' take risks with men's lives merely to provide better clothes; why not use wool and cotton, as our grandparents did?

However, the 'fatal accident content' of wool and cotton is higher than that of man-made fibres. The price of any article is the price of the labour used to make it, capital being other people's labour. Agriculture is a low wage, high accident industry so there will be more labour and more accidents in a cotton shirt than a nylon one. Flixborough was the price of nylon but the price of wool and cotton is higher.

These examples and others[2] show that operations research and hazard analysis do not necessarily involve complex mathematics. The calculations are often very simple. The essence is the scientific approach. See *Fermi estimates*.

1. P M S Blackett, *Studies of War*, Oliver and Boyd, Edinburgh, 1962, pp. 169, 173 and 210
2. T A Kletz, *Loss Prevention in the Process Industries*, Institution of Chemical Engineers Symposium Series No. 34, 1971, p. 34

Optional extras

Some people divide safety equipment and procedures into essential ones, that should never be omitted or neglected, and 'optional extras', luxuries that can be ignored when we are busy or under stress. The optional extras are not usually the ones which, after careful consideration, someone has decided can be omitted, but the newer equipment or procedures, recently introduced. The feeling seems to be, 'We managed without *nitrogen blanketing*, *trip testing* or a *modification* control system for a long time, so it won't matter if we do without it now'. In fact, since a deliberate decision has been taken to introduce new equipment or procedures, the case for them is probably stronger than for many others which have been merely hallowed by time, *spark-resistant tools*, for example.

When we introduce new safety equipment or procedures we have to sell then to those who will have to use them. This selling takes longer and requires more effort than most managers realize. Most of us are trams rather than buses and like to follow familar tracks, so regular checking is needed to make sure that the new equipment is being used and the new procedures are being followed or people will revert to old habits. See *LFA*, Chapter 6.

Other industries

Whatever industry you work in, you will find it helpful to read reports of accidents in other industries. Many of the lessons to be learnt will apply to all industries and as you are not involved you will see them more clearly; and it will be recreation rather than work. (See *anaesthetics*.) Railway accidents are particularly recommended as they been occurring for a long time and have been thoroughly investigated. Errors by drivers and signalmen have caused many accidents; slowly people came to realize that prosecution or exhortation had no effect and that we have to accept an occasional accident or change designs or methods of working so that errors are less likely to occur or the consequences are less serious. See *alertness, human failing* and *EVHE*, Chapter 2.

I recommend the following books on railway accidents. Not all are in print at the time of writing but they will be available from libraries.

- A Schneider and A Masé, *Railway Accidents of Great Britain and Europe*, David and Charles, Newton Abbott, Devon, 1968
- L T C Rolt, *Red for Danger*, David and Charles, 4th edition, 1987
- J A B Hamilton, *Trains to Nowhere*, Allen and Unwin, Shepperton, Surrey, 1981
- M Herald and J A B Hamilton, *Rails to Disaster*, Allen and Unwin, 1984
- O S Nock and B K Cooper, *Historic Railway Disasters*, Allen and Unwin, 4th edition, 1982
- S Hall, *Danger Signals*, Allen and Unwin, 1987
- R C Reid, *Train Wrecks*, Bonanza Books, New York, 1968 (for US accidents)

Hall's book discusses recent accidents and current problems.

We should not forget William McGonagall (the poet laureate of loss prevention?) who described many accidents in terrible verse which nevertheless has a certain charm[1], for example:

So the train moved slowly across the bridge of Tay,
Until it was about midway,
Then the central girders with a crash gave way,
And down went the train and the passengers into the Tay!

See *Tay Bridge*.

1. W McGonagall, *The Railway Bridge of the Silvery Tay and Other Disasters*, Sphere Books, London, 1972

Overreaction

I am thinking of overreaction by people, not chemical *reactions* that go too far.

Often, after an accident, people overreact and spend money extravagantly. 'It must never happen again', is often said. This is understandable but if the chance that it will recur is small and the consequences are unlikely to be serious the right thing to do may be nothing. Resources are not unlimited and every time we spend money preventing an unlikely or trivial accident there is less money available for preventing more serious accidents.

While standing on a ladder carrying out a minor repair a man burnt his wrist, not seriously. The safety officer reported a few days later that the line had now been insulated. The line was normally out-of-reach. It was doubtful if anyone had gone near it since the plant was built or would do so again; insulating the pipe was an overreaction.

Even if the accident was serious, and we reduce the chance that it will happen again, we cannot make the chance zero. Safety is often approached asymptotically. (See *asymptote*.) After a chlorine leak the directors said, 'It must never happen again'. They had to be told that this was impossible but that the chance could be made as low as they wished. As a result a target was drawn up and a procedure devised for meeting it. See *chlorine*.

After a runaway reaction on a batch reactor a director said, '*Hazard and operability studies* must be carried out on all processes.' He had to be told that this was impossible on any realistic timescale and as a result a rapid method of ranking the processes, to decide which ones should be studied, was devised.

Of course, after an accident it may be necessary to spend more and do more than is technically necessary in order to reassure employees and the public. This expenditure can be looked on as a fine for letting the accident occur rather than as effective expenditure on prevention.

'The Amercican way is to wait until it's an emergency and then overreact[1].'

When there has been no accident but a hazard has come to light in other ways we do not get overreaction but its opposite, *cognitive dissonance*.

1. G B Craig, quoted in *The Sciences*, Vol. 28, No. 1, January/February 1988, p. 2

Over-reporting

See *under-reporting*.

Oxidation

Fires and *explosions* are much more frequent in oxidation plants than in other plants of comparable size[1]. This is not surprising when we remember that:

- We are deliberately adding *air* or *oxygen* to flammable liquids or gases, so errors in *design, maintenance* or operation of the oxygen control or distribution systems may produce an explosive mixture.
- The *reactions* are often slow, low conversion processes, so inventories are large. (Many of the reactions are not really slow – the apparent slowness may be due to poor mixing – and the conversion may be kept low because with poor mixing, higher conversion will give unwanted side reactions.) See *Flixborough*.
- Some of the products of oxidation are unstable and may decompose violently.

To reduce the incidence of fires and explosions we should:

- Operate as far as possible from the flammable range or from conditions under which explosive decomposition may occur.

- Design the oxygen control system so that it will achieve an agreed hazard rate, *test* it regularly and do not allow unauthorized changes to it.
- Prevent overheating or *contamination* of unstable reaction products.
- Prevent *reverse flow* from *reactors* into storage vessels or air (or oxygen) lines.
- If possible, develop smaller, better mixed reactors which will produce the same output from a lower inventory. See *intensification*.

See *LFA*, Chapter 8

1. *Plant/Operations Progress*, Vol. 7, No. 4, October 1988, p. 230

Oxygen

Oxygen (and *nitrogen*) illustrate the saying that if something is good, more of it may not be better. Many people have been killed or injured because they did not realize that in a high oxygen concentration clothing ignites readily and burns easily.

High oxygen concentrations may be due to *leaks* from oxygen *cylinders* or *hoses* or may occur because the oxygen supplied to a torch is not all consumed. For example, welding was taking place inside a tank. The cylinders were outside and hoses led to the welding set. One of the welders lit a cigarette and noticed that his lighter flame was longer than usual and that the cigarette burned more quickly than usual, but he did not realize what this meant. When he started to weld a spark fell on another man's pullover. It caught fire at once; the flame spread to his entire clothing and he died from his injuries. The oxygen hose was in poor condition and had leaked[1]. See *alertness* for a similar incident ten years later[2].

Oxygen should be used only in well-ventilated situations, the oxygen content should be measured with a portable analyser and everyone present should be taught to recognize the signs of high oxygen concentration. Cylinders should be isolated when work stops.

Other accidents have occurred because grease left in valves caught fire, because oxygen was used to sweeten a foul atmosphere and because oxygen was used instead of compressed air, for spraying paint, operating tools, clearing chokes or starting diesel engines[3].

Several tanks and tankers containing liquid oxygen have exploded; some of the incidents were due to *contamination* and at least one was due to overheating of a submerged pump. Another possible mechanism is reaction of oxygen with the fresh aluminium surface inside a crack. Normally the aluminium is protected by an oxide coating[4].

1. *Accidents*, October 1968
2. *The Fire on HMS Glasgow, 23 September 1976*, HMSO, London, 1977
3. *Fires and Explosions due to the Misuse of Oxygen*, Health and Safety Executive, London, 1984
4. S M Lainoff, *Advances in Cryogenic Engineering*, Vol. 27, 1982, p. 953

Packaged deals

This term is used to describe a unit such as a *boiler* or refrigeration plant which is sold complete with all its ancillary equipment, such as *instruments* and *relief devices*, ready for connection to the main plant. Many companies do not check to see that package deals comply with their usual safety standards, or even with acceptable standards, or that relief valve sizes have been estimated correctly. Here are some incidents that have occurred as a result:

- A reciprocating compressor was started up in error with the delivery valve closed. The relief valve was too small and the packing round the cylinder rod was blown out. Unknown to the users, the relief valve was merely a 'sentinel' valve to warn the operator. The compressor had been in use for ten years.
- A reciprocating pump was ordered capable of delivering $2\,m^3/hr$. The manufacturer supplied his nearest standard size, which was capable of delivering $3\,m^3/hr$, but sized the relief valve for $2\,m^3/hr$. When the pump was operated against a restricted delivery the coupling rod was bent; fortunately, it was the weakest part of the system.
- An under-pressure connection was being made to a pipeline by a specialist *contractor* when a ¼ inch branch was knocked off by a scaffolding plank. The company did not allow ¼ inch connections on process lines – all branches up to the first isolation valve were 1 inch minimum – but they did not check the contractor's equipment.
- The support legs on a tank trailer, used to support the tank when it is not connected to a tractor, were designed so that they could not be lubricated adequately; several failures occurred.
- Figure 62 shows a relief valve which was supplied with a compressor; it was of an unsuitable type, was mounted horizontally and vibrated excessively. Relief valves should be mounted vertically so that any condensation or dirt which collects in them has the maximum chance of falling out.

Packaged equipment may not use the same *threads* as the main plant.

Panic

The press sometimes gives the impression that people panic when there is a *fire* but, according to an article in *Fire Prevention*[1], there is little evidence that they do. In fact the opposite often occurs. People take their time, tidy their desk before they leave the office and then leave by the exit they usually use, instead of the nearest one. Another feature of fire behaviour is that if people decide to fight a fire they may run past fire extinguishers two or three times, while filling buckets of water. It seems that because the extinguishers are not used regularly they are not noticed and are forgotten.

1. *Fire Prevention*, No. 145, November 1977, p. 31

Figure 62 This relief valve, supplied with a compressor, vibrated excessively

Patience

A child plants an acorn and expects to see a tree the next day. There are people like that in industry. A manager or *safety professional*, new to a works, sees equipment or a practice that is obviously unsafe. The equipment will take time to put right – *modifications* have to be designed and made – but why not *change* the practice right away? If we do not, someone may be injured. Unfortunately, rapid change is impracticable, unless the procedure is very simple. If we find, for example, that the systems for preparing equipment for *maintenance* or *entry* are unsatisfactory it is no use issuing an edict for immediate change. We have to persuade people that change is necessary, involve them in drawing up the new procedures, explain them and the reasons for them and discuss objections. All this takes time. We cannot rule out the possibility that an accident may occur during the period of change but the fact that change is in the air may make people more aware of the hazards.

Most people do not succumb on the spot; they have to be seduced by degrees.

However, after a serious accident people do accept change by decree.

Penny-pinching

We are often amused by the petty economies of wealthy people who re-use old envelopes and plastic bags. In industry similar penny-pinching can be tragic, for example, when old tab washers, split pins and pipes are re-used.

The piston of a reciprocating engine was secured to the piston rod by a nut, which was locked in position by a tab washer. When the compressor was overhauled the tightness of this nut was checked; to do this the tab on the washer had to be knocked down and then knocked up again. This weakened the washer so that the tab snapped off in service, the nut worked loose and the piston hit the end of the cylinder, fracturing the piston rod.

The load on a 30 tonne hoist slipped, fortunately without injuring anyone. It was then found that a fulcrum pin in the brake mechanism had worked loose as the split pin holding it in position had fractured and fallen. The bits of the pin were found on the floor.

Split pins and tab washers should not be re-used but replaced every time they are disturbed. Perhaps it is not penny-pinching but lack of spares that prevents us doing so. Perhaps we cannot be bothered to go to the store; perhaps there are none in the store.

Similarly, if we are altering or extending the plant it seems sensible to re-use old pipe which shows no signs of *corrosion* and looks fit for re-use. Unfortunately not, as the pipe may be affected in ways that are hard to detect even by expert examination. If it has been used hot, it may have used up some of its creep life and may fail in service under conditions in which new pipe would not fail. If the old pipe is made from stainless steel and has been in contact with chloride it may have been affected by stress corrosion cracking. The amount of chloride in town water may be sufficient. Old pipe should never be re-used unless we know its history and can take advice from a materials specialist.

To quote a Chinese proverb, 'What use going to bed early to save candles, if the result is twins?'

See *extravagance, old equipment* and *Tay Bridge*.

Perception of risk

Most people's perception of risk does not agree with that of experts: '... the persons taking the risk and having limited control over it, tend to see it as being more severe than those who control it but are not physically or immediately threatened by it[1]'. Many people think, for example, that *nuclear power* is more dangerous than *smoking*. In earlier times the experts would simply have told the public that they were wrong. Today we live in the era of the common man and such elitist action is out of keeping with the spirit of the age. The Royal Society, in their 1983 report on *Risk Assessment*[2], did not simply give the facts, as they would have done in an earlier age, but described the public's views, the reasons for them and ways of measuring them. They ascribed a validity to the public's views at least equal to those of the experts though they conceded that reliance on the public's views 'should not totally replace the objective statistical risk approach'.

There are three steps in *hazard analysis*, the application of quantitative methods to safety problems :

- Estimating how often the hazard will occur.
- Estimating the consequences.
- Deciding what action to take (if any).

The first two steps should be matters for expert calculation, or expert judgement, if calculation is not possible. The third step is a matter for public decision. The expert has no more right than anyone else to decide what are *acceptable risks* and which risks should be reduced or removed: that is, how society should spend its resources. The Royal Society, however, went further than this. They gave the public's opinions on the first two steps a validity equal to that of the expert.

In a democracy, Government departments, industry and experts are all servants of the public, and should do what the public wants. But a good servant does not unhesitatingly obey his master or mistress. Like Jeeves in the P G Wodehouse stories, he speaks up when he thinks they are wrong. If we think the public is wrong we should say so, and why we think so, in words they can understand. If they still wish to follow the policies we think are wrong, they have a right to do so.

It is important to know public perceptions of reality but it is also important to know reality itself.

Most experts believe that, if they patiently explain the facts, in the end most people will accept them. Thus most people have come to accept that smoking is harmful, though it has taken about 30 years for them to do so. However, many advertising agents believe that images are more important than facts. Do not, they say, waste time telling people the facts about the chemical or nuclear industries. The facts are boring, and if they contradict existing beliefs they will be tuned out. (See *cognitive dissonance*.) Instead try to associate these industries in people's minds with things they find pleasant.

See *aversion to risk, false alarms, perspective, risk compensation, selling safety, 'they'* and *water*.

1. E Falconer, *Health and Safety at Work*, Vol. 11, No. 5, May 1989, p. 72
2. *Risk Assessment*, The Royal Society, London, 1983

Permits-to-work

See *maintenance* and *short cuts*.

Persistence

We ask for money for what we think is a necessary safety measure. Our proposal is returned with questions attached. 'What do other companies do? What alternatives have been considered? Is there a cheaper solution?' I have known people say, '*They* don't want to do it. I shall not bother any more'.

Expenditure proposals should be thoroughly probed, especially when there is no financial return. Just because safety is involved, we should not ask for blank cheques. There is usually more than one solution to a problem. Many *safety professionals* waste money on the obvious way of removing a hazard when a little thought might produce a better way. We should answer the questions patiently and in the end, when we have shown that the hazard is not trivial and that our case is sound, we will get the money.

Persistence is also necessary when persuading people to carry out jobs which do not cost money but need time or effort. If you keep on reminding people, in a good-natured way, they may finally realize that as they are going to do what you want in the end it might be easier to do it sooner rather than later. Only on very rare occasions should the safety professional have to ask a senior manager for support. See *persuation*.

Persistence of a different sort was described by Charles Darwin in *The Origin of Species*:

> But on looking closely between the stems of the heath, I found a multitude of seedlings and little trees which had been perpetually browsed down by the cattle. In one square yard . . . I counted thirty-two little trees; and one of them with twenty-six years of growth, had, during many years, tried to raise its head above the stems of the heath, and had failed.

Some safety professionals show the same persistence as Darwin's little tree, trying year after year and never giving up. Admirable, but it might be better to try a different method if the original one fails to succeed. The roots of a tree provide a better analogy. If they cannot find a way through, they try one direction and then another until they find the tiniest cracks.

Perspective

Watching the Cup Final on TV, I wondered what would happen if a jumbo jet from Heathrow Airport, not all that far away, crashed into the crowd. Thousands of people might be killed. Had emergency plans been prepared? Should we close the Airport when Wembley Stadium is in use? Since we can watch the match in comfort on TV, is there any need to attend in person? Why take unnecessary risks?

When I talk like this people think I am being facetious but I am not. I am trying to get them to keep risks in perspective. The risk of a *Chernobyl*-type accident in the UK and the risk of an aeroplane crashing on a crowded football stadium are probably comparable, but we get worked up about one and ignore the other.

So far no aeroplane has crashed on a football crowd, but that is just good luck; our *nuclear power* stations are under our own control but we have little control over the planes that use our airports.

Another example: The most hazardous of all man-made objects (except weapons), judged by their record, are dams. Their collapse killed 2200 people in the US in 1889, 1900 in Italy in 1963, several thousand in India in 1979 and 2500 in Sri Lanka in 1986. The last incident killed about as many as the toxic gas escape at *Bhopal* in 1984 but *The Daily Telegraph* decided that it was worth only two column-inches. No one is asking for dams to be 'phased out', new construction stopped or enquiries held into their design and operation. Of course, everyone understands how a dam works (or thinks they do – dams are actually high technology) while few understand chemical plants or nuclear power stations.

There is, however, a difference between nuclear power stations, aeroplanes and dams. All can fail catastrophically but in addition nuclear power stations emit radiation all the time. Not very much, about a hundredth of the radiation we get from natural sources, but we cannot do anything about natural radiation so why accept any increase however small?

We can reduce our exposure to natural radiation if we want to do so. Nearly half of it comes from radon, a radioactive gas given off by the ground and by building materials, which accumulates in houses and other buildings. In Cornwall the average concentration of radon is about three times the UK average and in 5% of the houses it is 30 times the average. Nuclear power is hardly worth worrying about compared with radon.

The risks from radon can be reduced by underfloor ventilation and by leaving windows open. However, before you rush to open your windows and catch pneumonia let me point out that the average risk is the equivalent of smoking about ten cigarettes per year. (See *radioactivity*.)

Similarly, in industry it is very easy to spend so much time dealing with small risks that we ignore larger ones. We should be concerned about all risks but we cannot do everything at once and so we should deal with the biggest risks first; if there is a large hole in our roof we repair that before the holes that are letting in drips.

See *aversion to risk, false alarms, hazard analysis, perception of risk* and *water*.

Persuation

Persuation is an essential skill of the *safety professional* as he has no line *management* authority and has to persuade others to carry out the actions he considers necessary. After an *accident* his task is usually easy and he may have to dissuade his colleagues from going too far. (See *overreaction*.) If he has recognized a hazard but no accident has occurred his task is more difficult but it can be done. (See *actions*.) People who have reluctantly admitted that there may be a problem may say that the proposed solution is too expensive or impracticable, will interfere with production, cause teething problems and may have unforeseen effects. (See *cognitive dissonance*.) With *patience* and *persistence* these objections can be overcome. Arguments should be put clearly, objections answered patiently. We should not give up just because success is not immediate. The pace of change may be slow but it is rapid by the standards of earlier generations.

The safety profession does not lack for ideas on how individuals can be protected or how mechanical processes should be guarded. What is not so clearly understood is the means by which we should manage our affairs to get the organization to do what we know must be done[1].

See *selling safety*.

1. W C Pope, *The Molding of a Movement*, National Safety Management Society, Oakland, California, 1977

Philosophers' stone

In the Middle Ages alchemists tried to find a philosophers' stone which would turn base metals into gold. Today some managers and *safety professionals* hope that they will find a wonder-working recipe, a star in the east, that will turn a poor safety record into a good one. There have been many false Messiahs – including *audits, incentive schemes, damage control, inspections*, safety sampling – but none of them did the trick.

All these techniques are useful and can play a part in a safety programme, if they are pursued with energy and enthusiasm, but the energy and enthusiasm matter more than the technique. In the 1960s one of my ICI colleagues, the late Brian Cornford, visited the US and was surprised to find that that these techniques were little used. To quote from his report: 'The stress lay on better *design* and *layout* coupled with *training* and supervision. In their words . . . such techniques are not useful to a well run outfit and, to improve a bad one, get better *management* rather than a technique.'

To quote J V *Grimaldi*[1]: 'Improvement in safety . . . is more certain when managers apply the rigorous and positive adminstrative persuasiveness that underlies success in any business function . . . outstanding safety performance occurs when the plant management does its job well. A low accident rate, like efficient production, is an implicit consequence of managerial control.'

I have heard senior managers say, '*Du Pont* have a fine accident record. Let's visit them and copy their techniques.' However, it is not Du Pont's techniques that give them their fine record but their commitment to safety at all levels.

See *remedies*.

1. *Management and Industrial Safety Achievement*, International Safety and Health Information Centre (CIS), Information Sheet No 13, 1966.

Photography

See *serendipity*.

Pipe failures

About half the large *leaks* that occur are due to pipe failures[1] and the biggest cause of pipe failures is the failure of *construction* teams to follow the *design* in detail, or to construct well, in accordance with *good* engineering *practice*, details that were left to their *discretion*. The most effective action we can take to prevent pipe failures is to specify the design in detail and then to inspect thoroughly, during and after construction, to make sure that the design has been followed and that details not specified in the design are in accordance with good engineering practice.

Other pipe failures can be prevented by better control of operations, better control of purchased material and, in a few cases, better monitoring for corrosion and better supply of information to the designers.

For examples and more detail see *alertness, (better) late than never, construction, dead-ends, leaks, penny-pinching* and *LFA*, Chapter 16.

1. K W Blything and S T Parry, *Pipework Failures – A Review of Historical Failures*, Report No. SRD R441, UK Atomic Energy Authority, Warrington, 1988

Piper Alpha

An *explosion* and *fire* on this North Sea oil platform in July 1988 caused the deaths of 167 people. At the time of writing, the report of the official inquiry has not been issued but the interim report of the technical investigation[1] is available. This states that the initial *leak* may have occurred because a pump relief valve was removed for test and the opening not properly blanked; oil then came out when the pump was started up. (See *isolation*.) Alternatively, liquid may have been carried over into a reciprocating compressor, causing damage.

The leak ignited and an explosion occurred in the poorly ventilated module in which the equipment was located and this explosion and the subsequent fire damaged the accommodation module which was located above.

Whatever the origin of the leak, the congested nature of the platform magnified its effects. A similar leak on shore would have caused few casualties. (See *domino effects*.)

In the months following the accident the need to carry out so many processing operations on platforms was questioned. Would it be possible, it was asked, to pump a mixture of oil and gas ashore instead of separating them on a platform?[2] Is underwater separation feasible?[3]

1. J R Petrie, *Piper Alpha Technical Investigation – Interim Report*, Department of Energy, London, September 1988
2. C Butcher, *The Chemical Engineer*, No. 443, December 1988, p. 31
3. P Varey, *The Chemical Engineer*, No. 444, January 1989, p. 87

Plastics

As *materials of construction* plastics have many advantages, most notably their resistance to *corrosion*, but they also have disadvantages and accidents have occurred because these were not taken into account.

Plastics are flammable and are not suitable for use where flammable liquids are handled (except for a few glass-reinforced plastics which burn slowly and may be suitable for special applications when metal is unsuitable).

Most plastics are electric insulators and care is needed that a section of plastic pipe does not leave equipment beyond it unearthed. See *static electricity*.

Some plastics are porous and contamination can result. Thus when plastic water pipes are run through oil-soaked ground the water may become contaminated with oil.

Many plastics are attacked by common solvents. In one plant a spillage of solvent attacked the plastic water main which disintegrated.

Many tanks made from glass-reinforced plastic have failed in service as the result of corrosion, lack of regular *inspection*, unsuitable choice of plastic or poor welding[1]. See *WWW*, §5.7.

As a rule, plastic equipment should not be heated electrically, since an electric fault may cause overheating and failure. Metal plugs should not be fitted in plastic equipment as the plastic may wear and the plug blow out.

Plastics have different mechanical properties to metal and have failed when metal designs were copied in plastic. For example, a plastic pipe was given insufficient support, it sagged and cracked. A *hose*, pushed over a plastic pipe and secured by a clamp, came off; the clamp would have secured the hose to a metal pipe but the friction between the plastic and the hose was less than between metal and the hose.

In general, when we use a new material of construction we should take advantage of its strengths and allow for its weaknesses, not just copy old designs in new materials.

1. T E Maddison, *Loss Prevention Bulletin*, No. 076, August 1987, p. 31

Platitudes

Safety gets more than its fair share of these. Talks and articles by senior managers and directors on safety are almost always full of phrases such as '. . . matter of major importance . . .', '. . . responsibility of all employees . . .', '. . . deserving of the highest priority . . .' There is no evidence that this sort of talk ever prevented a single accident but we can read it in every company newspaper. It suggests that the speakers are not really interested and have not bothered to understand and address the current problems. Unfortunately, this is usually obvious to their audience and the message is 'tuned out'. (See the quotation from Swain under *human failing*.) The speakers may wish to give the impression that they consider safety important but just saying so, without discussing specific problems, can have the opposite effect. If the same people were discussing output, efficiency, costs or product quality, they would not talk generalities but would identify and draw attention to the technical problems that were preventing achievement of their objectives and would make regular checks on progress.

Lord Sieff, former chairman of Marks & Spencer, writes[1],

> It is no use the Chief Executive saying this is the policy unless he or she follows it up consistently, congratulating those who are successful in carrying out the policy.

I do not wish to suggest that managers who produce platitudes on safety are incompetent; like the rest of us they trapped by *culture*, circumstances and habit.

1. 1. M. Sieff, Don't Ask The Price, Collins, London, 1988

Plugs

Construction teams often make openings in equipment so that water can be drained after pressure testing; they then fit screwed plugs in them. Such plugs have corroded and blown out many years later. On other occasions they have been found to be holding by only one or two threads. Unless the plugs are needed for future pressure testing they should be welded up. Do not weld over an ordinary screwed plug; if the thread corrodes the full pressure will be applied to the weld. Use a specially designed plug with a full strength weld.

Unwelded plugs should be avoided whenever possible. If they have to be used, they should be locked in position so that they cannot vibrate loose.

Plugs of another sort are used to seal leaking heat exchanger tubes. When fitting them remember that one day the plugs may have to be removed for retubing. Use a design of plug which is easy to remove. Do not leave problems for those who will come after you.

Poisons

See *toxicity*.

Policy

In theory the big men at the top lay down the policy and the rest of us follow it. In practice it does not work like that. We deal with problems as well as we can, subject to various constraints. Looking back we see a common pattern; that is our policy. Statements of policy often put into 'statute law' what is already the 'common law' of the company. Action does not follow policy; policy follows action. Policy statements which are counter to the 'common law' – to *'custom and practice'* – are often ignored.

We see this more clearly if we look at the past. Early men did not start to domesticate plants and animals because a ruler decided on a change in policy. They did so 'not really because they opted for a technological solution to the dilemma of too many mouths and too few resources, but rather because they felt subconsciously forced to do so by pressures they did not understand[1]'.

If I want to know a company's policy on canteens I do not ask for a statement of their policy. A meal in the canteen will tell me more. Similarly, if you want to know a company's policy on safety, look around. Policy is what we do, not what we say. Policy statements are required by the UK *Health and Safety at Work Act* but I regard them as one of the least important parts of the Act.

Of course, looking back, we may not always see consistency. Our actions may be confused and contradictory. They are more likely to be consistent if there is a generally accepted *culture*.

If we want to change a policy we should not start by persuading the board to issue a directive. Instead we should persuade individual managers, the lowest level that has the power to make the changes we want. Gradually more and more of them will do what we want and then it will become the policy. It is then that a policy statement can be issued to tidy things up and try to bring into line the 10% or so who will never be persuaded.

A similar view has been put forward in another context by Roger Fisher[2]:

> ... we should stop thinking in terms of producing something mysterious called 'policy' and think instead of the people we are trying to influence, and the decisions we want them to make. We do not ask a motor mechanic for a consistent statement of his *attitude* towards diferent kinds of car. We ask him to repair the vehicle . . . we should ask those we are trying to influence for a definite decision.

In the state of shock following a serious accident people will accept a change imposed from above by decree, but not at other times. They have to be persuaded that the change is necessary and reasonable.

See *detail* and *variety*.

1. T C Young in *Man in Nature*, edited by L D Levine, Royal Ontario Museum, Toronto, 1975, p. 9
2. *Financial Times*, 25 March 1971

Polymerization

See *(accidental) purification*.

Poor equipment

Poor safety equipment may be worse than no equipment at all.

A belt drive was unfenced so that the operator could stop the machine by throwing off the belt. No accident had occurred in living memory. A *Factory Inspector* objected and the belt was enclosed in a wire mesh guard. Soon afterwards someone leaned against the mesh, his finger went through the mesh and his hand was so severely injured that he lost it[1].

The lesson to be learnt is not, of course, that the machine should have been left unguarded but that the guard should have been stronger. (At one stage in my career I was building instruments for use on the plant. I soon learnt that thay had to be made strong enough for someone to stand on them.) See *fencing* and *modifications*.

Similarly, an unreliable *alarm* is worse than no alarm as operators rely on it and do not watch the plant as closely as they would watch it if there was no alarm.

1. A H Little, *Chemistry in Britain*, Vol. 23, No. 4, April 1987, p. 323

Posters

Many companies display safety posters, their own or those sold by safety organizations, and many of them are well designed. But they have two disadvantages:

1. The message is very general, such as, 'We all shoulder the responsibility for safety'
2. After a while nobody notices them. Perhaps this is as well as they become very tatty.

For these reasons, safety posters should give a specific technical message, rather than a general one, and they should be in the form of a calendar, so that they are changed every month. Most of the Figures in this book were originally calendar posters.

A way of displaying small safety posters is to stick them on the covers of books of permits-to-work, or other forms that are used frequently. A new poster comes with each new book.

Powders

See *dusts*.

Practicable

See *'reasonably practicable'*.

Practice

See *custom and practice* and *good practice*.

Pressure

Pressure and *force* are often confused, especially when pressure is measured in pounds per square inch (or kilograms per square centimetre). Because '30 pounds per square inch' is rather a mouthful we say that the pressure is '30 pounds'. Since 30 pounds force is not a large force we then assume that 30 pounds pressure is not a large pressure, forgetting that a force of 30 pounds is exerted on every square inch of the vessel wall. When vessels have burst as the result of excessive pressure, operators have often expressed surprise that a pressure of 'only 30 pounds' could cause so much damage. On several occasions explosion experts have had to be brought in to convince them that there had not been a chemical explosion. See *EVHE*, § 2.2.1.

In the SI system quite different units are used for force (newtons) and pressure (pascals or bars). Their increasing use may help to avoid confusion and operators may come to realize that 2 bar (30 pounds per square inch) is a big enough pressure to kill someone.

See *trapped pressure*.

Pressure tests

Equipment is usually pressure tested when new and at intervals during its life, to prove and confirm that it is fit for service. The practice seems to date back to 1817 when the report of a Parliamentary Select Committee on the explosion of the *boiler* of a steam packet recommended that boilers should be inspected by a 'skilful engineer . . . who should ascertain by trial the strength of such boiler'. However, as described in the item on boilers, it was many years before action was taken.

The following points on pressure testing should be noted:

- Equipment may fail when tested – if we were sure it would not fail, we would not need to *test* it – so make sure that failure will not cause injury or, as far as possible, damage to other equipment.
- Make sure before hydraulic testing that the equipment and its foundations can withstand the weight of water; many large distillation columns will not.
- Hydraulic testing is normal but sometimes pneumatic testing is necessary, for example, if the vessel cannot withstand the weight of water. Failure of a vessel during pneumatic testing can cause much more damage than failure during hydraulic testing, as air is more compressible than water, and greater precautions are necessary. If pipework has to be tested pneumatically it should be carried out in sections so that the product of test pressure and volume does not exceed 100 bar/m^3. This is the energy released by the explosion of 0.7 kg TNT.
- Make sure you know the test pressure of the equipment. 1.5 times design is common but some codes ask for 1.3 or 1.25 times design.
- Do not pressure test below the 'minimum permitted test temperature' or brittle failure may occur. The test water may have to be heated. Spectacular failures have occurred because the water was too cold. For example, a large pressure *vessel* (16 m long by 1.7 m diameter) failed at a pressure of 345 bar when tested at the manufacturers. Four large pieces were flung from the vessel and one, weighing 2 tonnes, went through the workshop wall and travelled nearly 50 m[1]. See *action replays*.
- When pressure testing pipework make sure that no sections are overlooked and that *non-return valves* do not prevent the pressure reaching the sections of pipework that lie beyond them.
- Many people believe that pressure testing of *steam* receivers is required by the UK Factories Act but in fact it requires thorough examination by a competent person every 26 months. The competent person has to decide the method of examination. At one time pressure testing was the only practicable method but is not so now that many methods of non-destructive testing are available. Pressure testing of other vessels and pipework is not specifically required by any UK law but is normal practice; again, other methods of assuring the integrity of the equipment can be used when pressure testing is difficult to carry out. Failure to assure the integrity of equipment is *prima facie* evidence that it has not been made safe, so far as is *'reasonably practicable'*, as required by the *Health and Safety at Work Act*.

At the time of writing, publication, in the UK, of new regulations on pressure systems, was imminent.

- As well as demonstrating the integrity of a vessel, pressure testing may result in 'shake-down', i.e, a redistribution of the stresses in a vessel.
- There is no advantage in pressure testing more often than necessary. It may raise the level of stress at defects such as slag inclusions and cracks. Leak testing, at or below design pressure, can of course, be carried out as often as we wish without risk and can be hydraulic or pneumatic.

While pressure vessels are usually tested at regular intervals, regular testing of pipework is much less common though large pipes may be bigger than small vessels. While it is not necesary to test all pipework at regular intervals, those pipes that carry hazardous materials at high pressures or temperatures should be scheduled for regular testing and/or *inspection*. The nature of the test or inspection should be specified.

1. *British Welding Research Association Bulletin*, Vol. 17, Part 6, June 1966, p. 149

Pressure vessels

See *vessels*.

Priorities

See *acceptable risk, choice of problems, hazard analysis, overreaction, perception of risk, perspective, production or safety?* and *water*.

Probabilities

See *hazard analysis, confidence limits* and *will and might*.

Probit

A probit is an equation of the form:

Probit = a + b ln V

where a and b are constants

V is a measure of something that causes injury, for example, explosion over-pressure or concentration of toxic gas.

The probit is a measure of the probability that death or a defined degree of injury will occur. *Lees*, §9.6 gives a table relating the value of the probit to the probability. Thus 3.72 corresponds to 10% probability, 5.00 to 50% and 6.28 to 90%. See *ammonia, chlorine* and *toxicity*.

Probits were originally used mainly by toxicologists but the method is now applied to *explosions* and *fires*. See *Lees*.

Problems

I am sometimes asked what were the biggest problems I had to tackle during my 14 years as a *safety professional* with *ICI*.
They were:

- The preparation of equipment for maintenance. The problem was threefold: agreeing better rules, persuading people to adopt them and checking from time to time to see that they were being followed.
- The control of *modifications*. This problem had the same three stages as above.
- Keeping alive the memory of past accidents. See *lost knowledge*.

Perhaps some readers expected me to say that the biggest problem of all was getting the money needed for improvements in safety, but this was not so. Many of the changes I suggested needed management effort only and grey hairs cost the company nothing. When expenditure was necessary, if a good case was made, the money was found. Questions had to be answered, *persistence* was necessary, but we got there in the end.
See *choice of problems*.

Procedures

See *exchanging one problem for another, methods of working* and *removing equipment and procedures*.

Process safety

See *loss prevention*.

Production or safety?

If we cannot have both, which should it be? In fact a poor safety record, like poor production or or poor efficiency, is evidence of poor *management*. Many studies have shown '. . . that there is a strong relationship between management's ability to control all contingency *loss* areas and its success in reducing severe preventable work accidents[1]'.

However, though this is true overall many people find themselves in a situation where the decision is less obvious. If there is a leak, do we shut down or finish the batch? Obviously we do not shut down immediately every time there is a small leak, but there are times when we should shut down; judgement is needed but guidance is often lacking. Sometimes people are given *contradictory instructions*: put safety first (general instruction) and finish the batch on time (specific instruction). They follow the specific instruction. (See *Chernobyl*.)

Nearly every manager claims that safety is the most important thing in his factory. What happens in practice usually makes it clear that managers do not set as high a priority on safety as they indicate.

Research might show that we could achieve better results by persuading people not to dramatically overstate their convictions about safety but to approach the subject in a less dramatic and more detailed way[2]. (See *attitude*.)

LFA, Chapter 3 describes an accident which occurred when a young, inexperienced graduate had to decide between production and safety.

1. J V Grimaldi, *Management and Industrial Safety Achievement*, International Occupational Safety and Health Information Centre (CIS), Information Sheet No. 13, 1966
2. Institution of Professional Civil Servants, *Evidence Submitted to the Committee on Safety and Health at Work*, 1972

Prohibition Notice

See *prosecution*.

(self-fulfilling) Prophecies

Some prophecies are sure to be proved correct, for example, the need for intermediate storage between two sections of a plant. It is essential, we are told, despite the hazard, as otherwise the whole plant will have to shut down when one section breaks down and much production will be lost. Computer calculations confirm the hunch. If the storage is provided it is used, and, looking back, we see that it prevented lost production.

However, if we did not have the intermediate storage more strenuous efforts would have been made to get the offline section repaired and back on line sooner.

Holusha, reviewing a book on Japanese industry writes[1]:

... the American practice of having buffer stocks of partly finished components all along the production line concealed problems. If a machine broke down, parts from a nearby buffer stock could keep the assembly line working until somebody got round to fixing the machine. But ... if the security of the buffers was removed, if the whole plant were in jeopardy of being shut down by one malfunctioning machine, the workers would be forced to tend their machines more carefully, so that all worked correctly all the time. The discipline this system imposed brought surprising improvements in productivity and quality.

In the process industries reduced intermediate stocks will not encourage the operators to treat the plant more carefully but they will encourage the maintenance team to get it back on line more quickly after an a breakdown.

Another example: we prophecy that a heat exchanger will foul and we over-design it. When it is first used the temperature is too high or the operator reduces the flow rate and this causes fouling. The prophecy is proved correct. If the heat exchanger had been designed on the assumption there was no (or less) fouling, fouling might not have occurred[2].

1. J Holusha, *New York Times*, 2 April 1986 (in a Review of M A Cusumano, *The Japanese Automobile Industry*, Harvard University Press, 1986)
2. J Redman, *The Chemical Engineer*, No. 452, September 1988, p. 12

Prosecution

Under the UK *Health and Safety at Work Act* (1974) and earlier legislation the Health and Safety Executive (HSE) can prosecute companies and employees (see *blind-eyes*) but there are only about 1200 prosecutions per year, most of them after an accident has occurred. Some trade unions and some safety magazines ask for more prosecutions. In a long interview with the Director General of the UK Health and Safety Executive, the interviewer repeatedly suggests that, to prevent accidents, we need more *Factory Inspectors* and more prosecutions; other ways of preventing accidents are hardly mentioned[1]. Prosecution is not a very effective means of preventing accidents as:

- Most offences are due to carelessness, oversight or inefficiency rather than criminal intent. See the quotation from the Robens report under *blame*.
- Responsibilty is spread over many people. See *chains*.
- It takes up a lot of time, which might be better spent.
- The fines imposed by the courts are usually very small (although in 1988 a UK company was fined £750 000).

The HSE prefers to issue Improvement and Prohibition Notices, about 8000 per year, so that hazards are removed before an accident occurs. Advice is even more effective, if people will listen to it, and this is usually the first approach. Prosecution is undertaken only 'where employers or others concerned appear deliberately to have disregarded the relevant regulations or where they have been reckless in exposing people to hazard or where there is a record of repeated infringement[2]'.

If there is a prosecution, it is usually the company that is prosecuted. Only a handful of individuals have been prosecuted, usually for wilful and deliberate actions, such as damaging safety equipment, or for gross negligence, such as failing to carry out duties which were clearly laid down as part of the job.

However, if a company ceases to exist before a prosecution can take place the HSE might prosecute a responsible official.

So far as I am aware no one has ever been prosecuted in the UK for an error of professional judgement. As long as one does one's best and exercises reasonable care, I do not think that there is any danger of prosecution. Your employer might be prosecuted for not employing a more competent or better trained person but that is another matter. 'The Courts must say firmly that, in a professional man, an error of judgement is not negligence[3].'

In some countries the authorities are much more ready to prosecute. They are confrontative while in the UK the HSE and its predecessors, the Factory, Explosives, Mines and Quarries and other Inspectorates, have always found it more effective to cooperate with industry.

1. D Gee, *Health and Safety at Work*, Vol. 11, No. 3, March 1989, p. 29
2. *The Leakage of Radioactive Liquor into the Ground, British Nuclear Fuels Ltd, Windscale, 15 March 1979*, Health and Safety Executive, 1980, §51
3. Lord Denning (in the Court of Appeal), *The Times*, 6 December 1979

Protective clothing

Protective clothing is widely supplied, when a hazard cannot be removed, but is often not worn. Many accidents occur as a result. The following is typical:
A man was employed to lift out molten metal with a ladle and pour it into a die. After some months he was given goggles and told to wear them. They misted up so he told his supervisor they were useless and stopped wearing them; nobody tried to make him wear them. After a year he was splashed with molten metal, losing the sight of one eye and part of the sight of the other. He claimed damages against his employer for negligence. The trial judge found that though the company had provided goggles they had failed in their duty of care to their employee because they did not take reasonable steps to ensure that the goggles were worn. This decision was upheld by the Court of Appeal. The courts decided that 40% of the blame for the injury was due to the injured man and therefore damages were reduced from £30 000 to £18 000 (1974 prices).

What are 'reasonable steps'? An employer is not expected to stand over a man all the time but he is expected, through his supervisors, to check up from time to time to see that protective clothing is being worn and they must not turn a *blind-eye* if they see it is not being worn.

See *adequacy, COSHH Regulations, eye protection, 'reasonably practicable'* and *EVHE*, §11.2.

Protective equipment

All protective equipment should be tested regularly or it may not work when required. (See *Bhopal*.) The test frequency depends on the failure rate. *Relief devices* are very reliable, they fail about once per hundred years on average, so testing every one or two years is adequate. Protective systems based on *instruments*, such as trips and *alarms*, fail more often, about once every couple of years on average, so more frequent testing is necessary, about once per month. See *tests*.

As well as relief valves, trips and alarms the following protective equipment should also be tested or inspected regularly, though they are often overlooked:

- Drain holes in relief valve tailpipes.
- Drain valves in tank bunds.
- *Emergency equipment* such as diesel-driven fire water pumps and generators.
- Earth connections, especially the moveable ones used for earthing road tankers.
- *Fire* and *smoke* detectors and fire-fighting equipment.
- *Flame traps*.
- *Hired equipment*.
- Labels are a sort of protective equipment. They vanish with remarkable speed, and regular checks should be made to make sure that they are still there. See *identification of equipment*.

- Mechanical protective equipment such as overspeed trips.
- *Nitrogen* blanketing.
- *Non-return valves*, if relief valves have been sized on the assumption that they will operate or if their failure can affect the safety and operability of the plant in other ways.
- Open vents. These are in effect relief devices, the simplest possible sort of relief device, and should be treated with the same respect.
- Passive protective equipment such as *insulation*.
- Spare *pumps*, especially those fitted with auto-starts.
- *Steam* traps.
- Trace heating (steam or electrical).
- Valves, remotely-operated and hand-operated, which have to be used in an emergency. See *emergency isolation valves*.
- *Ventilation* equipment.
- Water sprays and steam curtains.

All protective equipment should be designed so that it can be tested. *Audits* should include a check that the tests are carried out and the results acted on.

Tests should be like 'real life'. A high temperature trip failed to work despite regular testing. It was removed from its case before testing so the test did not disclose that the pointer rubbed against the case. This prevented it indicating a high temperature.

Test results should be displayed in the control room. It is good practice to list all the protective equipment on a board, showing the dates on which tests are due, and the test results. Everyone can then see when testing is overdue.

Operators sometimes regard testing as a nuisance. (See *optional extras*.) 'Why are the men in the instrument section always wanting to test their trips?' Such operators fail to realise that the trips are there for their protection and that they should own them. See *redundancy*.

It is easy to buy protective equipment. All we need is money. It is much more difficult to make sure it is tested regularly and kept in working order.

See *methods of working, useless equipment and procedures*, *WWW*, Chapter 14 and *LFA*, Chapters 2 and 10.

Publication

We are publishing fewer *accident reports* and less information on the action needed to prevent the accidents happening again than we were doing a few years ago, for the following reasons:

- Time: with reductions in staff people have more to do and actions which do not have to be completed by a set date get repeatedly postponed.
- Lawyers, especially in the US: they fear that publication may affect claims for compensation or make prosecution more likely.
- Fear of adverse publicity.
- Fear that secrets will be given away. (See below.)
- Cumbersome company procedures for authorizing publication.

We should encourage publication, for four reasons:

- Moral: if we have information that can prevent an accident we have a duty to make it known. See *morality*.
- Pragmatic: we may learn about other accidents in return. See *debts* and *failure*.
- Economic: we should encourage other companies to spend as much on safety as we do.
- The industry is one: when one company has an accident all suffer from loss of public esteem and Government interference. 'We are hostages to the weakest performer[1].'

Gilinsky has described the effect on the US *nuclear power* programme of a failure to share information[2]:

> ... 115 or so nuclear units were in the hands of about 55 companies with widely varying ability. They had all ordered and gone ahead more or less in parallel. That had meant about 50 learning curves, instead of one or a few as in other countries, so the same mistakes had been made in utility after utility.

I suggest that companies should:

- Encourage their staff to publish.
- Remove any suggestion of *blame* from accident reports and, if they wish, publish then anonymously in, for example, the Institution of Chemical Engineers' *Loss Prevention Bulletin*.
- Publish reports on *dangerous occurrences*, if they are unwilling to publish reports on accidents in which people were injured.
- Circulate the information within an industry club, if they are unwilling to publish in the open literature.
- Describe the action they took as the result of an accident, if they cannot describe the accident itself.

At conferences many papers describe company procedures for ensuring that plants are designed and operated safely. (If the procedures were always followed there would be few accidents.) Accounts of accidents that have happened have far more impact and teach us far more.

Regarding secrecy, accident reports can be written so that they describe the general lessons to be learnt without giving away the details of individual processes, as George E Davis, one of the founders of chemical engineering, made clear in 1888. After a lecture someone accused him, as a consultant, of giving away the secrets of the companies he visited. He replied, 'The science of chemical engineering does not consist in hawking about trade secrets... Chemical engineering has higher aims, it endeavours to work out the application of machinery and plant to the utilisation of chemical action on the large scale[3].'

See reference 4.

1. Lord Marshall, quoted by A Conway, *Atom*, No. 389, March 1989, p. 2
2. Quoted by A Conway, *Atom*, No. 389, March 1989, p. 2
3. J F Donnelly, *Annals of Science*, Vol. 45, 1988, p. 555
4. *Plant/Operations Progress*, Vol. 7, No. 3, July 1988, p. 145

Pumps

To the loss prevention engineer, pumps are devices with glands that leak. Packed glands are cheap and simple and rarely fail catastrophically but there is usually a slight leak (see *fugitive emissions*) and they need regular adjustment. Mechanical seals are therefore widely used, often with a second seal to contain any leakage. They require skilled assembly and accurate shaft alignment. Canned motors and magnetic drives eliminate the need for seals but their initial cost is high[1].

When experience shows that pumps handling hazardous liquids are particularly liable to leak, for example, those handling very hot or cold liquids, remotely-operated *emergency isolation valves* should be fitted in the suction lines so that leaks can be stopped from a safe distance[2]. It is not usually possible to reach the normal hand-operated isolation valves. A *non-return valve* can be fitted in the delivery line instead of another emergency isolation valve. Emergency isolation valves, and non-return valves used as emergency isolation valves, should be *tested* regularly.

All glands may fail catastrophically if there is a bearing failure. If the quantity of liquid in the suction vessel is large, say, 20 tonnes or more, and hazardous, then emergency isolation valves should be installed even though the pumps have no history of failure. Otherwise, if failure should occur, a large quantity of liquid will be spilt.

See *WWW*, Section 7.2.

1. T Newby, *The Chemical Engineer*, No. 447, April 1988, p. 33
2. *Chemical Engineering Progress*, Vol. 71, No. 9, September 1975, p. 63

(accidental) Purification

Contamination causes accidents. The materials we handle may be safe but contaminants may make them unsafe. The opposite effect also occurs, though less often. Accidental removal of an inhibitor may cause an accident.

The liquid in a tank was inhibited to prevent polymerization. The vapour that condensed on the roof was not inhibited; it polymerized at the base of the vent and nearly blocked it. The vent was inspected regularly but the polymer was not noticed. When inspecting a vent it is not sufficient to look through it; its diameter should be checked by passing a rod through it. (The top of the rod should be bent or enlarged so that it cannot be dropped into the tank.) See *relief devices*.

Acrylic acid is usually inhibited to prevent polymerization. Nevertheless a road tanker containing the acid ruptured violently while it was being heated with a mixture of steam and water though the temperature was controlled to prevent excessive heating. It is believed that some of the acrylic acid had solidified and the inhibitor separated by fractional crystallization; after melting, the acid polymerized explosively[1].

A similar incident occurred on a ship. The inhibitor used was effective only in the presence of oxygen and a small leak of liquid from the next tank, through a crack, may have reacted with the oxygen and removed it[1].

1. J J Kurland and D R Bryant, *Plant/Operations Progress*, Vol. 6, No. 4, October 1987, p. 203

Qualitative

Qualitative expressions mean different things to different people and are therefore best avoided. To one person a leak is 'likely' if it occurs once per month, to another if it occurs once during the lifetime of the plant. To one person a 'large' leak is over 100 tonnes, to another anything over a kg is 'large'.

Of course, if everyone agrees on the meaning of 'likely' or 'large' there is no harm in using them, but then the words are no longer qualitative.

Another qualitative, and therefore largely meaningless, word that is often used in loss prevention is 'probable'. Insurers often talk of 'maximum probable loss'.

'Reasonably practicable' is a qualitative phrase that has served us well but the need for greater precision led to the development of *hazard analysis*.

Quantitative

See *hazard analysis*.

Radioactivity

The accidents at *Chernobyl* and *Three Mile Island* hit the headlines but misuse of sealed sources used for measurement, radiography and radiotherapy has caused more harm than Three Mile Island. One of the worst incidents occurred at Goiânia in Brazil in 1985[1,2]. When a radiotherapy institute moved to new premises a powerful caesium 137 source was left behind. The building was partly demolished. Two years later two men removed the assembly containing the source to a nearby house. One man became ill and his doctor diagnosed food poisoning. The other man broke open the assembly. The radioactive material glowed in the dark and bits were given to friends and neighbours; many became ill and several died.

Once radioactivity had been recognized decontamination was a major problem; seven houses had to be demolished and tonnes of soil had to be removed from gardens and yards.

Altogether mishandling of radioactive materials has killed about 60 people since 1945; a third were members of the public killed by lost sources[1].

Nevertheless, the risks from radioactivity are far smaller than the public believe them to be. (See *false alarms, nuclear power, perception of risk* and *perspective.*) In fact, there is growing evidence that small doses of radiation, like small doses of *arsenic*, ultra-violet light and vitamins, are beneficial[3,4].

We usually estimate the effects of small doses of radioactivity (and toxic chemicals) by *extrapolation* from the effects of large doses but this may give an incorrect result. The body's repair system may be able to cope with small doses (if they are not beneficial) and harm may occur only when the repair system is overwhelmed.

Even though a little of something is good for us, a lot may be harmful. Conversely, if a lot of something is harmful, a little of it may not be proportionately harmful.

1. *Atom*, No. 388, February 1989, pp. 15 and 26
2. *The Radiological Accident in Goiânia*, International Atomic Energy Agency, Vienna, 1988
3. K Brown, *Atom*, No. 378. April 1988, p. 26
4. J H Fremlin, *Atom*, No. 380, April 1989, p. 4

Railways

See *alarms, alertness, modifications after an accident, other industries, stress, Tay Bridge* and *team working.*

Random accidents

See *accident-prone*

Reactions

It is impossible to cover reactions adequately in a book of this nature and all I can do is draw attention to a few points that are often overlooked.

Today it is generally accepted that we cannot say after an accident, 'I did not know that the reaction would runaway if it got too hot/stood for too long/the reactants were added in the wrong order/too quickly/too slowly/too much reactant was added'. We should ask if these deviations, and others, could occur, assess their consequences and, if these are hazardous, decide what action to take. *Hazard and operability studies* can help us make sure that all possible deviations are considered and there are methods available for measuring reaction rate, for example, Accelerating Rate Calorimetry[1], and for estimating the size of *relief device* required[2,3].

Runaways are often due to side reactions which become important when conditions change.

Contamination of a reactant may affect reaction rate[4].

Runaway reactions have occurred in scrubbers where reactor exit gases were scrubbed[5].

As well as the reactants we should consider auxiliary materials. Silica was used to remove water evolved in a reaction. Whenever a new batch of drying agent was installed there was a rise in temperature. The silica was reacting with one of the reactants, a fluorine compound. The operators never reported the rise in temperature until one day conditions were such that the rise got out of hand. The incident shows the importance of responding to unusual observations. See *alertness*.

Plant instructions gave the time over which a reactant should be added. An operator started to add the reactant too slowly and then added the rest of it too quickly; a runaway occurred. Instructions may have to specify the rate of addition of a reactant as well as the total time to be taken.

The designers of a hydrogenation plant considered what would happen if liquid flow was lost and if hydrogen flow was lost but did not consider the possibility that both might be lost at the same time. When this occurred the gas and liquid in the reactor reacted together, the temperature reached over 600°C and the reactor was damaged. No temperature points were fitted to the reactor as hydrogenation was considered a mild reaction that could not runaway. See *wolves in sheep's clothing*.

Road tankers were filled with waste sludge containing aluminium particles. One day the tanker contained some caustic soda left over from the previous load. The soda reacted with the aluminium and the increase in pressure blew open an inspection port and knocked an operator onto the ground; he was seriously injured.

Handbook of Reactive Chemical Hazards[6] by *Bretherick* lists all known hazardous chemical reactions.

See *(better) late than never*.

1. T Yoshida, *Safety of Reactive Chemicals*, Elsevier, Amsterdam, 1987
2. Design Institute for Emergency Relief Systems (DIERS) reports, American Institute of Chemical Engineers, New York
3. M F Pantony, N F Scilly and J A Barton, *Plant/Operations Progress*, Vol. 8, No. 2, April 1989, p. 113
4. *Loss Prevention Bulletin*, No. 003, June 1975
5. *Chemical Safety Summary*, Vol. 55, No. 218, 1984, p. 34
6. L Bretherick, *Handbook of Reactive Chemical Hazards*, Butterworths, 3rd edition, 1985

Reactors

From the *loss prevention* point of view reactors should be designed to fulfil three requirements. They should be:

- As small as possible.
- Tubular rather than pot.
- Vapour phase rather than liquid phase[1].

Reactors should be as small as possible as 'what you don't have, can't leak'. However, reactors are often large because the reaction is said to be slow but often it is the mixing rather than the chemistry that is slow. Sometimes reactors are large because conversion is low and large quantities of raw material get a 'free ride' and have to be recycled. Conversion is often kept low to minimize by-product formation but better mixing can sometimes allow conversion to be increased. See *intensification* and *oxidation*.

Long thin tubular reactors are safer than large pot reactors as the leak rate is limited by the cross-section of the tube and can be stopped by closing a valve in the line. If we wish to manufacture 20 000 tonnes per year of a product then in theory all of it can pass through a line 2 inches bore (assuming a flow rate of 1 m/sec). If we wish to manufacture 1000 tonnes per year a line 1 cm diameter should be sufficient. Any larger line is a reflection on our ability as engineers.

Vapour phase reactors are safer than liquid phase reactors as the mass flow through a hole of given size in a vapour phase reactor is many times, typically 50 times, smaller.

1. T A Kletz, *Cheaper, Safer Plants – Notes on Inherently Safer and Simpler Plants*, Institution of Chemical Engineers, 2nd edition, 1984

Figure 63 Reactors   **Learn beforehand – before it gets out of hand**

'Reasonable care'

The law requires employers and their 'servants' to take 'reasonable care' for the safety of employees and the public but what degree of care is considered reasonable? This was discussed by the UK Court of Appeal in 1987[1]. The captain of the Herald of Free Enterprise (see *Zeebrugge*) was suspended for one year by a Wreck Commissioner for allowing the ferry boat to sail without checking that the bow doors were closed. The captain appealed against the sentence.

In dismissing the appeal the judge agreed that the system adopted by the captain was the same as that adopted by the masters of the other Spirit class vessels. General practice was usually evidence of reasonable care but in this case the requirement to check that the doors were shut before putting to sea was so absolute that it was folly to ignore it. The practice of other masters did not evidence reasonable care but a general and culpable complacency, born perhaps of repetitive routine and fostered on the part of owners and managers.

One wonders why all the masters were not suspended. The captain of the boat which sank was no more guilty than the others, just unlucky.

See *blame*.

1. *Guardian*, Law Report, 21 December 1987

'Reasonably practicable'

The *Health and Safety at Work Act* requires employers to provide a safe plant and system of work so far as is 'reasonably practicable'. Other legislation requires *leaks* of hazardous material to be 'as low as reasonably practicable' (ALARP) or 'as low as reasonably achievable' (ALARA). Although the *Factories Act* contains many *absolute requirements*, much of it is qualified by the phrase 'reasonably practicable'. What do these words mean?

> 'Reasonably practicable' . . . implies that a computation must be made in which the *quantum* of risk is placed in one scale and the sacrifice, whether in money, time or trouble, involved in the measures necessary to avert the risk is placed in the other; and that, if it be shown that there is a gross disproportion between them, the risk being insignificant in relation to the sacrifice, the person on whom the duty is laid discharges the burden of proving that compliance is not 'reasonably practicable'.
>
> Where an obligation is qualified by the term 'practicable' a stricter standard is imposed . . . but, nonetheless, 'practicable' means something other than physically possible. The measures must be possible in the light of current knowledge and invention[1].

Even when the law lays down an absolute requirement, UK Factory Inspectors seem to act as if the words 'reasonably practicable' were there.

Many other countries have an entirely different approach. Commenting on an announcement from the US Occupational Safety and Health Administration (OSHA), Alexander writes:

Absent from the OSHA's preamble is much indication that the agency gives a hoot for how much its standards would cost or for the benefits of substances it might effectively be banning. It gives no hint as to whether, in picking a suitable 'substitute' or a 'feasible' exposure level, any weight would be given to economic considerations as well as technological ones[2].

See *asbestos, honesty* and *'reasonable care'*.

1. *Redgraves's Health and Safety in Factories*, edited by I Fife and A E Machin, Butterworths, 2nd edition, 1982, p. 15
2. T Alexander, *Fortune*, 3 July 1978, p. 86

Recognition

See *perception of risk*.

Redundancy

If the *reliability* of a component is not as good as we would like it to be and it cannot easily be improved, then we may install two or more in parallel. This is known as redundancy[1]. For example, suppose a plant is fitted with a high temperature trip, to switch off a heater when the temperature reaches a pre-set level, and we require a very high reliability. We may install two (or more) independent trips, or perhaps just two independent temperature measurements which operate the same switch.

Replication is a better term than redundancy as the latter implies more than is necesary.

Redundant items are never completely independent. They may suffer from the same design or manufacturing faults; they may be affected by power failure or *contamination* of the process stream or instrument *air*; they may all be affected by fire or mechanical damage; or they may be maintained by the same man who makes the same mistakes. These are known as common mode failures. For this reason, whenever possible, we should use different types of equipment, made by different companies, or better still, measure a different physical property. If a rise in temperature is accompanied by a rise in pressure we could measure pressure and temperature, instead of measuring temperature twice. This is known as diversity.

A disadvantage of redundancy is that false alarms or trips occur more often. If there are two temperature measurements instead of one we shall get false high temperature readings twice as often. These are often called fail-safe faults but they may not really be safe. Shutting off the heat unnecessarily may upset the plant. These faults may be made less frequent by voting systems. Three temperature measurements are fitted instead of two and two out of the three must indicate a high temperature before the heat supply is shut off.

Another weakness of redundancy (and of *defence in depth*) is that people may not bother to repair the second trip as they have still got another to use. The reasons for installing redundancy should be explained to operators and regular checks should be made (as with all *protective equipment*) to see that the redundant equipment is in working order.

Redundancy does not work with people. If they know that someone else is checking their actions they may rely on the checker and work less reliably. A man plus a checker may be less reliable than a man alone.

See *useless equipment and practices*.

1. T A Kletz, *Hazop and Hazan – Notes on the Identification and Assessment of Hazards*, Institution of Chemical Engineers, Rugby, 2nd edition, 1986, § 3.5.10 and 3.6.4/5.

Recognition

See *perception of risk*.

Regulations

See *absolute requirements, CIMAH Regulations, COSHH Regulations, Factories Acts, Health and Safety at Work Act* and *rules and regulations*.

Reliability

The difference between *loss prevention* and the traditional safety approach is illustrated by their treatment of reliability. To the *old-timer*, equipment was reliable or it was not. The loss prevention engineer, however, does not asks if equipment, or people, will fail, but how often? He realizes that no person or equipment is 100% reliable and he tries to estimate the reliability that will give an *acceptable risk*. When the consequences of failure are not serious, or there is duplication of equipment and/or frequent testing, quite low reliabilities may be acceptable. When we are dealing with hazardous materials, however, the consequences of failure may be serious and high reliability is usually necessary. With equipment this may be obtained by *redundancy*.

Men are usually very reliable but there are many opportunities for error in the course of a day's work and the best that a man can do may not be good enough. (See *human failing*.) Automatic protection is therefore often necessary. (Of course, whenever possible we should design *friendly plants* in which equipment failure or human error does not have serious effects on output or safety.)

In loss prevention we usually express reliability as the fractional dead time, the fraction of the time that a protective device is not in working order, that is, the probability of failure on demand. (See *tests*.)

Fractional dead time and failure rate are often confused; fractional dead time is a dimensionless number while failure rate is expressed as the number of failures per year (or some other period of time). Nonsense answers have been produced because beginners got these quantities confused. As in any calculation we should always make our units clear at each stage[1].

A review of published papers on the reliability of chemical plants shows that the number grew from three in 1962-6 to 68 in 1982-6[2]. See *tests*, and *useless equipment and procedures*. For more informatiom on reliability, including its history, see Lees, Chapter 7.

1. T A Kletz, *Hazop and Hazan – Notes on the Identification and Assessment of Hazards*, Institution of Chemical Engineers, 2nd edition, 1985, Chapter 3
2. B S Dhillon and S N Rayapati, *IEEE Transactions on Reliability*, Vol. 57, No. 2, June 1988, p. 199

Reliability data

Much information on failure rates is scattered throughout the subject literature and some of it has been summarized by Lees, Appendix A9. The best source of such data is the data bank operated by the Systems Reliability Service of the UK Atomic Energy Authority, Culcheth, Warrington, UK. Subscribing companies are expected to contribute data and they in return have access to the data in the bank.

Many companies have their own books of data but they are often misused. Instead of looking in the book for any data on, say, relief valves, and then consulting the original reference for details, people often just copy a figure from the book and miss important qualifications.

Data on events, such as fires, explosions and toxic gas releases are stored in the MHIDAS data base, also operated by the UK Atomic Energy Authority. The *Loss Prevention Bulletin*, published by the Institution of Chemical Engineers, Rugby, UK prints an annual summary of such incidents.

In the early 1970s many companies started to collect data on equipment reliability. These would be useful, they thought, not only for estimating failure rates but also for estimating plant availability and maintenance requirements and for deciding which design was the 'best buy'[1]. Most of these systems soon fell into disuse. The maintenance workers, who had to record the causes of failure when they repaired a piece of equipment, did not like filling in a form that, so far as they were concerned, served no useful purpose. They did so if at all in an inaccurate or perfunctory manner. Reliability engineers decided that if they wanted data on the failure rate of non-return valves, for example, it was better to wait until they needed the data and then spend some time on the plant collecting the information themselves. Describing a successful data collection project, Stevens says[2], 'The AMT (advanced machine tools) project had succeeded because the people on the shop floor had been approached as equals, and had been convinced that their intelligent co-operation would help them gain pay rather than lose their jobs.'

1. T A Kletz, *Process Technology International*, Vol. 18, No. 3, March 1973, p. 111
2. B Stevens, quoted by A Conway, *Atom*, No. 386, December 1988, p. 10

Relief devices

For a relief device (relief valve, bursting disc or open vent; see Figure 64) to function correctly it has to be correctly sized, chosen, installed and maintained.

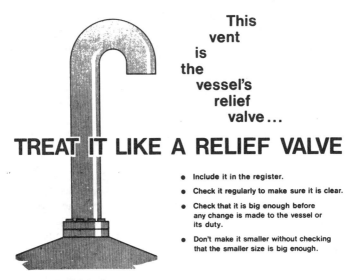

This vent is the vessel's relief valve...

TREAT IT LIKE A RELIEF VALVE

- Include it in the register.
- Check it regularly to make sure it is clear.
- Check that it is big enough before any change is made to the vessel or its duty.
- Don't make it smaller without checking that the smaller size is big enough.

Figure 64 Relief devices

Recognized codes give formulae for sizing relief devices. Fitt has described some common pitfalls[1]. At one time it was difficult to size relief valves to control runaway *reactions*, particularly when the flow was two-phase, but this subject is now much better understood as a result of the work carried out by the Design Institute for Emergency Relief Systems (DIERS) set up by the American Institute of Chemical Engineers[2].

Once the relief device has been sized, choosing suitable hardware is not usually a problem but sometimes it is installed incorrectly. Relief valves should be installed vertically (see *packaged deals*) and the tail pipes should be secure. (See *WWW*, §10.4.) Bursting discs should be installed the right way round. (See *WWW*, §3.2.1.) Relief devices should, whenever possible, be installed directly on the vessels they are protecting, not at the end of long lines; if they have to be installed at the end of a line the pressure drop in the line should be taken into account.

Most companies do not allow isolation valves to be fitted below relief devices (unless there are two devices with interlocked valves). If isolation valves are fitted, so that relief devices can be maintained with the plant on line, then rigorous control and inspection procedures are necessary to make sure that the isolation valves are not left closed any longer than necessary.

Installing a grossly oversized relief valve may not make for increased safety as the relief valve may chatter, that is, repeatedly and rapidly open and close: such vibration has caused leaks.

Relief valves are usually tested regularly and correctly maintained (but see *WWW*, § 10.4.5). Drain holes in tail pipes are often overlooked; they should be checked regularly to make sure they are clear.

A common failing is not recognizing that an open vent is a relief valve and should be treated like one. Its size should not be changed unless we

Figure 65 Relief devices

have gone through the same procedure as we would go through before changing the size of a relief valve. It should be inspected regularly to make sure that it is clear and that its size has not been changed. *LFA*, Chapter 7 describes how a vessel was ruptured and two men killed because the size of a vent was changed without authority. There was no procedure for regular inspection and the vent was often choked. (See Figure 65.)

Relief devices should not discharge to atmosphere unless the material discharged is harmless (for example, steam or compressed air), the probability that a discharge will occur is low (for example, because a trip system has been installed to isolate the source of pressure before the relief valve lifts and it will lift only when the trip fails) or the quantity discharged is small. At one time flammable gases such as ethylene and propylene were often discharged to atmosphere when the exit velocity was high enough to produce jet mixing. Although a flammable cloud cannot be formed, and this practice is not unsafe, it is now discouraged or forbidden for environmental reasons.

If a relief device discharges into a closed system, such as a flare system, catchpot or scrubbing system, then the pressure drop in the connecting lines should be taken into account. There should be no dips in which liquid can accumulate.

See *boilers, tests* and *useless equipment and procedures*.

1. J S Fitt, *Loss Prevention and Safety Promotion in the Process Industries*, Elsevier, Amsterdam, 1974, p. 317
2. Reports available from the American Institute of Chemical Engineers

Remedies

One of the pioneers of loss prevention, the late Bill Doyle, used to say that for every complex problem there is a simple, plausible, wrong solution. Here are some of the simple, plausible, wrong solutions suggested to prevent accidents:

Money

Open a safety magazine and someone will be complaining that the money allocated to safety is insufficient. It is true that safety by design should be our aim. Whenever possible we should design our plants so as to remove opportunities for accidents, but unfortunately this is not always possible. For about half the accidents that occur there is no *'reasonably practicable'* way of modifying the hardware; all we can do is to change the software, that is, the method of working, training, inspections and so on. We cannot spend our way out of every problem[1].

Some *safety professionals* waste their companies' money on the obvious ways of controlling a hazard when a little thought, and involvement in the early stages of a project, would produce ways of avoiding the hazard. See *inherently safer design*.

More Factory Inspectors

Another favourite remedy of safety magazines. Too many inspectors can be counterproductive. They may give the impression that the *identification of hazards* and deciding what action should be taken can be left to them. In fact, those who produce the hazards know more about them than anyone else and it is their responsibility to remove or control them. The inspector is there to monitor and advise and, occasionally, to get tough with those who are neglecting their responsibilities, not to act as a factory safety officer.

More care

At one time most accident reports told the injured man to take more care and in many companies they still do. However, it is now widely realized that there is little we can do to prevent people making occasional mistakes and that we should therefore design our plants and methods of working so that there are fewer opportunities for error or so that those errors that do occur do not result in accidents. See *friendly plants, human failing* and *EVHE*.

Writing about the extinct Maya civilization of Central America, Pendergast asks what we can learn from their fate[2]:

> . . .it may be that a society that seizes upon a single technique to solve all problems may well have made a fatal mistake . . . The Maya chose religion and building as the means of providing for all their wants and needs, and ensuring that their way of life would survive . . . They sought

no alternative means of providing what they required, and they perished, or at least their way of life disappeared.

See *philosophers' stone* and *prosecution*.

1. T A Kletz, *Plant/Operations Progress*, Vol. 3, No. 4, October 1984, p. 210
2. D M Pendergast in *Man in Nature*, edited by L D Levine, Royal Ontario Museum, Toronto, 1975, p. 111

Removing equipment or procedures

Equipment that is no longer used should be removed or the plant will resemble a scrap dump. But before you remove equipment, or stop using it, ask why it was put there.

Many years ago, as a young manager, I had a salutary experience. Any spillages in a pumphouse drained out into an open trench through pipes fitted with U-bends. The U-bends did not seem to fulfil any useful purpose, and rubbish collected in them, so I had them removed. One of the shift foremen was on holiday at the time. When he returned he told me that the U-bends had been installed about ten years earlier following a fatal accident. A leak of flammable gas had entered the pumphouse through the drain lines and had exploded. The U-bends had been installed to prevent this happening again. The other foremen either never knew about this explosion or had forgotten.

During the intervening ten years the pumphouse walls had been removed, it became no more than a roof supported on four poles (see *ventilation*), so I was right, the U-bends no longer served any purpose. But nevertheless the incident taught me a useful lesson: never remove anything unless you know why it was put there in the first place.

Similarly, we should never stop carrying out a procedure just because we do not know the reason for it. For example, many accidents have occurred in the chemical industry because *maintenance* workers opened up equipment which contained flammable or corrosive chemicals. Sometimes the equipment had not been properly isolated; sometimes they opened up the wrong equipment. After these incidents rules are written to say that equipment under repair must be isolated by locked valves; in addition, slip-plates should be inserted (unless the job is quick one) and a numbered label should be fixed to the equipment (unless it has a permanent number). (See *identification of equipment* and *isolation*.)

After a few years the reasons for these rules are forgotten and a yuppy in a hurry, keen to save time and reduce costs (both admirable things to do) asks why we are following these time-consuming procedures. No one can remember, and the procedures are scrapped. A few years later the accident happens again.

What can we do to keep memories alive? I suggest that we should:

• Include in our rules the reasons for them and remind people of the accidents that happened before we had them. Most rules are like the ten commandments. They tell us what to do but not why.

- Remind people of old accidents in our safety bulletins and by regular talks and discussions. Looking at slides of past accidents and talking about them is much more effective than reading a book of rules.
- Keep a *black book*.

Reorganization

Reorganization can result, and has resulted, in safety prodedures being discontinued, not because someone took a considered decision to stop them but because they were overlooked when the new organization was set up and no one was made responsible for them.

For example, some years ago, when I was in industry, *hazard and operability studies* (hazops) were carried out by a management services department. When this department was disbanded the hazop expert was transferred to another department that knew nothing about them, was not interested in them and wanted to transfer the expert to other work. I had to do a lot of lobbying to prevent hazops, generally agreed to be a useful technique, coming to an end.

On another occasion, the job of shift general foreman in a works was abolished. It was thought that all their duties had been allocated to others but after a minor explosion had occurred in the *drains* it was discovered that one of their duties, never written down, was to monitor the drains for flammable vapour which, incidentally, can be done with a flammable gas detector.

Many companies have moved from a functional organization, in which there are departments for research, design, production, marketing etc., to a business-centred organization in which separate departments for each product (or group of products) carry out all (or most of) the functions. With this organization it is not always clear who is responsible for technological innovation. The business is managed but is the technology? To overcome this problem in *ICI*, 'We have a process by which scientists can talk about science strategy which cuts across business boundaries. We have seven science groups covering the whole of ICI's scientific ambitions. Each of these scientific areas might serve up to five or six different business strategies[1]'.

See *amalgamation* and *innovation*.

1. C Reece, quoted by R Stevenson, *Chemistry in Britain*, Vol. 22, No. 8, August 1986, p. 695

Repeated accidents

Early societies saw history as cyclical, like the seasons. Everything comes round again in time.

The Israelites were the first society, the Bible the first book, to see history as linear, as progress towards a goal, and in the West we have come to expect this as normal, or at least desirable. The cyclical view, however, must correspond to something deep in human experience or we would not have believed in it for so long.

In safety and loss prevention, although there has undoubtedly been much progress, old problems come round again and old views reappear. Some examples are described in the item on *lost knowledge*. Other accidents that keep recurring involve *entry* to vessels, asphyxiation by *nitrogen*, fires in oil-soaked *insulation*, poor *ventilation* (because flammable gases are handled in closed buildings), water hammer (because the *steam* traps on steam mains are insufficient or not in working order) and *instrument* failure or neglect.

Some old views that keep reappearing are described under *human failing* and *sleeping beauties*.

Old accident reports are, to quote the title of a book by Barbara Tuchman, 'A Distant Mirror', reflecting the same lessons as those of today. (See *bible*.)

Research

There are two ways in which research should be discussed and both are covered here

Research in safety

Although there is much we would like to know, lack of knowledge is not a major problem. Accidents do not occur because we do not know how to prevent them but because we lack the necessary energy, drive and commitment. We know what to do – or someone does if we do not – but we lack the will to get on with it. On several occasions I have made myself unpopular, at a meeting called to discuss proposals for research into an aspect of safety, by asking, 'How will you persuade people to use the new knowledge you are going to give them when they are not using the knowledge they have already?' I do not wish to discourage research in safety but I suggest that an attempt is made to answer this question. If specific detailed knowledge is required by a designer, say, on the yield strength of a new grade of steel, the knowledge is likely to be used. I am less sure about the results of research on, say, human behaviour.

Howard[1] suggests that research is needed in the following areas:

- Testing of short-cut design procedures for vent sizing for runaway *reactions*.
- Vent sizing for gas *explosions* when the initial pressure is above atmospheric and when the gas is turbulent.
- Short-term variations in concentration in gas clouds.
- Better methods for measuring pH.
- Evaluation of research on unconfined vapour cloud explosions.

Roberts[2] has summarized the Health and Safety Executive's research programme.

Safety in research

Laboratory safety is discussed in more detail by the Royal Society of Chemistry[3,4].

In experimental and other small-scale plants we should aim for the same standard of safety as on a full-scale plant but the methods we use to achieve that standard may be different. The reasons are discussed under *scale-up*.

When carrying out research, especially on large units, we should list possible outcomes and decide on the the action to take in each case. See *Chernobyl*.

1. W B Howard, *Chemical Engineering Progress*, Vol. 84, No. 9, September 1988, p. 25
2. A F Roberts, *Chemical Engineering Research and Design*, Vol. 65, No. 4, July 1987, p. 291
3. L Bretherick, *Hazards in the Chemical Laboratory*, Royal Society of Chemistry, 4th edition, 1987 (618 pages)
4. *Guide to Safe Practices in Chemical Laboratories*, Royal Society of Chemistry, 1987 (40 pages)

Responsibility

See *employers' responsibility*.

(legal) Responsibility

Under the *Health and Safety at Work Act* individuals can be prosecuted as well as companies (see *blind-eyes*) but *prosecutions* of individuals are rare; there are only a handful per year. Official reports rarely criticize individuals (but see *Zeebrugge*). In the past courts of enquiry were more willing to blame individuals than they are today. For example, the report on the collapse of the *Tay Bridge* in 1879 said, 'We find that the bridge was badly designed, badly constructed and badly maintained and that its downfall was due to inherent defects in the structure . . . For these defects in design, construction and maintenance Sir Thomas Bouch is in our opinion mainly to blame'[1].

The report of an enquiry into a shipyard explosion in 1929 said, 'We hold the Manchester Steam Users Association (the insurance company) primarily to blame for the explosion . . . With regard to Mr Petrie, DSc, Chief Engineer to the Association, it is difficult to know what to say. We cannot relieve him of responsibility because he has undertaken to carry out duties which he is wholly incompetent to fulfil . . . The explosion . . . caused the death of two men whose lives would have been saved but for the gross neglect of the Association and the two officers to whom we have referred'[1].

See *knowledge of what we don't know* and *law*.

1. Quoted by R Booth, inaugural lecture, University of Aston, 1979

(managerial) Responsibility

The first lost-time accident in which I was involved, nearly 40 years ago, as a young manager, taught me several lessons.

A *research* worker wanted to attach an *instrument* to a branch on a caustic soda pipeline to carry out some measurements. The foreman and I thought that the instrument looked too fragile to withstand the *pressure* but the research worker assured us that he had used it before. I let him go ahead but decided to watch, with the foreman, from a safe distance. The instrument leaked, spraying caustic soda over a visiting workman who, like the postman in one of G K Chesterton's Father Brown stories, had not been noticed by the foreman or myself.

The most important lesson I learnt from this incident is that a manager should not be afraid to back his own judgement against the expert, especially when, as in this case, the expert says it is safe and the manager is doubtful. Obviously expert advice should not be disregarded lightly, especially when the reputation of the expert is high. But one of the skills a manager has to acquire before he earns his spurs is the ability to assess the mass of expert and often conflicting advice that he constantly receives.

A manager is responsible for everything that goes on on his plant. He cannot say, 'It is not my fault; the expert said it was OK'. It is the responsibility of the manager to decide whether or not to accept the expert's advice and to specify suitable precautions.

The official report on the accident tended to blame the expert but I felt responsible.

Just as a manager cannot hide behind an expert so he cannot hide behind his boss. Napoleon said[1]:

A commander-in-chief cannot take as an excuse for his mistakes in warfare an order given by his minister or sovereign, when the person giving the order is absent from the field of operations and is imperfectly aware or wholly unaware of the latest state of affairs. It follows that any commander-in-chief who undertakes to carry out a plan which he considers defective is at fault; he must put forward his reasons, insist on the plan being changed, and finally tender his resignation rather than be the instrument of his army's downfall.

See *Summerland*.

Other lessons that come out from my first lost-time accident are:

- All *modifications*, including temporary ones, should be systematically assessed, using a formal procedure, before they are authorized. In this case the procedure should have made me ask:
  - What is the pressure in the pipeline?
  - What pressure will the instrument withstand? How do we know? Has it been tested? When and by whom?
  - What is the material of construction? Will it withstand caustic soda of the concentration used?
  - If the instrument fails, will it do so suddenly or will it leak before failure?

- If we wish to restrict access to an area it should be roped off. It is not sufficient to look out for anyone approaching. People are part of the scene and it is easy not to notice them.

Fortunately, the workman's injuries were not serious and he was absent from work for only two or three days. But the incident showed me that there are lessons to be learnt from all accidents, not just the really serious ones. See *dangerous occurrences*.

For the responsibilities of *safety professionals*, see that item.

1. Napoleon, *Military Maxims and Thoughts*

(personal) Responsibility

A former colleague of mine had a frightening experience as a young engineer. Soon after joining his first company he was asked to inspect the inside of a large pipe, 1.5 m diameter. He had to wear breathing apparatus supplied from the compressed *air* main. He had walked 60 m down the pipe when his face mask started to fill with water. Fortunately he stayed calm, pulled off his mask, held his breath and walked quickly out of the pipe.

He then discovered that the breathing apparatus has been connected to the bottom of the compressed air main instead of the top so that a slug of water in the main had entered the breathing apparatus. (See Figure 58, p. 221.)

As a young engineer with little experience he had assumed that the 'safety people' and the company procedures would take care of his safety; he did not carry out any checks himself or ask any questions and he entered the pipe without hesitation.

It is, of course, the responsibilty of the process team, before issuing an *entry* permit, to make sure that the job has been properly prepared and that all necessary precautions have been taken. Nevertheless, the men who are going to enter a vessel should be encouraged to carry out their own checks. If they do so, they are not being awkward; they are making a real contribution to safety.

Today, compressed air for breathing apparatus is usually supplied from *cylinders*, not the mains.

The company my colleague worked for normally had high standards but even in the best organizations things can go wrong from time to time. The men whose lives are at risk therefore have the right, and should be encouraged, to carry out their own checks, not to decrease the responsibility of those primarily responsible but to provide some additional or bonus safety.

Similarly, before opening up equipment, *maintenance* workers should be encouraged to check that it is isolated and drained. Some maintenance workers start off making their own checks but after a few years, if they have never found anything wrong, they stop doing so. They should remember that once in a lifetime is too often to open up equipment that is full of chemicals. (See *isolation*.)

If maintenance men check the precautions, is there a danger that those who prepare the equipment will make less care? I do not think so, as they will not want to be caught out.

Increasingly, we are designing *friendly plants* which contain fewer opportunities for *human failing* or can withstand its effects. Instead, should we not expect people to take more care, to accept more responsibility?

At the personal level we should try to do our best, follow the rules, never take short cuts. However, at the social level we must expect people to continue to make the sort of mistakes that they have made in the past. When we are dealing with hazardous materials only a very low level of error can be tolerated, a level so low that very often it is unrealistic to expect people to achieve it. We should therefore design our plants so that they are tolerant of human error. To quote B Inglis: 'Personal responsibility is a noble ideal, a necessary individual aim, but it is no use basing social expectations on it; they will prove to be illusions[1].'

1. B Inglis, *Private Conscience – Public Morality*, 1964, p. 138

Retribution

At one time retribution used to be taken against equipment as well as people. In his *Annual Report for 1971*, the UK Chief Inspector of Factories quoted the following description of a fatal accident which occurred in 1540[1].

A young childe . . standinge neere to the whele of a horse myll . . . was by some myshap come within the swepe or compass of the cogge whele and therewith was torne in peces and killed. And, upon inquisition taken, it was founde that the whele was the cause of the childes dethe, whereupon the myll was forthwith defaced and pulled down.

The Chief Inspector added that many no doubt feel that this sort of decisive action against bad conditions in industry should be widely applied today!

Factory Inspectors have the power to seize hazardous equipment, but the power is rarely used. They would have a problem dealing with most of the equipment used by the process industries.

See *blame*.

1. UK Chief Inspector of Factories, *Annual Report for 1971*, HMSO, London, 1972, p. xv

Reverse flow

Of all the deviations from design intention considered during a *hazard and operability study* reverse flow seems to be the one that is most often overlooked. Many designers fail to foresee that gas or liquid may travel along a pipeline in the opposite direction to that intended. Spectacular incidents have occurred when the contents of a *reactor* have travelled back up a feed line into one of the raw material storage tanks and reacted explosively with the raw material. A very high reliability *protective system*, far more reliable than a *non-return* (check) *valve*, may be needed to prevent this occurring; alternatively a break tank can be used.

Reverse flow has often occurred from process equipment into *service lines*, from *drains* into process equipment or onto the ground and from blowdown lines into process equipment that has been depressurized for *maintenance*. Pumps have been damaged by reverse flow.

See *WWW*, Chapter 18 and reference 1.

Sometimes our thoughts flow the wrong way. Instead of looking for the evidence and then forming opinions, we often form opinions and then look for the evidence.

However, ideas are sometimes more acceptable if they are put in reverse. Agnostics who will not accept that God is good may agree that goodness is godly. Those who react to the phrase, 'humans make mistakes' by saying they should not, may accept that 'mistakes are human', that is, are inevitable when humans are involved.

See *(better) late than never*.

1. T A Kletz, *Hydrocarbon Processing*, Vol. 55, No. 3, March 1976, p. 187

Risk

See *aversion to risk, hazard and risk* and *perception of risk*.

Risk compensation

I advocate safety by *design* (see *human failing* and *friendly plants*) but it has its limitations. If something is redesigned to make it safer, people may find a way to keep the level of safety the same and achieve another objective instead. This is called risk compensation. For example:

- If roads are straightened to remove dangerous bends, drivers may compensate by going faster.
- If chainsaws are made safer, do-it-yourself enthusiasts may start using them.
- If a *tank* is fitted with a high level *alarm*, the operator may stop watching the level and wait until the alarm sounds. After a year or two the alarm will fail and the tank will overflow.

When people compensate for reduced risks in this way, they are saying that they are satisfied with the existing level of risk and would rather have something else in place of reduced risk[1]. Up to a point, this is reasonable. We do not want absolute safety, at work or anywhere else. Unfortunately, most people's *perception of risk* is poor; they do not know which risks are large and which are small. See *false alarms*.

When safety is obtained by adding on *protective equipment*, risk compensation is often easy; we can switch off the protective equipment or put too much reliance on it (as in the example quoted). *Friendly plants* and *inherently safer designs* make risk compensation less easy.

1. A R Hale and A I Glendon, *Individual Behaviour and the Control of Risk*, Elsevier, Amsterdam, 1987

Risk criteria

See *acceptable risk*.

Risk management

Risk management in its widest sense includes activities such as:

- Avoiding the risk (see *friendly plants*).
- Controlling the risk, by adding on *protective equipment* or by *methods of working*.
- Protecting people from the effects. (See *layout and location*.)
- Controlling the effects by fire-fighting, emergency planning etc.
- Cleaning up after an accident.
- Taking out *insurance* against the effects of the risk.

In practice, the term risk management usually describes an activity in which the last item is emphasized. If a company has a risk management department it is probably concerned mainly with insurance (not just placing the insurance but deciding when to insure and when not to) while the other activities listed above are the concern of a (process) safety or *loss prevention* department.

Risk perception

See *perception of risk*.

Rotameters

See *level glasses*.

Rules and regulations

If I follow all the rules and regulations, will I have a safe plant? The answer is not necessarily, as the rules may be out of date.

Some years ago a hydraulic *crane* collapsed on to a petrochemical plant, fortunately without causing any serious damage. (See Figure 66.) It collapsed because the driver was trying to lift too much weight for for the particular length and inclination of the jib, but the alarm bell did not sound. The crane was, however, fitted with all the alarms required by the then current *code of practice* and they were all in working order.

A mechanical strut crane has three degrees of freedom. The angle of the jib can be altered, the jib can be rotated and the load can be raised and lowered. An alarm sounds if the driver tries to lift too big a weight for the particular jib angle. A hydraulic crane has an extra degree of freedom. The length of the jib can be altered.

In the incident I have described the driver set the angle of the jib and then extended the jib to its maximum length. The crane was then so unstable that it tipped up as soon as a small weight was lifted. Because the hydraulic crane has an extra degree of freedom, it needs an extra alarm, but it was not fitted with one, only with the minimum number of alarms

282 Runaway reactions

Figure 66 Rules and regulations

required by the code of practice which had been written before hydraulic cranes came into general use.

So do not relax because you are following the rules or codes; they may be out of date. A horse and an 'iron-horse' need different codes.

If you have new knowledge, not previously available, then you may have to take additional precautions, whatever the rules say, or you may not have a safe plant and system of work, as required by the *Health and Safety at Work Act*.

See *breaking the rules, CIMAH Regulations, COSHH Regulations* and *instructions*.

Runaway reactions

See *reactions*.

Sabotage

See *arson* and *security*.

Safety

Safety, from the Latin *salvus*, meaning uninjured, entire, healthy, is defined by the *Shorter Oxford Dictionary* as 'the state of being safe, that is, unhurt, uninjured, unharmed; having escaped some real or apprehended danger'.

In everyday life the word often suggests a negative, almost cowardly, approach to life, as in such phrases as 'play for safety', 'be on the safe side' and *'safety first'*. Most people instinctively prefer the philosophy of 'nothing ventured, nothing gained' (see *manliness*) or that of James Graham:

> He either fears his fate too much,
> or his deserts are small;
> That puts it not unto the touch,
> to win or lose it all.

As children we were stirred by tales of derring-do and our heroes and adventurers did not pause to weigh up the hazards or to take out a permit-to-work. In fact it is not like this in real life. Successful adventurers accept risk only for good reason and plan carefully to avoid accidents. See, for example, *the Ascent of Everest* by Sir John Hunt[1] (it reads more like an *ICI* report than a boy's adventure story) or the words of Admiral Byrd:

> In all my travels and adventures in the interest of science and discovery I have never taken an unnecessary risk. Only the best and safest equipment was selected for planes and ships. Everything was safe that could be made safe.
> By careful planning and by taking no unnecessary chances my men and I have lived to enjoy the hazards and thrills of adventure and discovery. We found adventure only by planning for safety as far as possible[2].

Sadly, it now seems that, in contrast, the arctic explorer, Captain Scott, was less prudent. His expedition to the South Pole in 1912 would not have failed if he had left more supplies in his depots, taken more than one spare cylinder for his motor sleds, trained the zoologist in the party to navigate, or not stopped to collect rocks when time was running out. 'Prudence dictated that Scott plan for a very wide margin of safety. He left none and thereby killed not only himself but four others[3].'

What can we do in industry to overcome the negative image of safety?

- Try to make it clear that safety is never the main business of an organization. Our aim is to make goods, provide services, carry out research. A ship is safest when it is in harbour, but that is not what ships are for.
- Try to make it clear that risk taking is OK if the gain is worth the risk. We go rock climbing or sailing because we think that the pleasure is worth the risk. We take jobs as airline pilots or soldiers or become

missionaries among cannibals because we think that the pay or the interest of the job or the benefit to others makes the risk worthwhile. However, in industry the risks that people take are almost always quite out of proportion to the gain. To save a little time we fail to isolate equipment for maintenance and risk that someone will be injured, or we fail to put on our goggles and get chemicals in our eyes.

- Some companies try to avoid the word 'safety', because it makes people switch off, and instead they talk about *accident* prevention and *hazard* reduction. They have a loss control, *loss prevention* or *risk management* department (or even a regulatory affairs department – see *law*) instead of a safety department; I doubt if this fools anyone. If safety has the wrong image, let us try and give it a better image. If the *safety professionals* in the safety department give useful and balanced advice and are seen to be competent people then the image of safety will improve. (Of course, if *loss prevention* and *risk management* are really being practised then it is fair to use these terms.)

1. Sir John Hunt, *The Ascent of Everest*, Hodder and Stoughton, London, 1953
2. Admiral Byrd, quoted in *Family Safety*, Summer 1972
3. J Diamond, *Discover*, Vol. 10, No. 4, April 1989, p. 73

'Safety first'

The origins of 'Safety First' were described in *Family Safety*, Summer 1972:

> The early slogan 'Safety First' which made safety sound unattractive because it seemed to put safety ahead of every other consideration in human affairs, grew from an unfortunate circumstance.
>
> At the turn of the century, a man named Henry C Frick, president of the Henry C Frick Coke Company, inaugurated the country's first industry-wide safety campaign with the slogan, 'Safety First, Quality Second, Cost Third'. 'Safety First' was lifted out of context and widely used by other companies. No longer was safety 'first' only to quality and cost; the slogan seemed to put it ahead of everything.
>
> The safety movement never advocated safety as the number one objective in life. It never asked Americans to live by a timid and faint-hearted code. It never opposed adventure, because adventure and discovery are the very essence of life.

I would go further. Safety cannot always come before cost. If we spend every penny necessary to remove every conceivable hazard, however trivial or unlikely, our companies will go bankrupt, we will be out of a job and the public will be deprived of the products we make. In the UK the law does not ask us to remove every conceivable hazard, only to do what is *'reasonably practicable'*. We have to decide how far to go and the methods of *hazard analysis* can help us.

Safety professionals

I have used this term for people who are variously described as safety officers, safety engineers, safety advisers, loss prevention advisers and so on. The item on the *old-timer* describes the safety professional of the past. Today, especially in the high technology industries, he needs to be:

- A technical man, with experience of design and/or operations in the industry in which he advises, or others will not listen to him.
- A numerate man, able to take a balanced view, balancing the needs of the business with the requirements of humanity and able to put hazards in order of priority. See *hazard analysis*.
- A communicator, able to give advice which his audience would rather not hear; good-natured persistence is needed. See *persuasion*.
- A monitor, surveying his parish on the lookout for procedures that have lapsed, equipment which is not being properly maintained and so on. See *audits* and *magic charm*[1].

The safety professional is an adviser – ultimate responsibility should remain with the line manager; see *(managerial) responsibility* – and advisers differ in the nature of their contribution. Some provide information on which the line manager can base his decisions; their influence is remote. Others display the alternatives; they contribute to the decision. Others say what they think should be done and share the responsibility. Safety professionals should belong to this third group and make clear recommendation for action. Only then can they expect to receive the same salary as those they advise. Compare Mrs Thatcher's remark about about Lord Young: 'Other people bring me problems. David brings me solutions.'
See *actions* and *time and money*.

1. *Safety and Loss Prevention*, Proceedings of the Jubilee Symposium, Institution of Chemical Engineers, Rugby, 1982, p. B1

Sampling

Sampling is a common cause of accidents in the chemical industry so it is worth listing the precautions necessary when sampling hazardous liquids:

- Do not take more samples than necessary. (Is on-line analysis possible?)
- Wear goggles and gloves.
- Do not hold the sample bottle in the hand but use a stand. For sampling hot liquids or corrosive liquids, such as strong acids, the stand should be enclosed.
- Do not stand close to the sample point (unless it is enclosed). The valve should be located so that the sampler has to stand at a safe distance. See Figure 67.
- Provide good access.
- Do not carry sample bottles in the hand; use a carrier. For small samples of safe materials, such as oils below their flash points, a plastic carrier of the type used for milk bottles and soft drinks may be used but a closed carrier should be used for hot, highly flammable or corrosive liquids. See Figure 68.

The need for these precautions is shown by the following incident. While an operator was sampling a molten salt it suddenly spurted out and in

Figure 67 Sampling

Never carry bottles in the hand
Use a sample carrier

For corrosive liquids the bottles should be fully enclosed

A winchester containing oleum slipped from the employee's grasp. The winchester smashed on the concrete floor causing the acid to splash on the operator's leg.

Figure 68 Sampling

jumping back to avoid being splashed he pulled the valve fully open. The flow increased and splashed him on the face and left arm. He tried to escape, bumped into a post on which a pump starter was mounted and fell over. His helmet and goggles came off and he was splashed on the head, neck, shoulder, back and face[1].

Although the report blamed the operator's inexperience, the post was moved to provide better access, the valve was moved so that the operator was more likely to close it than open it, if he moved suddenly, and samples were taken less often, only when needed, instead of routinely.

1. *Petroleum Review*, July 1983, p. 26

Scale-up

Scale-up – increase in plant size from an experimental unit to a full-scale one or from a small plant to a larger one – can introduce hazards. The best known is the effect of size. If the volume of a piece of equipment is increased eight times, and the shape is the same, the surface area is increased only four times. Heat loss will be less and extra cooling may be necessary. Heat loss to the walls as well as through the walls may be important. Other effects of scale-up are:

- On a small unit the inventory may be so small that it does not matter if it all leaks out but on a larger unit it may be sufficient to cause a serious fire, explosion or toxic incident. See *change – new problems*.
- On an experimental unit it may be possible to disperse leaks by good artificial *ventilation* or to protect against the effects of an *explosion* by blast walls. On full-scale plants the former is difficult and the latter is usually impossible.
- Experimental plants are usually supervised by more highly qualified people than full-scale plants.
- It may be uneconomical to add a complex instrumented *protective system* to an experimental plant, and the information needed to design the system may not be available. The experimental plant may have been built to acquire that knowledge. 'If we knew what we were doing, it would not be *research*[1]'.

1. *Chemical Technology*, March 1980

Scrap

Many companies allow employees to take scrap material home, either free or for a nominal charge. Good, but take care that you do not take home, or allow anyone else to take home, equipment which is so old or damaged that it could be dangerous.

An electrician went to a friend's home to help him with a wiring job in his workshop. When he was there he happened to see a bench grinder fitted with a grinding wheel that had broken in two and been repaired with two metal strips and four nuts and bolts. The friend had found the wheel, in two pieces, in a waste bin and had 'repaired' it himself. The grinder was all set and ready for use. The electrician managed to persuade his friend that he should put the wheel back where it ought to be – in a waste bin[1].

In contrast, at lunch one day, when I worked in industry, I heard an accountant ask an engineer if he could let him have a length of scrap rope, so that he could make a swing for his daughter. A week later the accountant reminded the engineer who said that, having thought about it, he would not provide the rope. He could find a length of scrap rope that looked OK but he would not know its history; he would not know if anything had been spilt on it. He told the accountant to buy a length of new rope.

Laboratory workers are often asked to oblige people by supplying small quantities of chemicals. There is obviously no harm in supplying a drop of demineralized water for topping up a battery, but where does it stop? A drop of solvent for getting your lunch off your tie before you see a customer? Probably OK, but watch that you do not end up supplying corrosive chemicals for clearing a blocked drain, or worse, explosives for removing a tree stump[2]. If you are senior to the laboratory staff do not embarrass then by asking them to supply chemicals that they should not let out of the laboratory. Do not ask someone to decant some weedkiller from a drum into a plain bottle or, worse still, an empty drink bottle. Many people have been poisoned because they mistook weedkiller for a soft drink.

Scrap should be freed from hazardous materials before it is put into a scrap bin or scrap bay. (See Figure 69.) It is not necessary to clean it as thoroughly as equipment sent for repair but gross quantities of hazardous materials should be removed.

1. *Safety Management (South Africa)*, September 1987, p. 61
2. M J Pitt, *The Chemical Engineer*, No. 442, November 1987, p. 38

Secrecy

See *publication*.

Security

Unauthorized visitors, or authorized visitors who wander into the wrong place, can be a hazard to themselves and others. Pitt[1] suggests the following:

- Walk round the outside of the works from time to time, preferably on a Sunday or at night, to see how easily someone might gain entry.
- Follow the routes taken by visitors to see if they are easy to follow and if the ignorant or foolish could stray into hazardous areas. Could a dishonest person slip something into his car or lorry?
- Lay out the site, with barriers if necessary, so that people are not tempted to take short cuts through hazardous areas.

Figure 69 Scrap

- Ask yourself what a disgruntled employee could do to cause maximum damage. While we cannot stop the determined saboteur we should not, for example, leave drums of acids and alkalis close together where an ignorant, negligent or disgruntled employee, or someone who is merely tired or under *stress*, can easily confuse them.

  See *arson*.

1. M J Pitt, *The Chemical Engineer*, No. 442, November 1987, p. 38

Selling safety

The difficulties encounted when selling safety can be illustrated by similar situations elsewhere:

> It's hard enough to sell someone a cemetary lot – it's what we call an intangible, like insurance . . . whatever your approach, you've got to make sure your product is perfect, without a flaw. The minute there's something wrong with your product and your prospect knows it, he grabs onto it and uses it against you[1].

Safety is also an intangible. People are reluctant to buy cemetary plots, insurance or safety because they remind them of unpleasant events, such as death, accidents and burglaries, that they would rather not think about. So, if we are going to sell safety the product has got to be good.

In addition, there are two golden rules that salesmen use to sell insurance and other intangibles[2]:

1. Make the intangible as tangible as possible. Don't try to sell a policy or technique; try to sell fewer accidents or fewer losses. Don't describe standards; take your prospect down to the plant and point out the faults on the ground. Show him pictures of fires or explosions on similar plants and point out on his plant the features that led to the fires or explosions.
2. Identify with the buyer. Try to give the impression that you have a similar background and had similar problems when you were in design or production. Don't come across as a head office smoothie out of touch with practical problems or, at the other extreme, as a country bumpkin, ignorant of the theory. *Safety professionals* who have never worked in the industry as managers or designers may have a problem; however good their advice, they lack credibility.

See *perception of risk*.

1. H Kemmelman, *Saturday the Rabbi went Hungry*, Penguin Books, 1969
2. B Matthysen, *Safety Management* (South Africa), December 1986, p. 7

Serendipity

Serendipity is named after the three princes of Serendip (Ceylon) who 'were aways making discoveries, by accidents and sagacity, of things they were not in quest of' (Horace Walpole). This can occur, in safety and loss prevention, as in everything else. Sometimes we have a problem and after thinking about it we find an answer. At other times we stumble across an answer to a problem we had not even recognized.

Many years ago, when I was an assistant works manager, not a safety adviser, I was looking through a very dull book on industrial safety by one of the United Nations organizations. I came across the suggestion that hazards might be photographed and shown to the staff. Everything else in the book was a *platitude*. On my next tour of the works I took my camera with me and photographed all the hazards I saw and then showed the slides at the next monthly meeting of all the managerial staff. They were shocked to see hazards they had passed every day without noticing. Of course, this technique shows up the physical hazards rather than errors in procedures or failures to follow them (though I did photograph untidy and neglected instructions).

Here a few more of the things seen and photographed on my first such tour:

- Wood fixed to steam line by *insulation*
- Open sample cock
- Cock handle too close to insulation
- Vent on condensate line pointing over structure
- Water from drain hole on shower falling on to stairway
- Open vent on *tank* containing flammable vapour
- Unsupported sample point
- Staging without toe-boards

- Old food tin containing lubricating oil
- Damaged *hose*
- Hole in wall of tank compound
- Broken handrail
- Empty goggles box
- Hoses left on floor forming tripping hazard
- One end of pipe supported by string and the other end by a lashing
- Open gate on accumulator compound
- Unregistered ladder
- Broken glass
- Loose pipe at head height
- Oil-soaked insulation
- Spillage of chemical
- Liquid overflowing from open end of pipe through leaking valve.

A colleague, a safety adviser on another works, was less successful with the technique. When he showed his slides the works manager, who was present, was so shocked that he lost his temper and played hell with the staff concerned. The safety adviser was not popular.

I nevertheless, recommend the use of a camera in this way (though some people prefer to use a Polaroid type and put the prints on the manager's or foreman's desk). There is no need to read the riot act; the photographs speak loudly enough.

Service lines

Contamination of service lines by process materials is common. A service (*steam*, compressed *air, nitrogen* or *water*) is connected to a process line or vessel, the service pressure falls (or the process pressure rises) and the process material enters the service line. I have known leaks of nitrogen and water to catch fire; a steam lance to feed a small oil fire instead of extinguish it; compressed air to set solid and even steam to freeze.

There should always be a *non-return* (check) *valve* in the connection to a service line, but these valves are not totally reliable so, if the service is used occasionally, it should be connected by a *hose* which is disconnected when not in use, or by double block and bleed valves. If the service is used for much of the time it may be permanently connected but there should be a low pressure *alarm* on the service line (or a high pressure alarm on the process line if the process pressure is liable to rise). In addition, we should always test service lines before welding on or near them. See Figure 70. On several occasions maintenance workers have been hurt while working on untested service lines; for example:

- A welder was killed when the water tank he was repairing exploded. The water was used for cooling a stream of flammable gas, there was a leak in the cooler and some gas entered the water lines.
- A fitter was affected by fumes while working on a steam drum. Live steam from the drum was injected into a distillation column which operated at a gauge pressure of 2 bar (30 p.s.i.). A valve in the steam line was closed but the valve was leaking and the line was not

An empty water line caught fire while a welder was working on it!

Test the **INSIDE** of water, air, steam, and nitrogen lines before welding on them in a plant area.

Figure 70 Service lines

slip-plated. When the pressure in the steam drum was blown off, vapour from the distillation column entered the drum. Slip-plates were normally used to isolate equipment under repair but one was not fitted on this occasion as it was 'only a steam line'.

See *drains, hazardous substances, temporary jobs* and *Three Mile Island*.

Seveso

Seveso is a village near Milan, Italy, where, in 1976, a discharge from a reactor bursting disc contaminated the neighbourhood with *dioxin*. No one was killed but 250 people developed chloracne, a skin disease, and the area was became uninhabitable. No accident, even *Bhopal*, has produced greater 'legislative fall-out'. It led to the enactment by the European Community of the Seveso Directive[1], which requires all companies which handle more than defined quantities of hazardous materials to demonstrate that they are capable of handling them safely. In the UK the Seveso Directive was implemented by the *CIMAH Regulations*.

The reactor at Seveso was heated by exhaust steam from a turbine and the steam was normally too cool to overheat the reactor. But the plant was shutting down for the weekend, the turbine was on reduced load, and the steam was hotter than usual. A *hazard and operability study* might have disclosed the hazard.

Italian law required the plant to shut down for the weekend, another example of well-meaning legislative interference producing a hazard. (See *Factories Acts*.)

There was no catchpot to collect the discharge from the bursting disc. It may have been omitted by the designers because they did not foresee that a runaway reaction might occur and they may have installed the bursting disc

to guard against other sources of over-pressure such as overfilling. However, there had been three runaways on earlier plants and it is not good practice to discharge any liquid to atmosphere.
See *LFA*, Chapter 9.

1. European Community: Council Directive of 24 June 1982 on the Major Accident Hazards of Certain Industrial Activities, *Official Journal of the European Communities, 1989*, No. L230, 5 August 1982, p. 1

Shipping

In popular imagination shipwrecks are due to bad weather, *fate*, and other *Acts of God*. In fact they can be prevented by better design and better management.
In 1959 2.8 ships in every thousand were lost; in 1979 it went up to 5.6. According to Douglas Foy[1] this worsening record is due to:

- Many more ships flying the flags of countries that are unwilling or unable to regulate them and leave the responsibility to classification societies that are ill-equipped to do so.
- An increase in the number of small owners who leave crew hire to agents.
- Larger ships without any corresponding improvement in manning or maintenance.
- Willingness of insurance companies to insure unseaworthy ships.
- Poor design, particularly of large ore carriers and roll-on/roll-off ferries. See *Zeebrugge*.

From the many examples given by Foy I quote only one: In 1979 two very large crude oil carriers, the *Atlantic Empress* and the *Aegean Captain*, collided in darkness and rain in the West Indies. The *Atlantic Empress* exploded and sank:

- The tanks on the *Empress* had not been inerted.
- The watch-keeping officers on both ships had failed to make proper use of their radars and had failed to follow the Collision Regulations.
- A lifeboat on the *Atlantic Empress* was mishandled and 26 people were drowned; the master had not held a boat drill during his 10 months on the ship.

Early in 1989 an oil tanker ran aground near the coast of Alaska, spilling crude oil. According to press reports, the captain was drunk and the third mate, who was on the bridge, was unqualified for Alaskan waters. This incident, and the others quoted, show that the shipping industry adopts standards of *management* quite different from those followed elsewhere. The attitude seems to be that a captain has to be left in sole charge of his ship and that auditing or managerial supervision is impracticable. This was true in the time of Captain Cook, but is not true today. It is as easy, or almost as easy, to *audit* a ship as a distant plant.
See *coincidences, lost knowledge, 'reasonable care'* and *sloppy thinking*.

1. D Foy, *Journal of the Royal Society of Arts*, Vol. 136, No. 5377, December 1987, p. 13

Short cuts

Some years ago, when mini skirts were popular, a newspaper asked some clergymen for their views; one of them replied that he could not comment as he never looked at girls' skirts and did not know how long they were.

This is similar to the attitude than many managers used to have (and some still have) towards permits-to-work (see *maintenance*), *entry* certificates and similar documents. They left them to the foremen, did not look at them and so did not know whether or not they covered everything that ought to be covered.

When an accident occurs and it is found that a short cut has been taken it is unlikely it was taken for the first time. It is more likely that short cutting has been going on for weeks or months, even years, not just by one person but by many. Regular inspections of procedures and the corresponding paperwork by managers could have shown up the irregularities and stopped them before the accident occurred. All managers should inspect a sample of permits-to-work every week. They should not just look at the paperwork but should inspect the job to make sure that the precautions detailed on the permit are being followed.

See *audits, blind-eyes* and *breaking the rules*.

Sight glasses

See *boilers* and *level glasses*.

Simple causes

On the subject of simple causes, Timpson says[1]:

> The chemical industry . . . continually comes up with accidents the causes of which are more the province of Dr Spock than of Dr Who.

Sometimes we look for highly technical *causes* of accidents and operating problems when the real causes are simple, so simple that we overlook them.

For example, a manufacturer had trouble making metal bellows of the quality required by a customer; the welds were too porous. The manufacturer checked the quality of the plate and welding rods used and analysed the shielding gas; all were OK. The customer's metallurgists were called in and got involved in discussions about the shape of the pores. Finally, they went to see the welding machine and discovered the simple cause. They found that the machine was protected only by thin plastic sheets supported on an angle-iron frame. The top was covered with dust, one side panel was missing and others were torn. On a windy day, when the workshop doors were opened, the plastic sheets flapped and dirt showered down. See *(visit accident) sites*.

Here are some more incidents with simple causes:

- A number of pump failures indicated excessive wear on the bearings. After various complex explanations had been considered, the trouble

was traced to the drums of lubricating oil which had been left standing in the open with the lids loose, so that rain water got in.
- During a plant shutdown a piece of wood was used as a temporary support for a new pipe. When the pipe was bolted in place it was hard to remove the wood, so it was left there. The pipe operated at 300°C and the wood caught fire.
- A reactor kept choking. Many explanations were dreamt up, but none stood up to test. The trouble was finally traced to the supply of hydrogen which was bought from the unit next door. They purified the hydrogen by passing it through charcoal. When a new supply of charcoal was needed the original grade was no longer available and a slightly finer grade was supplied instead. It passed through the holes in the support plate and was carried along to the reactor in the consuming unit.
- Aircraft have crashed because the hull corroded near the toilets, due to the well-known inability of men to aim straight.

For other incidents with simple causes, see *identification of equipment*.
An insurer once said that many of his clients could not be faulted in areas of highly complex technology – they did difficult things well – but they were vulnerable in more mundane areas.

1. T Timpson, *Acceptable Risk, Who says so?*, British Safety Council, London, 1978, p. 25

Simplicity

Engineering was once defined as 'the art of doing that well with one dollar, which any bungler can do with two after a fashion[1]'. Looking at today's complex designs this definition hardly seems correct; and complexity not only costs money but provides more opportunities for *human failing* and equipment failure. Simpler plants are cheaper and safer; they are *friendly plants*[2]. Before we can design simpler plants we need to consider the reasons for complexity:

- Our plants contain large quantities of hazardous materials that are kept under control by much added-on instrumentation and other protective equipment. *Inherently safer designs*, which contain lower inventories of hazardous materials, are therefore also simpler.
- Many operating and safety problems are not identified until late in design when it is too late to make fundamental changes and avoid them. All we can do is to add on protective equipment to control the problems. To design simpler plants (and inherently safer plants) we need to carry out safety studies much earlier in design, at the conceptual stage, when we are deciding what process to use, and at the flowsheet stage. We need systematic, structured studies similar to *hazard and operabililty studies*.
- *Codes of practice*, standards and *custom and practice* are often still followed when they are no longer appropriate.
- Desire for flexibility, particularly on multistream plants where cross-overs are often installed so that each item of equipment can be used on any stream.
- Lavish provision of installed spares.

- A feeling that it is OK to spend money on complexity but not on simplicity. Simpler plants are usually, but not always, cheaper than complex plants.
- In some cases there is an excessive aversion to risk. We may go too far in adding on equipment to guard against trivial or unlikely hazards. See *acceptable risk* and *'reasonably practicable'*.

See *stacks*

1. A M Wellington (1847-1895), *The Economic Theory of Railway Location: Introduction*, Wiley, 6th edition, 1900
2. T A Kletz, *Cheaper, Safer Plants – Notes on Inherently Safer and Simpler Plants*, Institution of Chemical Engineers, Rugby, 2nd edition, 1985

Single-stream plants

See *large plants*.

(visit accident) Sites

Many companies investigate all accidents and dangerous occurrences but sometimes the investigation takes place in the office and the manager in charge does not visit the site. Instead he relies on written or verbal reports from those who were there at the time. These can mislead, as the following incidents show:

- While repairing a flowmeter in a 30 inch (0.75 m) diameter pipe a man hurt his back climbing over the pipe. When the investigator visited the site he found that the man was expected to use a footbridge 50 m away. He had to cross from one side of the pipe to the other six times so he would have had to walk 600 m in total. Not surprisingly, he preferred to climb over the pipe. Most of us would have done the same.
- A man slipped on a staircase, twisted his ankle and was absent for three weeks. His boots and the stairs were in good condition; was it just one of those things we can do nothing about? A visit to the site showed that the handrails were covered with plastic so that anyone using them collected a charge of *static electricity* and got a mild electric shock when he touched the metal of the plant. The shock was not dangerous, but unpleasant, so people tended not to use the handrails.
- While running a temporary cable to a *construction* site an electrician stumbled on a piece of wire mesh, lost his balance and hurt his foot on a metal bar. His foreman told him to take more care while crossing ground used by contractors. A site visit showed that the ground was covered in weeds 2 feet tall which hid many tripping hazards. The junk had been there for years and had not been disturbed by the contractors. The ground should have been bulldozed before work started.
- A fitter hurt his hand while using a hand-operated machine for clamping *hoses* on to pipes. A visit to the workshop showed that the machine was mounted parallel to the bench instead of at right angles to it and this made it awkward to use.

These incidents show the importance of visiting the site of an accident. See also *(treating) symptoms instead of diseases*. They also show that something can be done to to prevent many simple, mechanical accidents which at first sight seem entirely due to the carelessness of the injured man. See *human failing* and *simple causes*.

The historian Barbara Tuchman has emphazied the importance of visiting historic sites before writing about them. 'On the terrain motives become clear, reasons and explanations and origins of things emerge that might otherwise have remained obscure. As a source of understanding, not to mention as a corrective for fixed ideas and mistaken notions, nothing is more valuable than knowing the scene in person[1].' (See also the quotation at the end of the item on *management*.)

1. Barbara Tuchman, *Practicing History*, Ballantine Books, New York, 1982, p. 61

Sleeping beauties

After I had given a talk to some of the staff of a large oil company, someone made a comment that used to be heard quite often. 'Most accidents', he said, 'are due to the carelessness or irresponsibility of the injured man, or one of his fellow-workers, to follow instructions'. He asked why I had said so much about the actions managers should take, thus implying that they were responsible and that they could prevent accidents.

When I went to the airport for my plane I felt I should check in for a journey on a time machine back to the 1980s.

My host explained that the questioner had recently returned to the UK after spending 20 years in an overseas refinery; he had been shielded from all the changes that had taken place during his absence.

Such people can be a menace if they come back to the UK on promotion, and take up a senior job.

See *human failing* and *old-timer*.

Slop-over

See *foam-over*.

Sloppy thinking

I use this phrase to describe the thoughts of people whose hearts are in the right place but whose proposals will never produce the results they want. It is not enough to want greater safety; we have to be reasonably sure that our actions are going to produce it and, unfortunately, sometimes they are not. Here are some examples.

1. Many people, understandably horrified by the slaughter on the roads, do not want to see bigger goods vehicles. But if vehicles are made bigger, we need fewer of them, and accidents should therefore be fewer.

2. Liquefied petroleum gas (LPG) is transported by road at atmospheric temperature and high pressure. If a *tanker* is involved in an accident and a *leak* occurs most of the contents will rapidly form a cloud of flammable vapour. Someone therefore suggested, a few years ago, that it should be transported refrigerated, at low pressure, so that less vapour will be produced in an accident. The tankers would have to be insulated, and this would reduce the payload, but this, it was said, would be a small price to pay for increased safety.

However, tankers are vary rarely punctured in an accident but ordinary road accidents are very common. If we use refrigerated tankers the payload will be less and we shall need more tankers on the road; there will be more accidents and consequently, more people will be killed; the action will have the opposite result of that intended. (The LPG tanker accident in Spain in 1978, which killed 200 people, could not have occurred in the UK as all LPG tankers are fitted with relief valves.)
3. A company wanted to construct a berth for *ammonia* ships on the bank of a river. The port authority was concerned that a passing ship might collide with them and cause a leak and so they suggested that a recess for the ammonia boats should be excavated out of the river bank, at right angles to it. Experience in other ports showed that the ammonia boats were much more likely to be damaged while manoeuvering into a confined space than by impact with passing ships. Building the recess would increase the chance of a leak, not decrease it. (This conclusion applied to a particular river; it may not be true everywhere.)
4. Cross-country underground pipelines are often inspected by helicopter. What is the chance that the helicopter pilot or the inspector will be killed in an accident? Is it less than the chance that someone's life will be saved as the result of an inspection? Calculation showed that it probably is, so we should carry on with the helicopter surveys, but the question was worth asking.

Calculations such as those I have described are sometimes called systems thinking or *operations research* but they do not really need a special name. They are just the opposite of sloppy thinking.

Emotion is valuable; it gives us the drive and energy to put things right. But emotion uncontrolled by thought is like a powerful car without a steering wheel. It may go off with great energy in the wrong direction.

Small companies

Which is the more dangerous: a small quantity of a hazardous chemical in the hands of a poor *management* or large quantity in the hands of a good management? The attention paid to major hazards since *Flixborough* suggests that most authorities think the latter is more dangerous (see *CIMAH Regulations*) but many accidents in small companies indicate a frightening ignorance of elementary *good practice*. I do not suggest for a moment that all small companies are incompetent, far from it, but a number seem to have low standards. For example, consider the following incidents:

- A small factory in a residential area recovered solvent by distillation. The cooling water supply to the condenser, after giving trouble for several weeks, finally failed and hot vapour was discharged from a vent inside a building. The vapour exploded, killing one man, injuring another and seriously damaging the factory. Some of the surrounding houses were slightly damaged and five *drums* landed outside the factory, one on a house[1].

  There were no operating or emergency instructions, no indication of cooling water flow and drums were stored too near buildings. However, by far the most serious error was allowing the vent pipe to discharge inside the building. If it had discharged outside the vapour would have dispersed harmlessly or, at the worse, there would have been a small fire on the end of the vent pipe. Vent pipes are designed to vent so this was not an unforeseen *leak*. (See *foresight*.) Was the vent pipe placed indoors to try to minimize smells which had caused some complaints? If so, it is another example of *exchanging one problem for another*.
- A year later a fire and explosion occurred in a warehouse where sodium chlorate was stored. To quote from the official report[2], 'The management . . . were not aware of the guidance that had been issued by the Health and Safety Executive about the storage of sodium chlorate . . . Although vandalism clearly played a part, the risk could have been significantly reduced had the sodium chlorate and other chemicals been stored in an appropriate manner'.

Small companies cannot employ the range of experts that large companies employ but expertise can be hired. Do they know when to hire it, do they have *knowledge of what we don't know*? There are many courses and training packages available which will not turn their staff into experts but will give them enough knowledge to recognize the need to call in an expert. Small companies should not handle hazardous materials unless their employees know enough to handle them safely.

See *COSHH Regulations*.

1. Health and Safety Executive, *The Explosion and Fire at Chemstar Limited on 6 September 1981*, HMSO, London, 1982
2. Health and Safety Executive, *The Fire and Explosion at B & R Hauliers, Salford on 25 September 1982*, HMSO, London, 1983

Smoke

Flash *fires* and *BLEVEs* kill people before the smoke has had time to affect them but when fires occur in buildings, aircraft and trains more people are killed by toxic chemicals in the smoke, such as carbon monoxide and hydrogen cyanide, than by the fire. Several companies are now producing hoods which allow people to leave smoke-filled buildings and aircraft in safety but some of the hoods do not provide protection against carbon monoxide.

Smoke can also cause damage comparable with or even exceeding that caused by the fire. Even when a fire is confined to one part of a building toxic and corrosive products may be carried with the smoke throughout the building and cause damage to computers and other electrical equipment beyond repair.

Smoke screens

Some *accident reports* try to distract attention from the *causes* which are the responsibility of the writer by drawing attention to other people's shortcomings.

For example, a gland leak on a liquefied flammable gas *pump* caught fire and caused considerable damage. The manager's report rightly drew attention to the congested layout, the amount of redundant equipment in the area, the fact that a gearbox casing had been made of aluminium, which melted, and several other unsatisfactory features. It did not stress that there had been a number of gland leaks on this pump over the years, that reliable glands are available for liquefied gases at ambient temperatures and therefore there was no need to have tolerated a leaky pump on this duty.

Smoke screens are laid by employees at all levels but the factory manager, being a more able man, is more likely to get away with them than an operator who tries to kid his foreman.

Smoking

Smoking causes about 100 000 deaths per year in the UK. The few thousand caused by accidents and industrial disease seem small by comparison. However, that is no excuse for not doing all we can to prevent industrial accidents and disease. If we spend less money or effort on their prevention there is no mechanism for using these resources to reduce smoking.

Non-smokers who have to share an office or workroom with smokers are exposed to a hazard greater than that presented by many chemicals. In many countries the authorities lay down the maximum concentrations of hazardous chemicals in the atmosphere to which employees may be exposed. (See *COSHH Regulations*.) Tobacco smoke should be one of those covered.

Smokers are more likely than non-smokers to contract some industrial diseases. See *asbestos* and *will and might*.

Smoking can be described as living twice as fast. The average loss of *expectation of life* from a cigarette is about equal to the time taken to smoke it. However, about four out of five smokers lose no life and the fifth loses five times as much life.

Software

See *audits, computers, methods of working* and *viruses*.

Solvents

The UK's fire brigades deal with about 5000 incidents per year involving the use of flammable solvents, such as petrol, used mainly for cleaning.

Petrol should never be used as a solvent, at home or at work. Paraffin is safer but still flammable. (See *cleanliness*.) Non-flammable solvents (or detergents) should be used whenever possible. The toxicity of many solvents is low but nevertheless they should be used only in well-ventilated situations and never used hot in the open. They should not be applied to the skin, as they will degrease it, and should not be used near flames, as they may form harmful products when they decompose.

The following incidents are typical of the many that have occurred:

- A man took his motorcycle to bits in the kitchen and cleaned the parts with petrol in an open pan a metre away from a lighted paraffin heater; his wife and baby were in the room. The petrol caught fire and the man and his wife were badly burned[1].
- A man cleaned a clock using petrol in a plastic bucket standing in the kitchen sink. The vapour was ignited by the gas cooker and the man then tried to put out the fire by pouring water over it. This spread the fire[1].
- An operator cleaned the inside of a cabinet with a rag soaked in acetone. He then placed the end of a non-flameproof vacuum cleaner in the cabinet and switched it on. Flames over a metre long came out of the other end of the cleaner.

1. *Annual Report of HM Inspectors of Explosives for 1973*, HMSO, London, 1984

Spark-resistant tools

These tools, formerly called non-sparking tools, seem to be regarded as a sort of *magic charm* to ward off accidents. Though many companies use them, since 1930 a series of reports have shown that their value is very limited.

Impact of steel hand tools on steel may ignite ethylene, hydrogen, acetylene and carbon disulphide but not most hydrocarbons. It may seem, therefore, that there is a case for using spark-resistant tools on plants which handle these substances but people should never be allowed to enter a *flammable atmosphere* in the first place, much less work in one. The most that should be allowed is putting the hands, suitably protected, into a small area of flammable mixture (say, up to 0.5 m across) immediately surrounding a leaking joint, in order to tighten it. If the leaking gas is one of those mentioned then a spark-resistant hammer may be used. Spark-resistant spanners are less effective than steel ones and there is no need to use them.

These comments apply only to mixtures of flammable vapour and air, not to solid explosives.

See *Myths*, §12 and reference 1.

1. F Powell, *Reprint No. 1110 from Gas, Water, Wastewater*, No. 6, Swiss Gas and Water Industry Association, Zurich, 1986

Stacks

Many *explosions* have ocurred in flare and vent stacks. There should be a continuous flow of fuel gas or *nitrogen* up the stack to prevent air diffusing down and forming an explosive mixture. The flows of fuel gas or nitrogen should be measured and, in addition, the atmosphere should be checked for oxygen, continuously on large flare stacks and from time to time on small vent stacks.

Vent stacks are sometimes ignited by lightning. This is not dangerous if:

- The gas mixture in the stack is not flammable so that the flame cannot travel down the stack.
- The flame does not impinge on overhead equipment. (It may lean at an angle of 45° in a wind).
- The flame can be extinguished by passing nitrogen or steam up the stack.

Stacks have been blocked by choked *flame traps* or molecular seals, frozen water seals, frozen contents or by debris from a ceramic lining on the tip. As far as possible, avoid flame traps, water seals and other devices which spoil the simplicity of a stack and increase the chance of obstruction.

See *chokes, expert systems* and *WWW*, Chapter 6.

Standards

See *codes of practice, good practice* and *old plants and modern standards*.

(five) Star Grading System

See *criteria*.

Statistics

See *accident statistics*.

Static electricity

Static electricity (static for short) has been quoted as the source of *ignition* responsible for many *fires* and *explosions*, sometimes correctly. At other times the investigator has failed to find a source of ignition and assumed it must be static although he cannot show precisely how a static charge could have been formed and discharged. See *stress*.

A static charge is formed whenever two surfaces are in relative motion, for example, when a liquid flows past the walls of a pipeline, droplets or particles move through the air or someone walks, gets up from a chair or removes clothing. One charge is formed on one surface and an equal and opposite charge on the other. If the charges are large enough they will discharge as sparks. Static charges can also be induced on neighbouring bodies by induction.

If equipment is earthed any static charges on it rapidly drain away to earth but, contrary to widespread belief, earthing does not always prevent static sparks. Suppose a liquid is flowing into a *tank*:

1. If the tank is not earthed it will become charged and a spark may pass between the tank and earth, igniting any vapour present outside the tank.
2. If the tank is earthed and the liquid is conducting, all the charges on the tank and the liquid will flow to earth and there will be no spark.
3. If the tank is earthed and the liquid is non-conducting then the charge on the tank will drain away to earth but a charge will remain on the liquid. A spark may pass between the liquid surface and the tank roof or wall, igniting the vapour present in the vapour space of the tank.

To prevent explosions in such tanks we should earth them and:

1. Prevent the formation of static by keeping pumping rates low (below 7 m/s for pure liquids but below 1 m/s if any water is present) and by leaving a length of straight pipe, for charges to dissipate, after restrictions such as filters and valves. Few, if any, companies rely on this method alone.
2. Prevent the formation of an explosive mixture in the vapour space by blanketing it with *nitrogen*.
3. Eliminate the vapour space by using a tank with a floating roof.
4. Make the liquid conducting by use of suitable additives.

For more information see *action replays, lost knowledge, (visit accident) sites, WWW*, Chapter 15, *Lees*, §16.5 and reference 1.

1. *Plant/Operations Progress*, Vol. 7, No. 1, January 1988

Steam

Steam, like *nitrogen*, is one of those *hazardous substances* that many people do not consider hazardous. It is treated with less respect than it deserves and is involved in more than its fair share of accidents. The hazards of steam are heat, pressure, *contamination* (see *service lines*) and water hammer. Five men were killed when a *plastic* hot water tank split along a seam[1].

WWW, § 9.1.5 and a report by S Mortimer[2] describe several incidents in which steam mains were broken by water hammer. Most of them occurred when a steam main was being brought into use and the amount of condensate formed was greater than usual. It moved along the main until it reached an obstruction, such as a change in direction or a restriction, which it struck with a hammer-like blow.

To prevent such accidents we should:

- Provide an adequate number of steam traps and inspect them regularly. If they are cold they are not working. See Figure 71.

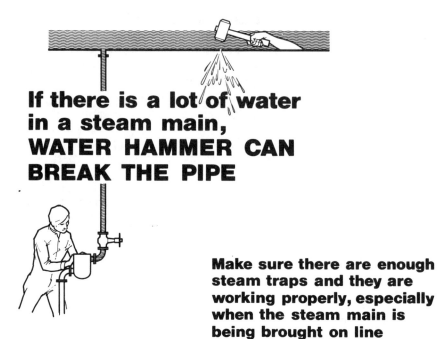

Figure 71 Steam

- Avoid dips in lines or weirs behind which condensate can collect, for example, globe *valves* in horizontal lines, and some types of strainer.
- Bring steam mains on line slowly.
- Avoid brittle valves such as those made from cast iron.

1. *Chemical Engineering*, 1 December 1977, p. 67
2. S Mortimer, *Avoiding Water Hammer in Steam Systems*, Specialist Inspector Report No. 5, Health and Safety Executive, Bootle, 1988

Storage

See *drums, dusts* and *tanks*.

Strength of equipment

See *(treating) symptoms instead of diseases*.

Stress

Stress and *distraction* (and *boredom*) can greatly increase the probability of those human errors that are due to a moment's forgetfulness. (See *human failing.*) Failure to allow for stress and distraction has made nonsense of many attempts to estimate the probability of error.

Stress is not just due to pressure of work but to more subtle factors such as relationships with supervisors and fellow workers. Wigglesworth writes, 'In any discussion of the behavioural environment, two factors stand out. The first is that such features as rest pauses, wage incentive schemes, and shorter hours are of minor significance compared with the major effect produced by constant consultation and communication with operators by management[1]'.

Apart from stress in the workplace, stress due to domestic and other outside problems may also increase the probability of error. One investigator found that train drivers are more likely to pass signals at danger when they are suffering from minor psychological disorders than at other times. However, '. . . the drivers he had investigated were volunteers anxious to rationalise their error by producing a psychological "excuse" for their "signal passed at danger" episode; on the other hand, the control group, presenting themselves for routine examination, had nothing to gain by revealing their psychological problems[2]'.

Stress is rather like *static electricity*; both cause accidents and both are blamed for more accidents than they cause. Static electricity will not cause a fire or explosion unless there is a *flammable atmosphere* present and stress will not cause an accident unless forgetting to close a valve or a similar error results in an accident. We try to avoid the formation of flammable atmospheres and similarly we should try to avoid situations in which a simple slip results in an accident. (See *friendly plants.*)

1. E C Wigglesworth, *Ergonomics*, March 1973
2. M Andrews, *Human Factors in Train Operation*, Institution of Railway Signal Engineers Meeting, 13 November 1979

Substitution

Substitution is one of the methods of achieving *inherently safer design*[1]. One of the most effective ways of preventing a leak of hazardous material is to use a safer material instead. For example, water (under pressure) can be used instead of flammable oil as a heat transfer medium (see *unexpected hazards*), other *solvents* can be used instead of *benzene*, or hypochlorite can be used instead of chlorine for disinfecting swimming pools. Alternative insecticides, which are safer to manufacture, can be used instead of carbaryl, which was manufactured at *Bhopal*, or carbaryl can be manufactured by a different route which does not involve the production of methyl isocyanate as an intermediate.

Other methods of achieving inherently safer design are *intensification* and *attenuation*.

See *LFA*, Chapter 10.

1. T A Kletz, *Cheaper, Safer Plants – Notes on Inherently Safer and Simpler Plants*, Institution of Chemical Engineers, Rugby, 2nd edition, 1985

Success

R L Miller says of success:

Our failures show very clearly but our successes never show; that is one of the problems. The technique we have been trying with some success in our reporting procedures is to emphasise success aspects of an incident. For example, a fire was controlled by a few sprinklers; what would the damage have been if the sprinkler system had not been commissioned and working properly? Or the Fire Brigade was on the job in two minutes and it was a ten thousand dollar loss. Without adequate control it could easily have been a two million dollar loss[1].

Here are a few more success stories, due to good procedures rather than good design. For other examples see *alertness*.

- A welding inspector was asked to check the radiographs of the pipework associated with a 15 000 tonne refrigerated *ammonia* storage *vessel*. He noticed that many of the radiographs, though satisfactory, seemed remarkably similar. Further investigation showed that they had all been taken from the same weld.

    The radiographer, a sub-contractor to the welding and pipe-fitting contractor, was found to be unqualified[2].

- On a new unit the project team had to order the initial stocks of materials. One member of the team, asked to order some TEA, ordered some drums of tri-ethylamine as he had previously worked on a plant where tri-ethylamine was used and was called TEA. The manager of the new unit ordered a continuing supply of drums of tri-ethanolamine, the material actually needed, and called TEA on the plant where he had previously worked. The confusion was discovered by an alert storeman who noticed that two different materials, with similar names, had been delivered for the same unit and he asked if both were really required.

- A Factory Inspector, visiting a chemical plant, saw 20 drums of highly flammable liquids in a workroom. He issued an Improvement Notice requiring the drums to be moved to a proper flammables store.

    Three months later there was a fire in the workroom. The flames did not spread rapidly and were soon extinguished by the Fire Brigade as there were no flammable liquids in the workroom[2].

Confucious is reported to have said, 'Think of the things you did not want to happen which did not happen.'

1. R L Miller, *Institution of Chemical Engineers Symposium Series No. 34*, 1971, p. 255
2. Health and Safety Executive, *Manufacturing and Service Industries 1981*, HMSO, London, 1982, p. 15

Summerland

The Summerland leisure complex in Douglas, Isle of Man was destroyed by *fire* in August 1973 with the loss of fifty lives. The causes, according to the Commission of Inquiry, were poor design and poor training[1,2].

Vandals set fire to the remains of a kiosk on a terrace. The wall near the kiosk was made from Galbestos, steel sheets coated with asbestos, bitumen and polyester, and lined on the inside with fibreboard. The bitumen and plastic did not ignite easily but once they were ignited the flames spread rapidly to the fibreboard lining and to the adjoining wall and roof which were made from Oroglas (polymethylmethacrylate). Means of escape were inadequate.

The underlying reason for the poor design was the lack of attention to safety, particularly fire safety, in the training of architects, and the Commission recommended that 'Architectural training should include a much extended study of fire protection and precautions'. The Commission also said that it 'was not impressed by repeated attempts by the designers to suggest that other people should have told them of any mistakes or inadequacies in the design'. (See *Health and Safety at Work Act* and *(managerial) responsibility*.) The Isle of Man Building Regulations lagged behind those of the mainland.

The poor design was compounded by the lack of training of the staff. For 20 minutes they tried to extinguish the fire in the kiosk and did not call the fire brigade. The general manager had never seen the 'excellent safety guide' published by the company, the loudspeakers were not used to broadcast instructions on evacuation and the emergency lighting generator failed to start. (See *emergency equipment*.)

Summerland, like *Aberfan*, is 'a terrifying tale of bungling ineptitude by many men charged with tasks for which they were totally unfitted . . .'.

1. *Report of the Summerland Fire Commission*, Isle of Man Government, 1974
2. *The Summerland Fire*, a summary of the official report published by the Fire Protection Association, London

Surprise

See *astonishment*.

Surveys

In a survey, a special sort of *audit*, instead of a quick look at everything we look at a few subjects in detail. If we wish to excavate a field for archaeological remains we could remove an equal depth of soil from the whole field or we could dig out a small area down to the underlying rock. If the area is well-chosen the latter will produce more results, especially if resources are limited, as they usually are.

Table 1 below shows some of the surveys that were carried out during my period as a safety adviser with *ICI*, mainly by people who spent most of their time on surveys and became expert in the subjects surveyed. The reasons why these subjects were chosen are also shown.

Note that in a survey we try to look at everything, not just a sample. When *tests* of trips and *alarms* were surveyed, the testing of every trip and alarm in the Petrochemicals Division was witnessed. When *sampling* was surveyed every sample point was inspected and most were seen in use.

However, when flameproof electrical equipment (see *electrical area classification*) was surveyed only every tenth item was inspected while every hundredth item was dismantled in the presence of the surveyor.

Table 1 ICI survey

Subject of survey	Reason why chosen
Equipment	
Testing of alarms and trips	Recognition of the importance of regular testing
Classified electrical equipment	Comments by a Factory Inspector
Sampling	Several serious accidents
Flame traps and open vents	Several accidents including one fatal
Equipment for handling liquefied flammable gases	The Feyzin fire (see *BLEVE*)
Foam-overs	A serious incident overseas and lesser incidents at home
Level glasses	Serious incidents elsewhere
Flammable *dust* hazards	*Reorganization* brought them into the division
Procedures	
Permits-to-work	A serious fire due to a poor permit system (see *isolation*)
Modification control	Flixborough
Registration and testing of *relief devices*	Recognition of the importance of regular testing
Testing of gas detectors	They do not always fail safe (see *analysis*)

From *Health and Safety at Work*, Vol. 4, No. 3, November 1981, p. 5

(treating) Symptoms instead of diseases

When something breaks repeatedly, do we ask why or do we just go on repairing it until someone is hurt or the damage is serious? Here are some incidents in which the latter happened.

- Twenty-seven times in nine years the cylinder lining of a compressor was found to be worn or cracked and had to be replaced. No one asked why it had to be replaced so often; they just went on replacing it. Finally, a bit of the lining got caught between the piston and the cylinder head and split the cylinder.
- To repair a piece of equipment a man had to remove a number of ¾ inch bolts. One of them broke. As it did so, he jerked backwards and hurt his back. During the investigation someone noticed several broken bolts on the floor. They had broken regularly, but no one had reported them; they just replaced them and carried on. Once it was known that the bolts were liable to break, it was possible to reduce the load on the bolts or use stronger ones.
- The pressure gauge on a steam locomotive was repeatedly changed as it was reading too high. No one considered that the gauges might be correct and the pressure too high. After the boiler had blown up, killing

the driver and fireman, the relief valve was found to be too small[1]. (See *instruments*.)
- A shearpin on a centrifuge failed about twice per day. The operator replaced it but did no more as he used the failure of the pin to tell him that the centrifuge was blocked. One day a pipe, weakened by vibration, fractured while he was changing the pin and he was sprayed with liquid[2].
- In 1814, in London, a wooden vat containing 600 m^3 of beer burst. Streets were deluged, two houses demolished and two people killed. The vat was held togther by 29 iron hoops. The lowest hoop fell off, the hoop above broke and the rest opened up one after the other. The lowest hoop had fallen off many times before but instead of telling anyone the operators just pushed it back into position, probably hooking the rivets back into place[3].

In the 1970s a colleague of mine visited the steel industry and found that their policy, when something broke, was simply to strengthen it.

When something goes wrong, particularly if it has gone wrong before, do not just repair it and put it back into use; ask why it went wrong.

1. C H Hewison, *Locomotive Boiler Explosions*, David and Charles, Newton Abbott, Devon, 1983, p. 119
2. Health and Safety Executive, *Dangerous Maintenance*, HMSO, London, 1987, p. 13
3. *Safety*, published by British Steel Corporation, August 1971

System accidents

See *'Normal accidents'*.

Systems of work

See *methods of working*.

Take-over

See *amalgamation*.

Talebearing

Talebearing (sometimes called whistle blowing) has always been unpopular. The Bible says, 'Do not go about as a talebearer among your countrymen' (Leviticus 19:16). But while tale-bearing is wrong, to stand idly by when there is a risk to another's life and limb is worse and to avoid the latter we should be prepared to speak out.

This does not mean that we should complain to the *Factory Inspector* whenever we see a hazard. We should complain within our company. Only when we are sure the hazard is serious and that the company is unwilling to do anything about it are we justified in complaining outside. There may be legitimate differences of opinion on the size of a hazard and the action required. We should not confuse these with a deliberate decision to ignore an undoubted hazard.

See *motivation*.

Tankers

Road tankers (tank trucks) and rail tankers (tank cars) are widely used for the conveyance of hazardous liquids and their record in the UK is remarkably good: an average of just over one person per year killed by the transport of chemicals and petrol. In other countries, including the US, the record is a good deal worse and in Spain, in 1978, over 200 people were killed when a tanker of propane ruptured and the contents ignited[1]. The good UK record is due to higher standards, plus a bit of luck as some of the overseas incidents could easily have occurred in the UK[2].

Most accidents involving tankers occur while they are being filled or emptied and the following are typical:

- A tanker is overfilled or it is emptied without a vent, such as a manhole, being open and is sucked in.
- A leak occurs because the wrong type of *hose* is used, or a damaged hose or the hose is not correctly fitted. See *threads*.
- A tanker is moved before the hose is disconnected.
- A tanker is splash filled with a flammable liquid such as petrol and the vapour is ignited by a discharge of *static electricity*. (See *(limitations of) experience*.) Sometimes a tanker is splash filled with a high boiling liquid such as paraffin and vapour in the tanker from the previous load is ignited (switch filling).
- The wrong liquid is put into a tanker or it is emptied into the wrong tank. Liquid *oxygen* and air have been delivered instead of *nitrogen* (see Figure 72) and acid has been discharged into alkali tanks. Liquid nitrogen should be tested before it is off-loaded.

Figure 72 Tankers

Figure 73 Tankers

- If a tank trailer is disconnected from its tractor, and the rear compartment is emptied first (or filled last), the tank trailer may tip up. See Figure 73.

See *force, sloppy thinking* and *WWW*, Chapter 13.

1. I Hymes, *Loss Prevention Bulletin*, No. 061, February 1985, p. 11 and H Ens, No. 068, April 1986, p. 17
2. T A Kletz, *Plant/Operations Progress*, Vol. 5, No. 3, July 1986, p. 160

Tanks

Large, low pressure storage tanks are the most fragile items of plant equipment in use. They are usually designed to withstand a gauge pressure of only 8 inches of water (0.3 p.s.i.) – they will burst at about three times this pressure – and a vacuum of only 2½ inches of water (0.1 p.s.i.). They will be sucked in if this vacuum is exceeded by more than a small amount. It is not surprising, therefore, that tanks are often damaged. The commonest reasons are described below.

Overpressuring

The tank is overfilled and the vent or overflow is too small. It may have been designed to pass the volume of air displaced but not the liquid rate; or gas is blown into the tank at a rate above the capacity of the vent. (See *extravagance*.) A tank may be overpressured if the liquid level is allowed to rise above the wall/roof seam into the roof space. See Figure 74 and *full-scale deflection*.

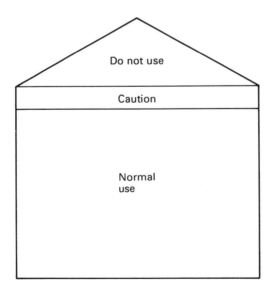

Figure 74 A tank may be overpressured and split if the liquid level is allowed to rise above the top of the walls (from reference 2)

Tanks 313

Figure 75 Tanks

Sucking in

The vent is too small for the rate of emptying or pump out or the vent is closed or blocked. A plastic bag tied over the vent may be sufficient to block it. (See Figure 75.) *WWW*, § 5.3 lists 13 ways in which tanks have been sucked in.

Explosions

If a tank is used to store a flammable liquid there may be an explosive mixture in the vapour space above the liquid. *Static electricity* or another source of *ignition* may then ignite it. Many companies store highly flammable liquids only in floating roof tanks or tanks that are blanketed with *nitrogen*. (Despite what I say in the items on *lost knowledge* and *repeated accidents*, we do improve. In 1928 93% of the tank fires started by lightning occurred on tanks with wooden roofs[1].) See *unexpected hazards*.

Tanks are usually designed so that the roof/wall weld will fail if they are overpressured and the contents of the tank will not spill. Any subsequent fire will be confined to the tank. However, the base/wall weld may fail first if it is weakened by *corrosion*.

Tanks are often overfilled without damage to the tank but there is, of course, a spillage of liquid. (See *alarms*.) Tanks containing hazardous chemicals are surrounded by bunds to confine the spillage but the drain valves in the bunds, fitted so that rain water can be removed, are often left open. Like all *protective equipment* they should be inspected regularly.

On many occasions men have been overcome by fumes during *entry* to a tank.

See *WWW*, Chapter 5 and *LFA*, Chapters 6 and 7.

1. C Dalley, *The Industrial Chemist*, March 1928, p. 117
2. D L Haines, R E Sanders, and J H Wood, AIChE Loss Prevention Symposium, 1989

Tay Bridge

The Tay Bridge disaster in 1879 was not the worst railway accident in British history, but it is probably the best remembered. (See *other industries*.) The bridge collapsed 6 months after it was opened and a train fell into the river. All the passengers and crew, 75 people, were drowned. The failure of a major engineering work so soon after opening caused much disquiet among the public and the engineering profession.

The failure was due to a change in the original *design*, poor *construction* and poor *inspection*, in short, poor *management*.

The original intention was to construct the piers entirely of brick and concrete and 14 out of 50 were built this way. However, the river bed was not as strong as expected so, to reduce the pressure on it and save cost, the designer, Sir Thomas Bouch, decided to build the piers of brick and concrete up to the high water mark and then construct the upper portions from cast iron pipes with cross-bracing. The brick piers still stand, many of the original girders which linked the tops of the piers were re-used and are in use today, but the cast iron piers collapsed in six months. They were not cheaper after all!

The casting and erection of the iron pipes were not properly supervised. The quality of the iron was poor, its thickness was unequal and many of the lugs that held the cross-bracing were not secure. The subsequent inspection of the bridge was entrusted to a man who was a good bricklayer but had little knowledge of ironwork. 'The most embarrassing of all the North British Railway witnesses was Henry Noble, the most honest and competent of men in his own limited sphere, but a bricklayer and not a man many railway companies would have chosen to take charge of the bridge[1].' (Compare *Flixborough* and *Aberfan*, where men were also asked to undertake tasks beyond their competence. They did not know that it was beyond their competence. See *knowledge of what we don't know*.)

After the inquiry Sir Thomas Bouch's reputation was finished. His health deteriorated and he died soon afterwards. He had made his name by building bridges cheaper than anyone had built then before, but with the Tay Bridge he went too far. In his desire to cut costs he went too far in economizing on the inspection of the ironwork both during and after construction. (See *penny-pinching* and *(legal) responsibility*.)

1. J Thomas, *The Tay Bridge Disaster*, David and Charles, 1981

Team working

The collapse of the Yarra Bridge in Melbourne in 1970, during *construction*, showed what can occur when all those concerned with a project do not work as a team. The bridge was one of the first box girder bridges to be built and no one told the construction team that more accurate construction than they had used in the past would be necessary.

It is traditional in the construction industry, if components do not fit, to force them into place. With a box girder bridge, however, parts have to be aligned accurately; if they do not fit, they should be modified until they do. The designers produced a design which took little account of construction problems, they did not tell the construction team that accurate assembly was necessary and they did not check to see that that it was taking place[1].

Discussing Ove Arup, Rice says, '. . . he knew that to design you had to know how to build[2]'.

Another example of different groups failing to work as a team occurs when a design department ignores running costs and aims for minimum capital cost rather than minimum lifetime cost.

It has become popular for plant operators to work as a team. Every man can carry out every operation and each task is done by whoever is available. The advantages are obvious but one day a control room operator asked the outside operators, over a loudspeaker, to shut down a pump. Both the outside operators were some way from the pump at the time and each assumed that the other would do it. Neither did it and the pump overheated and caught fire. A system in which everyone can do a job can easily degrade into a system in which no one does it.

A similar incident occurred on the railways in 1894. Both drivers on a double-headed train assumed that the other was in charge and was watching the signals. Neither was and the train ran into the train in front[3]!

1. *Report of the Royal Commission on the Failure of the West Gate Bridge*, State of Victoria Government Printer, Melbourne, Australia, 1971
2. P Rice, *Journal of the Royal Society of Arts*, Vol. 137, No. 5395, June 1989, p. 425
3. R O T Povey, *The Bedside Book of Railway History*, Dalesman Books, Keighley, West Yorkshire, 1974, p. 31

Temporary jobs

There are times when we all feel that there is nothing so permanent as a temporary building or a temporary job but in industry temporary repairs or lash-ups that are left too long can become hazardous. Scaffolding is not intended for permanent use but is often left for months. There should be a procedure for drawing management's attention to any that has been left for longer than, say, 3 months. Similarly, lashings may be used to support a new pipe while a permanent support is made but should not be left in position for more than a few weeks. If they are left they may corrode and the pipe may collapse. *Hoses* are OK as a temporary connection between two pipes but they are not intended for permanent use; a fixed pipe is safer, and cheaper!

A large bolted joint in a steam line was leaking so a temporary clamp was fitted round it. The leak stopped and everyone forget about it – until, 3½ years later, the joint came apart with explosive violence. Other equipment was damaged and there was a leak of oil which caught fire, causing £1M damage. The bolts in the joint were not normally in contact with the steam but when the clamp was fitted they became exposed to the steam and impurities in it caused corrosion.[1]

Expert advice should be taken before clamps are fitted to pipelines. The maximum time that the temporary repair can be left should be stated; the existence of the clamp should be registered and a permanent repair should be made as soon as possible.

The incident occurred in a refinery. Note that it was not flammable oil or gas that caused the initial failure but *steam*. In refineries and chemical plants steam, *nitrogen, water* and compressed *air* cause a disproportionate amount of trouble. Everyone realizes that flammable substances are hazardous and treats them with repect but *service lines* are often considered safe; they are not. Nitrogen has probably killed more people than any other substance except methyl isocyanate, the chemical that leaked at *Bhopal*. In one of the largest chemical companies in the US the worst accident of the last decade was the failure of a *plastic* hot water tank which killed five people.

Some temporary jobs are hazardous from the start, for example, the temporary pipe at *Flixborough*, the temporary cover described in the item on *nothing* and the equipment described in the following two incidents[2]:

1. The manhole on the top of a silo was removed for modification and a fitter was asked to cover the opening with a sheet of stainless steel 0.7 mm (1/36 inch) thick. While he was hammering down the edges of the sheet he put his hand on the centre. The sheet gave way under his weight and he fell 11 m into the silo and was killed.
2. A *tank* of phthalic anhydride, a corrosive liquid, contained a submerged pump. It was removed for repair by withdrawing it through a manhole on the top of the tank and disconnecting the electric cables. To stop fumes coming out of the tank the open manhole was then covered with a piece of jointing and a piece of plywood. An electrician who was repairing the cables stepped on to the plywood. It broke and he fell through the manhole. Fortunately he spread out his arms and stopped himself falling right in.

See *cranes* and *drums*.

1. *Loss Prevention Bulletin*, No. 069, June 1986, p. 17
2. J Bond, *Loss Prevention Bulletin*, No. 078, December 1987, p. 15

Tests

The items on *instruments, inspection* and *methods of working* have stressed the importance of testing or inspecting all *protective equipment* at regular intervals, or it may not work when required. The more reliable the equipment, the less often it needs testing. Relief valves are very reliable; a typical valve develops a fault that will prevent it lifting about once in a hundred years.

Suppose we test it every two years. It will develop a fault, on average, halfway between tests and will then be inactive or 'dead' until the next test, that is, for about one year in every hundred years or for 1% of the time. We say that its fractional dead time (FDT), the probability that it will fail to operate when required, is 0.01.

Suppose the valve is required to lift once per year, that is, its demand rate is 1/year. A demand will coincide with a dead period, the relief valve will fail to lift when required and the equipment will be overpressured about once in a hundred years. The hazard rate, the rate at which dangerous incidents occur, is therefore 0.01/year.

Note that the equipment may not burst when it is overpressured. It may withstand the excess pressure, or the relief valve may lift at a higher pressure than its set pressure. The hazard rate for bursting of equipment is lower, usually much lower, than the hazard rate for overpressuring.

Instruments are less reliable than relief valves and a typical failure rate for a trip is once every two years. To achieve an FDT of 0.01 we have to test it every 2 weeks (0.04 year). Trips should therefore be tested once a month. More frequent testing is not recommended as the time required for testing becomes significant.

See *redundancy* and *reliability*.

Suppose a relief valve has to be kept hot by steam or electric heating. Such heating systems are usually not very reliable. Someone may turn off the heating, a steam trap may fail, a fuse may blow or the insulation may be damaged. If the heating fails the relief valve will not operate. The relief valve should therefore be checked frequently, say, every day or every few days, to make sure that the heating is still in operation.

The item on *protective equipment* includes a list of equipment that should be tested regularly and stresses the importance of making sure that the tests simulate 'real life' conditions. If the *time of response* is important it should be checked during the test. When testing trips and alarms check that they operate at the correct set-point.

Emergency procedures should be tested as well as *emergency equipment*.

WWW, Chapter 10 and *LFA*, Chapters 2 and 6 describe accidents which occurred because protective equipment was not tested.

The testing of pressure vessels and pipework to prove their integrity is described under *pressure tests*.

For useless tests see *useless equipment and procedures*.

'They'

'Why don't they do more about safety?' is a question often asked. If we are talking to an operator, 'they' means the foremen or managers. If we are talking to a junior manager, 'they' means the senior managers. If we are talking to a senior manager, 'they' means the directors.

A possible answer to the question is to say that it is your job to 'do something' as much as theirs. You may not be able to do everything that needs doing but there is a lot you can do. Why not make a start on that? (See *(safety) books*.)

Perhaps you can do more than you think. In some companies the limits of people's authority are defined but in many, such as *ICI*, they are vague. The job can grow; you can do more than you did last week. Responsibility is left lying around for anyone who wants to pick it up.

If we are talking to the public, 'they' are more difficult to define. An objector to a proposal to store nuclear waste underground said on TV, 'You told us that pneumonconiosis was not a danger to the health of miners; so you should not be believed when you say that this method of storage is safe'. The speaker probably realized that that the expert on waste disposal knew nothing about pneumoconiosis. She meant that people who wore similar clothes and spoke in a similar way to him had said similar things on other occasions[1].

To answer the question from the public, 'Why don't they do more about safety and stop accidents?', we have to try to explain that:

- Money is not unlimited and so the more we spend on removing one hazard the less there is available for removing others.
- The expenditure is met in the end by the public, not by the Government or industry.
- Being human, we will never foresee all hazards.
- Complete safety is approached asymptotically. See *asymptote*.

See *false alarms, perception of risk* and *perspective*.

1. D E Broadbent in *Risk: Man-made Hazards to Man*, edited by M G Cooper, Clarendon Press, Oxford, 1985, p. 31

Thinking

See *sloppy thinking*.

Threads

A new concert hall collapsed during a rehearsal when the trumpets played. (The collapse of the walls of Jericho, it seems, should be included in my *repeated accidents*.) The roof trusses had been bolted together at mid-span using 20 mm diameter bolts with ¾ inch (19 mm) nuts. The nuts appeared to screw on satisfactorily but they rested only on the tops of the threads. The nuts stripped off along the bolts and the roof trusses collapsed[1].

In the UK several different types of thread are in use: the traditional Whitworth system, the Unified system, introduced in the 1940s, and the Metric system which is gradually replacing the older systems. These systems differ in thread angle and number of threads per inch. As a result it is possible to mix nuts and bolts from different systems. Thread engagement can vary from full but loose engagement to a tight fit of only one or two threads.

Eyebolts, which are screwed into equipment so that it can be lifted, are easily mismatched[2].

All those who handle nuts and bolts, from storekeepers to mechanics, should be aware of the problem and trained to distinguish the different types of thread. Ideally only one system should be used but this is not always possible as purchased equipment may use a different system. (See *packaged deals*.)

Mismatch can occur even within the same system. An eyebolt with a metric thread will engage in the next size up, giving a loose, very weak fit. A factory used breathing apparatus with ½ inch *hoses*. One department ordered equipment with 9/16 inch hoses. Inevitably, the two sorts of fittings became mixed and a ½ inch hose was secured to a face-piece with a 9/16 inch fixing ring. The hose came off while the wearer was working in a confined space; afterwards all the 9/16 inch equipment was withdrawn.

Sometimes threads of the correct type are not fully engaged. To stop a slight leak from the hose connection to a road tanker the driver inserted three washers. This left only ⅜ inch (10 mm) of screwed thread available for use. In addition the driver did not have the correct tool for tightening the connection. The leak which occurred was much larger than the one the washers were intended to stop.

On another occasion a hose came off a tanker while it was being loaded with liquefied flammable gas. The female thread on the hose was so worn that only a third of the original depth was left.

Left hand threads are sometimes used to prevent wrong connections. For example, *cylinder* valve connections may have left hand threads so that they cannot be fitted to the wrong service lines. However, such threads can be easily loosened by someone who intended to tighten them; a leak of pyrophoric gas and a fire started in this way. A contributing factor was a cylinder valve capable of passing a far higher gas rate than that needed[3].

1. *Journal of Occupational Accidents*, Vol. 3, No. 3
2. Health and Safety Executive, *Eyebolts – Guidance Note No. PM 16*, HMSO, London, 1980
3. T Bielli, *A Positive Approach to Compressed Gas Safety*, American Institute of Chemical Engineers Loss Prevention Symposium, 1989

Three Mile Island

The accident at Three Mile Island *nuclear power* station in Pennsylvania in 1979 has many lessons for other industries.

The trouble started with a choke in a resin polisher unit which removed impurities from the water used for steam raising. The design of this service unit had not received the same attention as the design of the main plant items. **When failure of ancillary equipment can shut down the main plant, its design should be given as much detailed attention as that of the main plant.**

To try to clear the choke the operators used the instrument *air* supply. Its pressure was lower than that of the water which entered the air lines. **Never connect** *service lines* **to equipment which is, or might be, at a higher pressure.**

The water in the air lines upset the instruments and the power station shut down automatically. Some heat was still produced by radioactive decay, the cooling water boiled and a relief valve lifted and stuck open.

The operators did not realize this as an indicator on the panel told then that it was shut. Unfortunately this indicator told them that a signal had been sent to the relief valve telling it to shut, not that it actually was shut; the operators did not realize this. *Instruments* **should measure directly what we need to know**

Other instruments were showing unusual readings but the operators did not realize their significance. Some faults may not be foreseen. *Instructions* cannot therefore cover every eventuality; so **operators should be trained in fault diagnosis.**

The managers and designers had assumed that, if the operators were trained to cope with major failures, minor failures would look after themselves. This is not true. Three Mile Island started as a minor incident but one that the operators had not been trained to handle. **Plan for all foreseeable failures, not just big ones.**

Another management error was to assume that if the *rules and regulations* were followed, the plant would be safe. This attitude is commoner in the US than in the UK. In the US the authorities believe that they should write books of regulations that look like telephone directories. **Authorities should lay down standards to be achieved and should check that they are achieved but should not try to say in detail how they are achieved.** See *Health and Safety at Work Act*.

For more information see *LFA*, Chapter 11.

Threshold limit values

See *COSHH Regulations*.

Tidiness

Junk left lying around is a major cause of industrial accidents, and according to some writers, **the major** cause. In the UK the Factories Act, Section 28 states that 'All floors . . . shall be kept free from any obstruction and from any substance likely to cause persons to slip'.

Untidiness can also contribute to accidents in more subtle ways. If the workplace is untidy then everyone will get the impression that tidiness is unimportant and may assume that slipshod and untidy work is acceptable. There may be less attention to detail and output and efficiency may suffer as well as safety.

A man went up a ladder on to a walkway and then, to retrieve a torch that he had dropped, climbed over the handrails on to a fragile roof which was clearly labelled. He tried to stand on the girders but slipped on to the roof sheets and fell into the room below. Figure 76 shows the foot of the ladder leading to the roof. As the man negotiated the pile of rubbish to get to the ladder the message he received was, 'In this place safety does not matter'. This may have encouraged him to take a chance and climb over the handrails.

It has been my privilege to visit many countries and I am sorry to have to say that UK plants are, on the whole, some of the most untidy that I have seen.

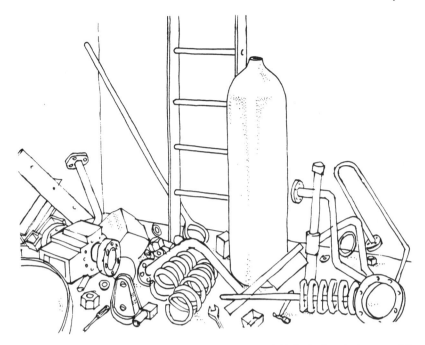

Figure 76 After going up this ladder a man acted unsafely. Did the junk at the bottom of the ladder tell him that safety was unimportant?

Tidiness, like everything else, can be carried too far; in one case, some *cylinders* were repainted in the wrong colour before an inspection. See *anaesthetics*.

See *cleanliness*.

Time and money

Companies sometimes draw attention to the amount of money they spend on safety (and the protection of the environment) (see *costs*) but money is not the only resource needed for safety or the only one that is in limited supply. Are companies equally willing to spend time and effort on safety and to assign to it some of their best technical staff? At one time safety was considered a suitable job for one of the less able employees. Those days have largely passed and the standard of *safety professionals* has improved greatly during the last 20 years but many companies still think that safety is a suitable home for those who have a few years to go before retirement. They may be able people but they are not able to bring to the job the continuity it needs.

Perhaps companies will allocate more of their best people to safety and loss prevention when they realize that the right sort of person will not just worry about hard hats and tripping hazards but will actively advocate designs that are cheaper and simpler as well as safer. See *friendly plants* and *inherently safer designs*.

Timebombs

'Timebombs' are faults in design or construction that may not become apparent until many years have passed. Here are some examples:

- A mobile crane was used to lower a load down a shaft. The cable had to be unwound to almost its full extent, an unusual occurrence. The load suddenly fell. It was then found that the end of the cable had never been fastened to the drum[1].
- While building a plant a *contractor* fitted a 1 inch screwed nipple in a pipeline so that he could fill it with water for pressure testing. He did not replace the nipple with a welded plug and it was not shown on any drawing. Twenty years later the nipple blew out and the plant was covered by a cloud of oil mist 30 m deep. It was sucked into the control room by the ventilating equipment making it difficult for the operators to shut the plant down. Fortunately, they did so before the mist caught fire 15 minutes later.
- During the War standards of plant *construction* were not always as good as they should have been. In 1978, 33 years after the end of the War, in Texas City, an 800 m^3 sphere, constructed during the War, and used to store liquefied petroleum gas, fractured along a faulty weld, causing a major *fire*. Other companies had checked the quality of wartime welding after the War, but not the company that owned this refinery[2].
- Many plants are now controlled by *computers*. Faults in the software – the instructions given to the computer – can lie like timebombs until a particular combination of circumstances arises, perhaps a particular fault condition at a particular stage of a batch. Thorough testing is needed to uncover these timebombs. Testing software can take twice as long as designing it.

Thorough checking is needed during and after construction to detect timebombs. If this was not done at the time then sample checking now may disclose defects that can then be investigated in full. Do you know the history of all your equipment and have you got design details? Are there any unused nipples or plugs? There is a list of about 25 points to look for in *LFA*, Chapter 16 and every experienced engineer will be able to add a few more.

1. R Stirzacker, *Health and Safety at Work*, Vol. 10, No. 8, August 1988, p. 17
2. *Loss Prevention Bulletin*, No. 077, October 1987, p. 11

Time of response

When designing a protective system we should not just ask what we want it to do but also how long it will take to do it[1]. When carrying out *tests* on *protective equipment*, if the time of response is important, it should always be measured. See Figure 77.

For example, machinery is often interlocked with guards so that if the guard is opened the machinery stops. Brakes are often fitted so that the

Figure 77 The time of response of an alarm or other protective system may be important

machinery stops quickly. The actual stopping time should be measured at regular intervals and compared with the design target.

Another example: a mixture of a solid and water had to be heated to 300 °C at a pressure of 70 bar before the solid would dissolve. The mixture was passed through the tubes of a heat exhanger while hot oil, at low pressure, was passed over the outside of the tubes. It was realized that if a tube burst, the water would come into direct contact with the hot oil and would turn to steam with explosive violence. An automatic system was therefore designed to measure any rise in the oil pressure and, if there was a rise, close four valves, in the water and oil inlet and exit lines. The heat exchanger was also fitted with a bursting disc which discharged into a catchpot. (See Figure 78.) The system was tested regularly but nevertheless when a tube actually burst the valves took too long to close and most of the oil was blown out of the system and caught fire. The valves had been designed to close quickly but had got sluggish; the time of response was not measured during the test so no one knew that they were not responding quickly enough[2].

Procedures, like equipment, also take time to operate. For example, how long does it take to empty your building when the fire alarm sounds? Is this quick enough? How long will it take the fire engines to arrive? If this

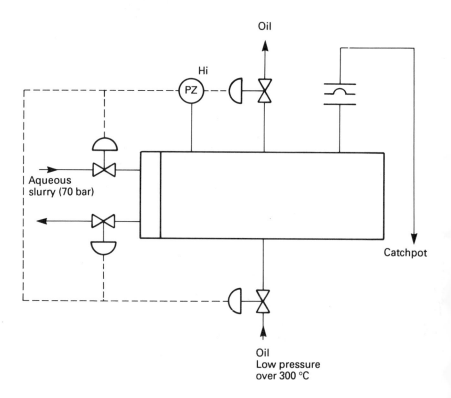

Figure 78 Protective system required when an aqueous slurry is heated by the heat transfer oil. The valves must be quick-acting

is too long we need to know the time they spend on the journey and also the time that elapses between an emergency telephone call and the departure of the fire engine. A study a few years ago showed that this time could be as long as the journey time[3].

One company was located near a fire station and trials showed that fire engines could be on site in a few minutes. When a fire actually occurred the fire engines took 20 minutes to arrive. The firemen had gone to another station for training. They had taken their engines with them but it still took them 20 minutes to get back.

Some companies have procedures for warning people who live nearby if there is a leak of toxic gas. I suspect that in many cases the gas cloud will have passed over before anyone could be told it is coming.

So, please remember, we need to know not just what protective systems will do but how long they will take to do it.

1. R Hill and D Kohan, *5th International Symposium – Loss Prevention and Safety Promotion in the Process Industries*, Société de Chimie Industrielle, Paris, 1986, p. 54–51
2. T A Kletz, *Cheaper, Safer Plants*, Institution of Chemical Engineers, 2nd edition, 1985, §3.5.3
3. *Fire Prevention*, No. 145, 1981, p. 31.

Time span of forgetfulness

Someone who needs instructions every hour is paid less than someone who sees his boss once per month. The period between instructions is the 'time span of discretion'.

For accidents there is a 'time span of forgetfulness', the period of time that has to elapse before an accident is forgotten and it happens again. For the sort of accident that causes minor injuries, or at the worst a few days absence, the time span is a couple of years. For example, a man is injured disconnecting a *hose*, as there is no way of venting the pressure. There is a campaign to fit vent valves. After a couple of years some of the vents have disappeared, some new hose points without vents have come into use and the accident happens again. See Figure 79.

For more serious accidents the time span is ten or more years. For example, after a serious fire in 1956 a company spent a considerable sum moving their equipment for handling liquefied petrolem gas into the open air. For the next ten years all similar units were built in the open. Then a unit was built indoors to reduce the noise level in a nearby workshop. Staff had changed, memories had faded and everyone had forgotten why open structures were preferred. (See *ventilation*.)

As it happened an explosion occurred in another part of the company inside a building and the walls of the new building were pulled down before the unit was commissioned.

For further examples see references 1 and 2.

Also see *lost knowledge*.

1. 'Accidents that will occur during the Coming Year', *Loss Prevention*, Vol. 10, 1976, p. 151
2. 'Organisations have no Memory', *Loss Prevention*, Vol. 13, 1980, p. 1

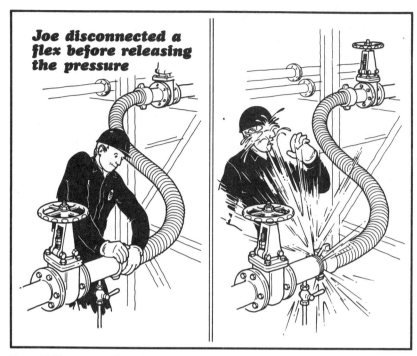

Figure 79 Time span of forgetfulness

Tolerable risk

See *acceptable risk*.

Tolerance

See *friendly plants*.

Tools

See *spark-resistant tools*.

Top event

See *defence in depth*.

Total loss control

See *damage control*.

Toxicity

Some toxic substances, such as *chlorine* and *ammonia*, produce immediate toxic effects. Some, such as *asbestos*, produce effects after a long period of time, typically several decades. Others, such as *arsenic, benzene* and *lead*, have both immediate effects, in high concentrations, and long-term effects, in low concentrations. For long-term effects see *COSHH Regulations, cancer, fugitive emissions* and *cause and effect*.

Short-term toxicities are usually expressed as the LC_{50}, the concentration that will kill half those exposed for a given period of time, or as the LD_{50}, the dose that will kill half those exposed. Concentrations or doses that will kill 1, 10 or some other percentage of those exposed are often of interest but such figures are rarely available. *Probits* can be used to calculate the figures but the results should be treated with caution as considerable *extrapolation* is involved.

The concentration, c, and time, t, that will produce similar effects are related by

$c^n t$ = constant.

For chlorine, values of n between 1.67 and 2.75 have been suggested[1].

The LC_{50} is not always the most appropriate criterion to use when comparing toxic gases or vapours as some gases and vapours spread more easily than others. An alternative criterion is to calculate the area that will be covered by a concentration equal to the LC_{50} (or some other figure) for a leak of a given size, say, 1 kg/s, dispersed by a wind of a given speed, say, 5 m/s.

Finally, remember that there is no sharp division between toxic and non-toxic substances. In the 16th Century Paracelcus said, 'All substances are poisons; only the dose makes it not a poison'.

See *ventilation*.

1. *Chlorine Toxicity Monograph*, Institution of Chemical Engineers, Rugby, 1987

Trade-off

See *alternatives*.

Tradition

See *old-timer* and *sleeping beauty*.

Training

Many accidents have occurred in the process industries because operators or supervisors or even managers did not understand the hazards of the technology. They did not know, for example, that addition of hot oil (above 100 °C) to a tank containing water will cause the water to turn to steam with explosive violence (see *foam-over*), that 30 'pounds' is not a small *pressure* because a *force* of 30 pounds is exerted on every square inch, or that mixtures of flammable vapour and air will sooner or later catch fire or explode even if everything possible is done to remove known sources of *ignition*.

Other accidents have occurred because operators, supervisors or managers did not understand the need for a permit-to-work system (see *maintenance*) or a *modification* control system or did not know the limitations of their equipment (see *tanks*) and allowed it to be taken up to too high (or low) a pressure or temperature. See *EVHE*, Chapter 2.

Training programmes are therefore needed, induction programmes for new employees and refresher programmes for old ones. Every company agrees that training is important but the practice is rather variable. Something better than an occasional talk or piece of paper is needed.

Discussions are better than talks. If an accident is being discussed the discussion leader should describe it briefly and then let the group question him to find out the rest of the facts. They should then say what they think should be done to prevent it happening again, not only on the plant where it occurred but on the plants which those present operate or are designing. Such discussions take much longer than a lecture but more is remembered and those present are more committed to the conclusions, as they are their conclusions.

The Institution of Chemical Engineers can supply sets of notes and slides for use in this way but it is better to discuss accidents that have occurred in your own company.

Many aids are now available to help the safety trainer: films, videos, slides and slide-tape sets. These aids are most effective if they are used as the starting point for a discussion (Are we doing this? If not, should we? How we do we start? and so on) and not just shown to the audience.

Most companies are better at devising good training programmes than at making sure that the programmes continue and everyone attends year after year.

Students of engineering need some training in loss prevention if they are to be capable of carrying the tasks that lie ahead of them in industry. In the UK the Institution of Chemical Engineers requires all undergraduates to receive some training before they can be considered for corporate membership but in other countries most undergraduates receive no such training. I have discussed elsewhere[1] the reasons why undergraduate chemical engineers should be trained in loss prevention and have suggested subjects that should be included in the syllabus.

1. T A Kletz, *Plant/Operations Progress*, Vol. 7, No. 2, April 1988, p. 95

Transport

See *airlines, caution, other industries, shipping* and *tankers*.

Trapped pressure

Trapped pressure is a major cause of accidents during *maintenance*. A man opening up equipment believes that the pressure has been blown off, but it has not and he is injured by escaping gas or liquid. The operator who prepared the equipment for maintenance may have forgotten to open the vents or drains but more often the vent or drain is choked or there is a *choke* somewhere else in the system. The choke may be due to dirt, to polymer, to liquids which set solid when heating is turned off or to foreign bodies.

Men breaking joints should assume that trapped pressure may be present and should loosen first the bolts furthest from them and then spring the joint before opening it fully. If any pressure is present it may be allowed to blow off or the joint can be remade. However, even after a joint has been fully broken, trapped pressure may clear suddenly.

The following incident is typical of many: an $8\,m^3$ vessel had been pressure-tested with nitrogen; a man was asked to blow off the pressure and remove the top cover. As the vessel was in a maintenance area, outside a workshop, and was no longer part of the plant, no permit-to-work was issued. He opened a valve until the pressure gauge read zero and he could not feel a blow of gas with his hand. He removed the eight nuts which secured the top cover and loosened it with a wedge. With a loud bang the cover blew 8 m into the air.

Several things were wrong:

- The vent line was only ⅝ inch (16 mm) diameter and was half full of debris.
- The pressure gauge read up to 250 bar and so a gauge pressure of 1 bar, sufficient to blow the cover into the air, would not show on it.
- The man doing the job should have loosened the cover before removing the nuts.
- Another person should have prepared the vesssel and issued a permit-to-work. Any system in which the same person prepares equipment and opens it up will sooner or later result in an accident. See *EVHE*, §2.2.1 and *WWW*, §17.1

Traps

Design and *construction* can leave traps into which others can fall. For example, an accident occurred while an old factory was being demolished. The floor boards rested on a series of joists which were supported by three walls. (See Figure 80.) The internal wall was not central so a beam A was installed to give additional support to the joists. One of the joists B was not quite long enough so a short joist C was used to bridge the gap between B and the nearest outside wall. B and C were not joined together, but they both rested on A.

Beam A and some of the floor boards had been removed but the end of B was still covered by a loose board. A man stepped on it, B broke at point D and the man fell to the floor below. The unsupported length of B was 5 m.

Joist B had been weakened by rot but the underlying cause of the accident was the unusual construction which was not obvious and set a trap for the *demolition* workers.

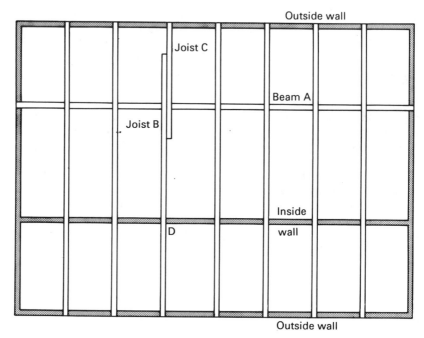

Figure 80 The builders left a trap for the men who would one day have to demolish the building

Other traps are more subtle. A plant consisted of two similar units. One control panel was a mirror image of the other. Men transferred from one unit to another made frequent mistakes.

See *flame traps* and *mechanical accidents*.

Triangles

According to Heinrich[1] for every major injury there are 29 minor injuries and 300 *dangerous occurrences* which do not cause injury. (See Figure 81.) It is a matter of chance whether an incident causes fatal injury, a major injury, a minor injury or no injury at all.

To prevent *accidents* we should base our actions on the study of a large sample of the accidents that occur. If we study only a small fraction picked at random, by the fact that they result in serious injury, then we may be wasting our money, time and effort on the wrong preventative measures. To decide on the right measures we should study a large number of accidents, including some which might have caused serious injury though, by good fortune, they did not.

It is possible to add another, smaller section at the top of the triangle to cover fatal accidents.

It follows from the Heinrich triangle that if we take action to halve (say) the number of fatal accidents we shall also halve the serious accidents, the minor accidents and the dangerous occurrences.

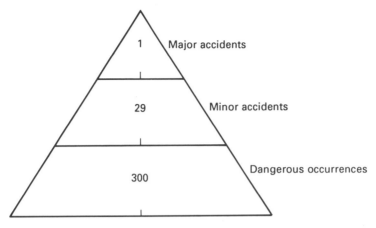

Figure 81 The Heinrich Triangle

Of course, the actual numbers in the Heinrich triangle will differ from one industry to another and from one type of accident to another, but the principle remains the same.

Senneck[2] has shown that for each type of accident the ratio of fatal to lost-time accidents is 'strikingly similar' for mines and factories. See Table 2.

Table 2 Accident ratios

Type of accident	Number of lost-time accidents per fatal accident	
	Mines	Factories
Transport	260	270
Machinery	450	550
Struck by falling objects	830	570
Loss of balance (excluding falls between levels)	17 000	4300
Use of tools	14 000	25 000
Handling	33 000	33 000
Others (including fire and explosion)	1400	820

1. H W Heinrich, *Industrial Accident Prevention*, McGraw-Hill, 4th edition, 1959, p. 27
2. C R Senneck, in *Proceedings of a Symposium on the Collection, Analysis and Interpretation of Accident Data*, edited by W D Hendry, Safety in Mines Research Establishment, Sheffield, 1975, Paper A1

Trips

See *alarms, instruments, protective equipment, tests* and *time of response*.

Underestimated hazards

When we start using a new substance or new equipment we are inclined to play down the hazards and say they are not really very serious. (See *cognitive dissonance*.) Here are some examples:

- In 1880 a spillage of oil from a tanker on the River Wear caught fire causing damage to several other ships. At the official inquiry several witnesses said that they did not think it was possible to ignite oil when it was floating on water and someone even carried out a demonstration in the courtroom to try to prove this point. However the Judge said that the defendants 'might bring all the chemists in the kingdom, but the fact remained that the liquid was on the water and the fire took place[1]'.
- In 1923 acetylene was recommended (by the carbide industry) for domestic lighting. An article recommending it[2] referred to 'its perfect safety'. It added '. . . there is no instance on record of an accident which was not directly due to foolishness or carelessness'. 'Table lamps can be . . . connected to the wall by flexible rubber hose.'
- In early domestic electric lighting installations electric shock was not considered a serious hazard and standards of insulation were poor; bare wires were concealed behind wooden panelling.

1. *Hazardous Cargo Bulletin*, March 1981
2. A Hoddle, *British Acetylene and Welding Handbook*, British Acetylene and Welding Association, 1923, p. 137

Under-reporting

In 1900 the UK Government reported that 100 men per year were killed by accidents at the docks. A trade union leader, Ben Tillett, wrote to *The Times* pointing out that over 100 died from accidents at one hospital alone[1]. I doubt if any accidental deaths in industry go unreported today in the UK but under-reporting of injuries and *dangerous occurrences* may still be a problem in some industries such as *construction*.

Under-reporting of industrial disease is a bigger problem, not because of a deliberate failure to report but because much industrial disease is unrecognized. In the UK only officially listed diseases have to be reported. It is generally agreed that many other diseases have both industrial (as well as non-industrial) causes but the actual number of cases due to occupational exposure is the subject of considerable dispute. In round figures, in the UK, about 1000 people per year die from listed industrial diseases, rather more than the number killed by industrial accidents. If we exclude pneumoconiosis, asbestosis and byssinosis, diseases restricted to specific industries and occupations, the number falls to a few hundred. Estimates of the total number killed by industrial diseases vary from about 1000 to 20000 from *cancer* alone. I have suggested elsewhere that the correct figure is probably near the lower end of this range[2]. The Health and Safety Executive estimate that there are about 2000 deaths per year from industrial disease. (See *COSHH Regulations* and *fugitive emissions*.)

There is probably no other industrial problem where such large sums are spent without any definite knowledge of the size of the problem. Before we spend resources on, say, improving output, efficiency or product quality or preventing accidents we ask first what is the present output, efficiency, quality or accident rate and what is the scope for improvement. When dealing with industrial disease we do not seem able to do this. Ways of improving reporting are discussed by the Health and Safety Executive[3].

Over-reporting is less common than under-reporting but has occurred. A works manager was puzzled by an increase in trivial accidents such as minor cuts and abrasions. The mystery was solved when he developed a headache, went to the first-aid room for an aspirin and found a very attractive nurse in charge.

See *accident statistics*.

1. Quoted in *The Second Cuckoo*, edited by K Gregory, Allen and Unwin, London, 1983, p. 15
2. T A Kletz in *Engineering Risk and Hazard Assessment*, edited by A Kandel and E Avni, CRC Press, Boca Raton, Florida, 1988, p. 1
3. *Collecting Information on Disease Caused by Work*, Health and Safety Executive, Bootle, 1988

Unexpected hazards

Sometimes we take precautions against obvious hazards, and then get hit by one that is not so obvious. For example:

1. Ethylene oxide is usually manufactured by the oxidation of ethylene with air or oxygen. For economic reasons it is necessary to operate close to the explosive limits and a complex protective system is necessary to make sure that the reaction mixture does not enter the explosive range[1]. Many *explosions* have occurred because the *protective equipment* was inadequate or was not kept in working order. However, because the *reaction* mixture is in the vapour phase these explosions have been localized and only rarely has anyone been killed[2].

    The reaction is exothermic and the catalyst tubes are surrounded by a coolant. Boiling oil under pressure is often used, hundreds of tonnes of it, enough for several *Flixboroughs*, and on many ethylene oxide plants it is a bigger hazard than the reaction mixture. Recent plants – and some older ones – use water under pressure as the coolant. (See *substitution*.)
2. Storage *tanks* containing petrol or other oils usually contain a layer of water. The oil is obviously hazardous but the water was the cause of one explosion. River water was used for the water layer in a kerosene tank. Bacterial action produced methane which exploded. The source of ignition was never identified with certainty but may have been pyrophoric iron sulphide formed by sulphate-reducing bacteria in the river water[3].
3. If welding has to be carried out on a storage tank that has contained petrol it is obviously necessary to clean the tank thoroughly and check that no petrol is left. Tanks that have contained *heavy oils* are expected to be less of a problem; not so, because it is almost impossible to remove all traces of the oil, which may be trapped behind rust or between plates. Welding then vaporizes the oil and ignites it. If welding has to be carried out on tanks that have contained heavy oils (or liquids that polymerize) the tanks should be inerted with nitrogen or fire fighting foam made from nitrogen (but not foam made from air). Water can reduce the volume to be inerted. See *WWW*, Chapters 12 and 18.

4. A factory manufactured chlorine and caustic soda from salt. The chlorine and soda were treated with great care but the salt was considered harmless. It was stored in cylindrical hoppers, 5 m diameter by 7 m tall, with a conical base. The salt flowed out by gravity but sometimes stuck to the sides. A man was lowered on a lifeline to the bottom of the hopper, well below the level of the salt which was stuck to the walls, to loosen it with a shovel and water hose. Suddenly a large quantity of salt slid down and buried him to a depth of 1.5 m; by the time he was rescued he was dead.

In the factory the employees were given considerable freedom in the way they did their job and no foreman or manager had approved the *method of working*. Although independence and initiative are good they should not be encouraged when *entry* to vessels is concerned. It should be permitted only after authorization by competent people. See *discretion*.

For other examples see *emergency equipment, nitrogen* and *service lines*.

1. R M Stewart, *Institute of Chemical Engineers Symposium Series No. 34*, 1971, p. 99
2. T A Kletz, *Plant/Operations Progress*, Vol. 7, No. 4, October 1988, p. 226
3. H E Watts, *Report on Explosion in Kerosene Tank at Killingholme, Lincs*, HMSO, London, 1938

Uniformity

See *variety*.

United Kingdom

In his annual report for 1971 the UK Chief Inspector of Factories wrote, '. . . Britain is one of the safest and healthiest countries in which to work[1]'. Ten years later (1981/2) the figures for fatal accidents at work per million population were:

Denmark	8
United Kingdom	9
Spain	17
France	25
West Germany	21
Italy	38

Quoting these figures, a UK Factory Inspector said, 'Inspectors in Britain had developed a professionalism in health and safety not found elsewhere in the European Community . . . and it means that British Inspectors' approach to health and safety problems is generally more effective than in other inspectorates[2]'.

See *Factories Acts, Health and Safety at Work Act, 'reasonably practicable'* and *United States*.

1. *Annual report of HM Chief Inspector of Factories for 1971*, HMSO, London, 1972, p. ix
2. S Campbell, quoted in *Health and Safety at Work*, Vol. 9, No. 10, October 1987, p. 8

United States

In 1932 *ICI*'s Chief Labour Officer visited the United States and found that the best US companies, such as *Du Pont*, had *lost-time accident rates* one-twentieth of ICI's. Their severity rates, the average number of days lost per accident, were also less. Some of the difference was due to the greater use of alternative work in the US but the major factors were good housekeeping and a belief that accidents are due to *management* failures. He wrote, '. . . the dispersal of noxious fumes is not left to the beneficient breezes of providence, and and a case of poisoning would not be explained by "failure of the prevailing wind"' and 'The chief difference between the British and American attitude towards accidents is one of mentality. Americans know what accidents really cost, are determined to stop them and act accordingly. British employers think that they have paid for their accidents when they have paid their premium to the *insurance* company, therefore, while paying lip service to safety, they are rather inclined to believe that too much emphasis interferes with production and after all, "accidents must happen".' (See *philosophers' stone*.)

Not all US companies are as good as the best and the difference between the UK and the US is now much less. Nevertheless, in 1987, when the UK Chemical Industry Safety, Health and Environment Council gave awards to companies with the best safety record in five consecutive years, all four awards were won by subsidiaries of US companies[1]. However, in *loss prevention* there is little to choose between the best US and UK companies.

If US management on the whole has a more enlightened attitude to safety than UK management, this does not apply to their legislation. See *Health and Safety at Work Act*.

1. *Health and Safety at Work*, Vol. 9, No. 11, November 1987, p. 8

Unlikely hazards

The White Knight in Lewis Carroll's *Through the Looking Glass* carried a mouse trap on his horse. He agreed it was not very likely that there would be any mice there but if they did come he preferred not to have them running about. 'It's as well to be provided for everything', he said. For the same reason his horse wore anklets to guard against the bites of sharks.

At one time many *safety professionals* were like the White Knight. If any hazard was identified, however unlikely, it must be removed. To do anything else would be to admit that we did not put *'safety first'*.

This view led to waste and inefficiency – the company's money was wasted on the removal of unlikely or trivial hazards while more serious ones went unrecognized. Some hazards should be put right immediately, some can wait until next year; if they are sufficiently unlikely, some can wait for ever. How do we decide what is 'sufficiently unlikely'? A number of attempts have been made to define levels of *acceptable risk*.

There is nothing callous or cold-blooded in this approach. The more we spend on unlikely risks the less we have available to spend on other risks.

The most effective humanitarian is the man who allocates his resources of time and money so that they bring the maximum benefit to his fellow men.
See *hazard analysis, perspective* and *'reasonably practicable'*.

Useless equipment and operations

We sometimes install equipment or carry out operations which serve no useful function. Here are some examples:

1. *Vessels* containing flammable gas are often fitted with fire relief valves to prevent the pressure exceeding the design pressure if the vessel is heated by fire. However, the metal will probably soften and the vessel burst before the relief valve lifts. It serves no useful purpose, though it is required by some of the codes. If the working pressure of the vessel and the set point of the relief valve are far below the design pressure, or the vessel is made of stainless steel, then the relief valve may lift and should certainly be installed.

   If a vessel contains liquid, or liquefied gas, then the liquid absorbs the heat and prevents the metal overheating. The metal will soften only if flames impinge on the vessel above the liquid level and obviously fire relief valves should be installed. (See *BLEVE*.) However, liquids close to their critical temperatures have a very low latent heat and may not absorb enough heat to prevent the vessel overheating.

   See *relief devices*.

2. Some years ago I visited a plant where the relief valves discharged to atmosphere. To prevent them lifting (or, to be more precise, to make the probability of them lifting very small) pressure switches were installed to isolate the source of pressure. As a high-reliability system was required three pressure switches, connected to a single isolation valve, were installed below each relief valve. The third pressure switch was useless. Once two pressure switches were installed the isolation valve became the weakest link in the system (the common mode – see *redundancy*) and it was a waste of money to install any more pressure switches.

   Wearing a second pair of braces, attached to the same buttons as the first pair, may reduce the chance that my trousers will fall down. The buttons are now the weak link (the common mode) and attaching a third pair of braces to the same buttons will produce no increase in safety. It would be better to spend the money on a second set of buttons.

   Of course, a third pressure switch would be useful if it formed part of a voting system to prevent spurious closing of the valve.

3. An example of a useless procedure is trying to increase the reliability of *protective equipment* (such as a high level trip on a *tank*) by testing it every day. The trip is out of action while it is being tested and a thorough *test* can easily take an hour or so. Testing then contributes more to the fractional dead time than accidental failure. The trip would be more reliable if it was tested less often.

4. Another example of useless testing is testing every month (say) a trip that operates every day. Any failure of the equipment is 30 times more

likely to be followed by a demand than by a test and the failure of the trip to respond to the demand will disclose the failure, that is, we will know the trip has failed when the tank overflows. Testing in these circumstances is a waste of time. If testing is to do any good the test frequency must be greater than the demand rate.

So what should we do if our trip operates every day? Testing every month, or even every week, is useless and daily testing gives a high fractional dead time. The answer is that we need two trips, or a better control system so that the trip does not need to operate every day.

Utilities

See *service lines*.

Vacuum

See *nothing*.

Valves

Isolation of equipment for maintenance should be based on slip-plates (*blinds*) rather than valves unless the job to be done is so quick that fitting slip-plates would take as long as the main job and be as dangerous.

Remotely-operated *emergency isolation valves* are widely used for isolating equipment which may leak and they have paid for themselves many times over. All valves, hand-operated and remotely-operated, which are not used regularly but may have to be used in an emergency should be exercised regularly or they may not work when required.

Whenever possible valves should be chosen so that we can see at a glance whether they are open or shut. (See *friendly plants*.) Valves with rising spindles are a good choice. Ball valves are OK if the handle cannot be fitted in the wrong position.

The *reliability* of valves is discussed by R J Aird and T J Moss[1]. Surprisingly, 20% of all incidents and 7% of the outage on French nuclear power stations are due to valve failure.

Old valves sold as new are a hazard on many chemical plants. Reference 2 describes ways of detecting such valves.

Many incidents have occurred because valves were assembled or dismantled incorrectly[3,4].

See *non-return valves* and *relief devices*.

1. R J Aird and T J Moss, *Mechanical Valve Reliabilty*, Seminar on Mechanical Reliability in the Process Industries, Institution of Mechanical Engineers, July 1984
2. *Chemical Engineering Progress*, 20 July 1987, p. 19
3. *WWW*, §1.5.4
4. *Loss Prevention Bulletin*, No. 066, December 1985, p. 23, No. 069, June 1986, p. 1 and No. 077, October 1987, p. 27

Variety

Some companies impose uniform standards, for safety and everything else, on their factories and divisions. Others, such as *ICI*, allow a good deal of variety within the organization. The reasons usually given for permitting this diversity are that:

- Factories making different products have different needs.
- Different systems having been developed in the past, the advantages of a common system do not justify the upheaval caused by a change.
- Each unit is committed to systems it has developed itself and might resent systems imposed by the centre.

Another reason has been suggested by the French anthropologist, C Levi-Strauss:

'Why was it that some societies had ossified and remained immobile for many hundreds of years, whereas others had been able to change and adapt?'

His answer was that civilized societies are pluralistic and always contain more than one viewpoint, more than one possibility of action, whereas primitive societies are monolithic, the existing viewpoint or the existing way of doing things, remains unquestioned. To put the matter another way, in a civilized society pluralism creates a tension and a debate which produces forward movement. In the primitive society, it seems as if the gods have given a once-for-all blueprint as to how things are to be done. In other words, the ideal has been laid down in the past and the necessity is to look backwards over one's shoulder and see what it consists of. In the civilized society heaven lies in the future – it is something to be reached out to. Heaven always lies ahead[1].

See *amalgamation*.

1. B Trapnell, *Daily Telegraph*, 15 December 1980

Vents

See *entry, explosion venting, hoses, (accidental) purification, stacks* and *relief devices*.

Ventilation

If flammable gases or highly flammable liquids are handled in closed buildings, small *leaks* can produce an explosive mixture of vapour and air. Many *explosions* have occurred in this way.

Ventilation machinery can be installed to remove leaking gases and vapours but it is expensive to install and operate and less efficient than natural ventilation. In the open air, even on a still day, there is enough air movement to disperse small leaks of gas or vapour, leaks that could build up to an explosive mixture indoors. In the open air several tonnes or tens of tonnes of vapour are usually necessary for an explosion but buildings have been destroyed by a few tens of kilogrammes.

Working out of doors in so-called temperate climates is not always pleasant (it may be impossible for part of the year in very cold climates) but the discomfort is a small price to pay for increased safety. There is no objection to a roof (with ridge ventilation) over the equipment but there should be no walls at equipment level. The wind should be free to blow through.

When there has been no explosion for some time, companies start putting equipment indoors, for the comfort of the workers or to reduce noise levels outside. When an explosion occurs there is a campaign to remove walls.

340 Ventilation

Figure 82 Ventilation

If toxic vapours are handled should we still put the equipment in the open? If a leak could spread so far that it affects the public (or untrained people on other plants), then it is better to put the equipment indoors, as this restricts the spread of the leak. The operators will be exposed but they can be supplied with breathing apparatus and trained in its use. For example, a hydrogen fluoride *tanker* offloading unit, including tanks and pumps, was enclosed, after a quantitative *hazard analysis* had shown that it was the biggest hazard on the plant[1]. If the leak could not affect the public (or untrained people on the neighbouring plants) then it may be better to locate the equipment out-of-doors.

The value of open construction has been known for many years. An article on an *ICI* factory published in 1936[2] said that workers handling nitrobenzene were affected by fumes, especially in warm weather. 'The problem was solved by grouping these processes into one shed . . . and removing as much of the shed walls as possible. The most undesirable operations are now actually carried out outside the shed.'

Cylinders of flammable gas should not be carried in closed vans, only in open trucks. See Figure 82.

See *dispersion, cognitive dissonance* and *cylinders*.

1. L C Schaller, *Off Site Risk Assessment and Risk Minimization Guidelines*, AIChE Summer Meeting, 1988
2. W G Hiscock, *Chemistry and Industry*, 20 March 1936, p. 222

Vessels

While *pipe failures* are the major cause of large *leaks*, failures of pressure vessels are rare. (*Tanks* designed for operation at atmospheric pressure are fragile and fail more often.) Most of the pressure vessel failures that have been reported occurred during pressure test or were cracks detected during routine *inspection*. Catastrophic ruptures in service are very rare. Few of those that have occurred could have been prevented by better design or construction (but see below). In most cases the vessel was subjected to conditions far outside the range it was designed to withstand. See *WWW*, Chapter 9.

It is generally accepted that, in estimating the probability of a major leak, pressure vessel failures can be ignored provided the vessels have been designed, constructed, operated and maintained according to recognized standards; these conditions do not always apply. A major *BLEVE* in Texas City in 1978 started when a wartime pressure vessel failed along a defective weld. See *time-bombs*.

A well-known and often-quoted report on vessel failures[1] states that the probability of catastrophic failure per year, for UK vessels, for the period 1962–1978, was 4.2×10^{-5}. However, 'catastrophic' is defined as 'destruction of the vessel or component, or a failure so severe as to necessitate major repair, or replacement'. The failure rate thus includes many failures which would not be considered catastrophic in the everyday sense of the word but the definition of 'catastrophic' is not always quoted – a good example of the errors that can be introduced by misquotation.

1. T A Smith and R G Warwick, *A Survey of Defects in Pressure Vessels in the UK for the Period 1962–78 and its Relevance to Nuclear Primary Circuits*, Report No. SRD R203, UK Atomic Energy Authority, Warrington, December 1981

Viruses

Much has been written about *computer* viruses – small pieces of code which have been maliciously inserted into floppy discs or other storage media to corrupt data – but at the time of writing there has been no study of viruses in real-time systems such as plant control programs. 'However, it seems that asserting the dangers of virus infection in such systems is not an idle concern. Any disruption in the operation of such systems can have consequences more serious than the loss of accounting or technical data[1].'

On the other hand, plant control software is harder to infect than other software as plant control computers are not usually connected to networks and, once set up, additional programs are added infrequently. For infection to occur, viruses would have to be present in the original software.

Computer viruses are like AIDS. Do not promiscuously share discs and data and you will not be infected[2].

Some control software contains codes which will prevent it being used after 6 months if the bill has not been paid.

1. W D Ehrenberger, *European Safety and Reliability Association Newsletter*, Vol. 5, No. 2, August 1988
2. M Becket, *Daily Telegraph*, 26 September 1989

Water

Water, like compressed *air*, is more hazardous than most people realize:

- Like other *service lines* water lines are often contaminated. Drinking water supplies should be connected to plants via a break tank so that *reverse flow* cannot occur. (See *WWW*, §18.2.)
- Lines have been broken by freezing water. Lines which are out of use should never be left full of water. *Dead-ends* in which water can accumulate should be avoided. Where stagnant water is unavoidable trace heating may be necessary. Like all *protective equipment* it should be checked regularly.
- If water comes into contact with hot oil it may vaporize with explosive violence. See *boil-over* and *foam-over*.
- Many people have been injured by hot water or water under pressure because they were treated with less respect than more obviously hazardous substances.

See *steam* and *unexpected hazards*.

The following article, written in 1976, has been quoted far more often than anything else I have written:

NEW FIRE-FIGHTING AGENT MEETS OPPOSITION
'COULD KILL MEN AS WELL AS FIRES'

ICI has announced the discovery of a new fire-fighting agent to add to their existing range. Known as WATER (Wonderful And Total Extinguishing Resource), it augments, rather than replaces existing agents such as dry powder and BCF which have been in use since time immemorial. It is particularly suitable for dealing with fires in buildings, timber yards and warehouses. Though required in large quantities, it is fairly cheap to produce and it is intended that quantities of about a million gallons should be stored in urban areas and near other installations of high risk ready for immediate use. BCF and dry powder are usually stored under pressure, but WATER will be stored in open ponds and reservoirs and conveyed to the scene of the fire by hoses and portable pumps.

ICI's proposals are already encountering strong opposition from safety and environmental groups. Professor Connie Barriner has pointed out that, if anyone immersed their head in a bucket of WATER, it would prove fatal in as little as 3 minutes. Each of ICI's proposed reservoirs will contain enough water to fill half a million two-gallon buckets. Each bucket-full could be used a hundred times so there is enough water in *one* reservoir to kill the entire population of the UK. Risks of this size, said Professor Barriner, should not be allowed whatever the gain. If the WATER were to get out of control the results of Flixborough and Seveso would pale into insignificance by comparison. What use was a fire-fighting agent that could kill men as well as fires?

A Local Authority spokesman said that he would strongly oppose planning permission for construction of a water reservoir in his area unless the most stringent precautions were followed. Open ponds were certainly not acceptable. What would prevent people falling in them? What would prevent the contents leaking out? At the very least the WATER would have to be contained in a steel pressure vessel surrounded by a leak-proof concrete wall.

A spokesman for the Fire Brigades said that he did not see the need for the new agent. Dry powder and BCF could cope with most fires. The new agent would bring with it risks, particularly to firemen, greater than any possible gain. Did we know what would happen to the new agent when it was exposed to intense heat? It had been reported that WATER was a constituent of beer. Did this mean that firemen would be intoxicated by the fumes?

The Friends of the World said that they had obtained a sample of water and found that it caused clothes to shrink. If it did this to cotton, what would it do to men?

In the House of Commons yesterday, the Home Secretary was asked if he would prohibit the sale of this lethal new material. The Home Secretary replied that, as it was clearly a major hazard, Local Authorities would have to take advice from the Health and Safety Executive before giving planning permission. A full investigation was needed and the Major Hazards Group would be asked to report.

WATER is being marketed in France under the name EAU (Element Anti-feu Universel).

See *aversion to risk, cause and effect, false alarms, perception of risk* and *perspective*.

Water hammer

See *steam*.

Welding

During electric welding the return lead (sometimes incorrectly called the earth lead) has to be attached to the object being welded. Failure to attach it correctly, that is, as close as possible to the weld and certainly no more than 2 m away, has caused several accidents. A welder was welding a bracket onto the outside of a tank containing caustic soda solution. He attached the return lead to a valve which was fixed to the tank. There was a high resistance between the valve and the tank so the return current travelled through the liquid and electrolysed it, producing hydrogen which exploded, killing one man and blinding another. (See Figure 83.)

A similar incident has been described by Nightingale[1].

We might feel that the hazards of a specialized technique such as welding can be left to the experts. If we suggest that something is wrong the experts may be offended; presumably they know their job. This incident (and others; see *alertness*) shows that we cannot assume that experts always know their job and that we should always speak up if we suspect that something may be wrong.

1. P J Nightingale, *Plant/Operations Progress*, Vol. 8, No. 1, January 1989, p. 29

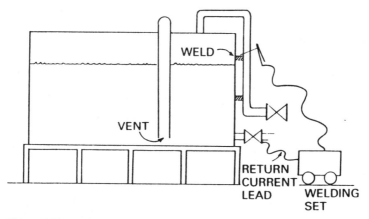

Figure 83 The welding return current electrolysed the liquid in the tank, producing hydrogen which exploded

Whistle blowing

See *talebearing*.

Will and might

Some *accidents* are deterministic – if we take (or fail to take) certain actions, an accident is inevitable. Others are probalilistic – if we take (or fail to take) certain actions, an accident might result.

An example of the first sort of accident is neglecting to clean the *flame trap* in the vent of a storage *tank*. Inevitably, after a few years, perhaps less, the flame trap will choke with dirt and the tank will be sucked in while it is being emptied. (See Figure 84.) If we fail to clean the flame trap, we are not taking a chance; we are awaiting an inevitable event. Perhaps we hope we will be promoted for reducing *maintenance* costs before the tank is sucked in. I have known one or two engineers who gained a reputation for saving on maintenance and left their successsors to pick up the tab.

An example of the second sort of accident is ignoring a *leak* of flammable gas or liquid until a more convenient time for repair. We may get away with it. If the leak is small and can be safely dispersed it may be reasonable to leave it for a while. The *law* does not require us to do everything possible to prevent every conceivable accident, only what is *'reasonably practicable'*, weighing in the balance the size of the risk and the cost, in money, time and trouble, of preventing it. Obviously, the more often we ignore leaks, the greater the chance that one will ignite. In one factory, where ethylene was handled at high pressure, so that leaks dispersed by jet mixing, hundreds of leaks per year were treated casually and often left until repair was convenient. For many years there was no ignition but in the end a serious explosion occurred. See *LFA*, Chapter 4.

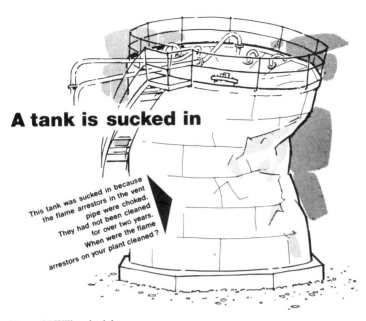

Figure 84 Will and might

If we consume insufficient Vitamin C the result is deterministic; we will inevitably get scurvy. If we smoke the result is probabilistic; we might get lung *cancer*, but most smokers do not. If we breath 400 p.p.m. of *chlorine* for 30 minutes (the so-called LC_{50}† concentration)[1] we might die; there is a 50% chance that we will do so. If we breath a much higher concentration the effects are deterministic; death is certain. The effects of small concentrations of toxic chemicals or radiation over long periods are probabilistic; some people will be affected but not all.

With deterministic accidents we can generalise from a single (or a few) incidents. If a heavy object is dropped from a height and hits the ground at a certain point, it will always do so. There is no need to carry out a hundred tests. With probabilistic accidents, we need much experience before we can generalize. Many years ago a leak of propylene exploded inside a building; it was a very mild *explosion*. Windows were broken but the glass fell as it broke and did not fly across the room. As a result one of the staff assumed that that this would occur every time propylene, or other hydrocarbons, exploded; unfortunately, this is not the case. The results of a vapour cloud explosion are probabilistic. (See *(limitations of) experience*.)

Probabilistic accidents are more common than deterministic ones but the latter are not scarce. Those who take a chance that a deterministic accident will not happen show not just a lack of safety awareness but a failure to understand the nature of the game.

1. *Chlorine Toxicity Monograph*, Institution of Chemical Engineers, Rugby, 1987

† Recent work suggests that the LC_{50} may be lower than the figure quoted.

Windows

Windows can be broken by mild *explosions* and more severe ones can send glass flying across the room. Window glass can, however, be protected cheaply and effectively by plastic film. It should be fitted whenever there is a significant risk that a peak incident pressure of more than about 1 p.s.i.g. (0.07 bar gauge) will be developed. At this pressure windows may be broken but the glass is unlikely to fly across the room. Explosion of a vapour cloud containing 50 tonnes of liquefied petroleum gas or similar materials will produce a pressure of up to 1 p.s.i.g. about 250 m away.

Of course, blast-resistant glass can be used instead of film but wired glass is not blast-resistant; a hand was nearly severed by wired glass after a *hydrogen* explosion[1].

Although the windows in a control building are usually the weakest point, many companies permit small windows, up to $1 m^2$ in area, so that operators can have a view of the plant.

1. J A MacDiarmid and G J T North, *Plant/Operations Progress*, Vol. 8, No. 2, April 1989, p. 96

Wolves in sheep's clothing

Some processes are believed to be safe and easy to control and then, after many years of safe operation, the 'shepherds' realize that they are dealing with a wolf, not a sheep.

For example, hydrogenation, particularly in the liquid phase, is often considered a mild *reaction*, unlikely to run away but *Bretherick*[1] lists ten incidents in which runaways occurred. Here is another.

Liquid, part fresh feed, part recycle, was hydrogenated in the plant shown in Figure 85. It had operated without serious incident for 14 years. The plant instructions detailed the action to be taken if liquid flow was lost (sweep out with *hydrogen*) and the action to be taken if hydrogen flow was lost (sweep out with liquid) but did not say what should be done if both flows were lost; one day this occurred. The liquid and gas in the reactor continued to react, the temperature rose well above the normal value of 235 °C, probably exceeding 450 °C, and the feed started to crack. The precise temperature reached is not known as there was no temperature point on the reactor, only on the exit line. When feed was restarted the fresh liquid fuelled the cracking reaction, the temperature rose above 600 °C and the reactor was damaged.

The designers of the *instruments* and *protective equipment* had assumed that the reaction was a gentle one that could not get out of hand.

The reason for the loss of hydrogen flow is interesting. The hydrogen was a by-product stream, contaminated with hydrocarbons, and usually had a density eight times that of pure hydrogen. On the day of the incident the hydrogen was purer than usual and its density was only twice that of pure hydrogen. The compressor was operating far from its design condition and this, combined with a zero error on the hydrogen flowmeter, made the hydrogen flow insignificant. If a *hazard and operability study* had been carried out, then someone would have asked what would happen if the

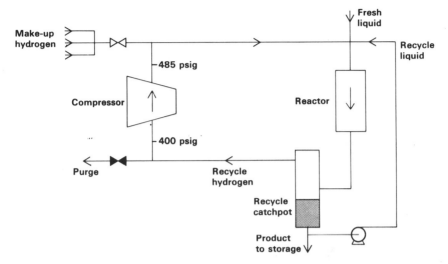

Figure 85 The operators were told what to do if the liquid flow failed (sweep out reactor with gas) and if the gas flow failed (sweep out reactor with liquid). What should they do if both fail?

hydrogen purity changed. After the incident the density was kept constant by injecting nitrogen.

The liquid flow was lost because the catchpot became empty, the recycle flow stopped and a trip stopped the flow of fresh liquid. (This trip was probably installed because 100 per cent fresh liquid would have been too reactive).

For other wolves in sheep's clothing see *(limitations of) experience, hazardous substances* and *unexpected hazards*.

1. L Bretherick, *A Handbook of Reactive Chemical Hazards*, Butterworths, 3rd edition, 1985

Working practice

See *custom and practice, good practice* and *methods of working*.

Zeebrugge

On 6 March 1987 the cross-channel roll-on/roll-off ferry *Herald of Free Enterprise* sank, with the loss of 186 passengers and crew, soon after leaving Zeebrugge in Belgium, en route for Dover. The vessel sank because the inner and outer bow doors had been left open; they had been left open because the assistant bosun, who should have closed them, was asleep in his cabin and did not hear an announcement on the loudspeakers that the ship was ready to sail. However, the underlying *causes* were weaknesses in design and poor management: 'From top to bottom the body corporate was infected with the disease of sloppiness.' This and other quotations are taken from the official report[1]. It provides an excellent illustration of the way in which layered *accident investigation* can look below the obvious recommendations and find ways of removing or reducing the hazards and improving the *management* system.

Removing or reducing the hazards

Because roll-on/roll-off ferries have large unbroken decks any water that gets on them spreads the full length of the ship and accumulates on one side, making the ship unstable. Despite the inconvenience of horizontal and vertical bulkheads, the report recommends that they are given further consideration. Water can get onto the vehicle deck through holes in the side, the result of collisions, and from above through openings, as well as through open doors.

Other recommendations made in the report are:

- Ways of providing additional bouyancy should be considered.
- Bigger pumps should be provided for removing water that gets aboard.
- Berths should be modified so that doors can be closed before ships leave.
- Warning lights and closed circuit TV should be used to tell the captain whether the doors are closed or open.
- The captain should know the draught and trim of the ship.

Though not mentioned in the report, the hazard will disappear (on this route) when the channel tunnel is completed.

Improving the management system

Sloppiness occurred at all levels. The officer in charge of loading did not check that the assistant bosun was on the job. He was unable to recognize him as the officers and crew worked different shift systems. There was pressure to keep to time and the boat was late. Before sailing the captain was not told that everything was OK; if no defects were reported, he assumed it was. There was no monitoring system. The industry apparently does not employ *safety professionals* and responsibility for safety was not clear. One director told the Court of Inquiry that he was responsible; another said no one was. '. . . those charged with the management of the Company's ro-ro fleet were not qualified to deal with many nautical matters and were unwilling to listen to their Masters, who were well-qualified.' In particular, requests for indicator lights to show that the doors were shut were not merely ignored but treated as absurd. '. . . the "Marine Department" did not listen to the complaints or suggestions or wishes of their Masters . . . the voice of the Masters fell on deaf ears ashore.' Complaints of overloading were brushed aside. Ships were known to have sailed before with open doors but nothing was done.

The Certificates of Competency of the captain and first officer were suspended but 'the underlying or cardinal faults lay higher up in the Company. The Board of Directors did not appreciate their responsibility for the safe management of their ships. . . They did not have any proper comprehension of what their duties were'. The Directors, however, seem to have continued in office. In fact, 6 months after the tragedy the Chairman of the holding company was quoted in the press as saying, 'Shore based management could not be blamed for duties not carried out at sea[2]'. He had spent most of his career in property management and may not have realized that managers in industry accept responsibility for everything that goes on, even though it is hundreds of miles from their offices. Sooner or later people fail to carry out routine tasks and so a monitoring or *audit* system should be established and, whenever possible, equipment and systems of work should be designed to tolerate errors without disaster. (See *friendly plants* and *human failing*.) Auditing may have been difficult on ships in the days of Captain Cook, when ships on long voyages were not seen for months, even years, but today a ferry on a short sea crossing is as easy to audit as a fixed plant. (See *shipping*, especially the last paragraph.)

The Zeebrugge report should be read by any director or senior manager who thinks that safety can be left to those on the job and that all they need do is produce a few expressions of goodwill. (See *platitudes*.)

Finally, there was a failure to learn the lessons of the past. Many other ro-ro ferries had capsized[3].

See *contradictory instructions, emergencies, 'reasonable care'* and *repeated accidents*.

1. Department of Transport, *MV Herald of Free Enterprise: Report of Court No 8074: Formal Investigation*, HMSO, London, 1987
2. *Daily Telegraph*, 10 October 1987
3. D Foy, *Journal of the Royal Society of Arts*, Vol. 136, No. 5377, December 1987, p. 13

RETURN CHEM.
TO ➡

SEP 19 1991